**Atomic Layer Deposition
in Energy Conversion
Applications**

Atomic Layer Deposition in Energy Conversion Applications

Julien Bachmann

Editor

Prof. Julien Bachmann
Friedrich-Alexander Universität
FAU Anorganische Chemie
Egerlandstr. 1
91058 Erlangen
Germany

Cover
Photograph in the background -
fotolia/danielschoenen

All books published by **Wiley-VCH** are carefully produced. Nevertheless, authors, editors, and publisher do not warrant the information contained in these books, including this book, to be free of errors. Readers are advised to keep in mind that statements, data, illustrations, procedural details or other items may inadvertently be inaccurate.

Library of Congress Card No.: applied for

British Library Cataloguing-in-Publication Data
A catalogue record for this book is available from the British Library.

Bibliographic information published by the Deutsche Nationalbibliothek
The Deutsche Nationalbibliothek lists this publication in the Deutsche Nationalbibliografie; detailed bibliographic data are available on the Internet at http://dnb.d-nb.de.

© 2017 Wiley-VCH Verlag GmbH & Co. KGaA, Boschstr. 12, 69469 Weinheim, Germany

All rights reserved (including those of translation into other languages). No part of this book may be reproduced in any form – by photoprinting, microfilm, or any other means – nor transmitted or translated into a machine language without written permission from the publishers. Registered names, trademarks, etc. used in this book, even when not specifically marked as such, are not to be considered unprotected by law.

Print ISBN: 978-3-527-33912-9
ePDF ISBN: 978-3-527-69481-5
ePub ISBN: 978-3-527-69483-9
Mobi ISBN: 978-3-527-69484-6
oBook ISBN: 978-3-527-69482-2

Cover Design Schulz Grafik-Design, Fußgönheim, Germany
Typesetting SPi Global, Chennai, India

Printed and bound by CPI Group (UK) Ltd, Croydon, CR0 4YY

Contents

Preface *xi*
Julien Bachmann
The Past of Energy Conversion *xi*
The Future of Energy Conversion *xi*
Technical Ingredients Needed *xiii*
Scope of This Book *xiv*
Photovoltaics: Strategies, Length Scales, and ALD *xv*
Electrochemical Energy Storage: Principles, Chemistries, and ALD *xvii*
Other Energy Conversion Strategies Based on Interfaces *xix*
References *xx*
List of Contributors *xxiii*

Part I Introduction to Atomic Layer Deposition *1*

1 Basics of Atomic Layer Deposition: Growth Characteristics and Conformality *3*
Jolien Dendooven and Christophe Detavernier
1.1 Atomic Layer Deposition *3*
1.1.1 Principle of ALD *3*
1.1.2 ALD Growth Characteristics – Linearity, Saturation, and ALD Window *5*
1.1.3 Plasma-Enhanced ALD *8*
1.1.3.1 Plasma Configurations for Plasma-Enhanced ALD *9*
1.1.3.2 Reactions in Plasma-Enhanced ALD *10*
1.1.3.3 Advantages and Challenges of Plasma-Enhanced ALD *10*
1.2 *In Situ* Characterization for Studying ALD Processes *11*
1.2.1 Quartz Crystal Microbalance *12*
1.2.2 Quadrupole Mass Spectrometry (QMS) *13*
1.2.3 Spectroscopic Ellipsometry *14*
1.2.4 Fourier Transform Infrared Spectroscopy *15*
1.2.5 Optical Emission Spectroscopy *15*

1.2.6	Other *In Situ* Techniques	*16*
1.3	Conformality of ALD Processes	*16*
1.3.1	Quantifying the Conformality of ALD Processes	*17*
1.3.2	Modeling the Conformality of ALD	*21*
1.3.3	The Conformality of Plasma-Enhanced ALD	*24*
1.3.4	Conformal Coating of Nanoporous Materials	*29*
	References	*34*

Part II Atomic Layer Deposition in Photovoltaic Devices *41*

2 Atomic Layer Deposition for High-Efficiency Crystalline Silicon Solar Cells *43*
Bart Macco, Bas W. H. van de Loo, and Wilhelmus M. M. Kessels

2.1	Introduction to High-Efficiency Crystalline Silicon Solar Cells	*43*
2.1.1	ALD for Si Homojunction Solar Cells	*44*
2.1.2	ALD for Si Heterojunction Solar Cells	*46*
2.1.3	Novel Passivating Contacts and ALD	*47*
2.1.4	Outline of this Chapter	*47*
2.2	Nanolayers for Surface Passivation of Si Homojunction Solar Cells	*48*
2.2.1	Basics of Surface Passivation	*48*
2.2.1.1	The Physics of Surface Recombination	*48*
2.2.1.2	Surface Passivation	*50*
2.2.1.3	Compatibility with Si Homojunction Solar Cells	*53*
2.2.2	Surface Passivation by ALD Al_2O_3	*54*
2.2.2.1	ALD of Al_2O_3 for Passivation	*55*
2.2.2.2	Hydrogenation of Interface Defects	*56*
2.2.2.3	Interface Engineering by Al_2O_3	*57*
2.2.2.4	Influence of the Surface Conditions on the Passivation Properties	*58*
2.2.3	ALD in Solar Cell Manufacturing	*59*
2.2.3.1	Requirements for Manufacturing in the PV Industry	*59*
2.2.3.2	High-Throughput ALD Reactors	*60*
2.2.3.3	ALD Al_2O_3 in the PV Industry	*60*
2.2.4	New Developments for ALD Passivation Schemes	*63*
2.2.4.1	ALD Stacks for the Passivation of n^+ Si or n^+ and p^+ Si surfaces	*63*
2.2.4.2	ALD for the Passivation of Surfaces with Demanding Topologies	*64*
2.2.4.3	Novel ALD-Based Passivation Schemes	*66*
2.3	Transparent Conductive Oxides for Si Heterojunction Solar Cells	*68*
2.3.1	Basics of TCOs in SHJ Solar Cells	*69*
2.3.1.1	Lateral Conductivity	*69*
2.3.1.2	Transparency	*71*
2.3.1.3	Compatibility with SHJ Solar Cells	*74*
2.3.2	ALD of Transparent Conductive Oxides	*74*
2.3.2.1	ALD of Doped ZnO	*74*
2.3.2.2	Beyond Al Doping: Doping by B, Ti, Ga, Hf, and H	*77*
2.3.2.3	ALD of In_2O_3	*78*
2.3.3	High-Volume Manufacturing of ALD TCOs	*79*

2.4	Prospects for ALD in Passivating Contacts	*80*
2.4.1	Basics of Passivating Contacts	*80*
2.4.1.1	How to Make a Passivating Contact?	*81*
2.4.1.2	Requirements of a Passivating Contact	*84*
2.4.2	ALD for Passivating Contacts	*86*
2.4.2.1	ALD for Tunneling Oxides	*86*
2.4.2.2	ALD for Electron-Selective Contacts	*87*
2.4.2.3	ALD for Hole-Selective Contacts	*89*
2.5	Conclusions and Outlook	*89*
	References	*90*
3	**ALD for Light Absorption**	*101*
	Alex Martinson	
3.1	Introduction to Solar Light Absorption	*101*
3.2	Why ALD for Solar Light Absorbers?	*104*
3.2.1	Uniformity and Precision of Large-Area Coatings	*104*
3.2.2	Orthogonalizing Light Harvesting and Charge Extraction	*105*
3.2.3	Pinhole-Free Ultrathin Films, ETA Cells	*107*
3.2.4	Chemical Control of Stoichiometry and Doping	*107*
3.2.5	Low-Temperature Epitaxy	*109*
3.3	ALD Processes for Visible and NIR Light Absorbers	*109*
3.3.1	ALD Metal Oxides for Light Absorption	*111*
3.3.2	ALD Metal Chalcogenides for Light Absorption	*111*
3.3.2.1	CIS	*112*
3.3.2.2	CZTS	*112*
3.3.2.3	Cu_2S	*112*
3.3.2.4	SnS	*113*
3.3.2.5	PbS	*113*
3.3.2.6	Sb_2S_3	*113*
3.3.2.7	CdS	*113*
3.3.2.8	In_2S_3	*114*
3.3.2.9	Bi_2S_3	*114*
3.3.3	Other ALD Materials for Light Absorption	*115*
3.4	Prospects and Future Challenges	*115*
	References	*115*
4	**Atomic Layer Deposition for Surface and Interface Engineering in Nanostructured Photovoltaic Devices**	*119*
	Carlos Guerra-Nuñez, Hyung Gyu Park, and Ivo Utke	
4.1	Introduction	*119*
4.2	ALD for Improved Nanostructured Solar Cells	*120*
4.2.1	Compact Layer: The TCO/Metal Oxide Interface	*121*
4.2.2	Blocking Layer: The Metal Oxide/Absorber Interface	*126*
4.2.3	Surface Passivation and Absorber Stabilization: The Absorber/HTM Interface	*130*
4.2.4	Atomic Layer Deposition on Quantum Dots	*132*

| Contents

4.2.5 ALD on Large-Surface-Area Current Collectors: Compact Blocking Layers *134*
4.3 ALD for Photoelectrochemical Devices for Water Splitting *138*
4.4 Prospects and Conclusions *142*
References *143*

Part III   ALD toward Electrochemical Energy Storage *149*

5   Atomic Layer Deposition of Electrocatalysts for Use in Fuel Cells and Electrolyzers *151*
Lifeng Liu
5.1 Introduction *151*
5.2 ALD of Pt-Group Metal and Alloy Electrocatalysts *153*
5.2.1 ALD of Pt Electrocatalysts *154*
5.2.1.1 Fabrication and Microstructure *154*
5.2.1.2 Electrochemical Performance *157*
5.2.2 ALD of Pd Electrocatalysts *168*
5.2.3 ALD of Pt-Based Alloy and Core/Shell Nanoparticle Electrocatalysts *169*
5.2.3.1 ALD of Pt Alloy Nanoparticle Electrocatalysts *170*
5.2.3.2 ALD of Core/Shell Nanoparticle Electrocatalysts *172*
5.3 ALD of Transition Metal Oxide Electrocatalysts *174*
5.4 Summary and Outlook *175*
Acknowledgment *178*
References *178*

6   Atomic Layer Deposition for Thin-Film Lithium-Ion Batteries *183*
Ola Nilsen, Knut B. Gandrud, Amund Ruud, and Helmer Fjellvåg
6.1 Introduction *183*
6.2 Coated Powder Battery Materials by ALD *184*
6.3 Li Chemistry for ALD *186*
6.4 Thin-Film Batteries *187*
6.5 ALD for Solid-State Electrolytes *189*
6.5.1 Li_2CO_3 *189*
6.5.2 Li–La–O *189*
6.5.3 LLT *189*
6.5.4 Li–Al–O ($LiAlO_2$) *190*
6.5.5 $Li_xSi_yO_z$ *191*
6.5.6 Li–Al–Si–O *191*
6.5.7 $LiNbO_3$ *192*
6.5.8 $LiTaO_3$ *192*
6.5.9 Li_3PO_4 *192*
6.5.10 Li_3N *192*
6.5.11 LiPON *193*

6.5.12	LiF *194*	
6.6	ALD for Cathode Materials *194*	
6.6.1	V_2O_5 *194*	
6.6.2	$LiCoO_2$ *195*	
6.6.3	$MnO_x/Li_2Mn_2O_4/LiMn_2O_4$ *196*	
6.6.4	Subsequent Lithiation *196*	
6.6.5	$LiFePO_4$ *197*	
6.6.6	Sulfides *198*	
6.7	ALD for Anode Materials *198*	
6.8	Outlook *199*	
	Acknowledgments *204*	
	References *204*	

7   ALD-Processed Oxides for High-Temperature Fuel Cells *209*
Michel Cassir, Arturo Meléndez-Ceballos, Marie-Hélène Chavanne, Dorra Dallel, and Armelle Ringuedé

7.1	Brief Description of High-Temperature Fuel Cells *209*
7.1.1	Solid Oxide Fuel Cells *209*
7.1.2	Molten Carbonate Fuel Cells *210*
7.2	Thin Layers in SOFC and MCFC Devices *210*
7.2.1	General Features *210*
7.2.2	Interest of ALD *212*
7.3	ALD for SOFC Materials *213*
7.3.1	Electrolytes and Interfaces *213*
7.3.1.1	Zirconia-Based Materials *213*
7.3.1.2	Ceria-Based Materials *214*
7.3.1.3	Gallate Materials *215*
7.3.2	Electrodes and Current Collectors *215*
7.3.2.1	Pt Deposits *215*
7.3.2.2	Anode *216*
7.3.2.3	Cathode *216*
7.4	Coatings for MCFC Cathodes and Bipolar Plates *216*
7.5	Conclusion and Emerging Topics *218*
	References *218*

Part IV   ALD in Photoelectrochemical and Thermoelectric Energy Conversion *223*

8   ALD for Photoelectrochemical Water Splitting *225*
Lionel Santinacci

8.1	Introduction *225*
8.2	Photoelectrochemical Cell: Principle, Materials, and Improvements *227*
8.2.1	Principle of the PEC *227*
8.2.2	Photoelectrode Materials *228*
8.2.2.1	Metal Oxides *229*

8.2.2.2	Elemental and Compound Semiconductors	*229*
8.2.2.3	Nitrides	*230*
8.2.3	Geometry of the Photoelectrodes: Micro- and Nanostructuring	*230*
8.2.4	Coating and Functionalization of the Photoelectrodes	*233*
8.3	Interest of ALD for PEC	*233*
8.3.1	Synthesis of Electrode Materials	*234*
8.3.2	Nanostructured Photoelectrodes	*235*
8.3.3	Catalyst Deposition	*239*
8.3.4	Passivation and Modification of the Junction	*240*
8.3.5	Photocorrosion Protection	*244*
8.3.5.1	Protection of Planar Photoanodes	*244*
8.3.5.2	Protection of Planar Photocathodes	*246*
8.3.5.3	Protection of Nanostructured Photoelectrodes	*246*
8.4	Conclusion and Outlook	*247*
	References	*247*

9	**Atomic Layer Deposition of Thermoelectric Materials**	*259*
	Maarit Karppinen and Antti J. Karttunen	
9.1	Introduction	*259*
9.1.1	Thermoelectric Energy Conversion and Cooling	*259*
9.1.2	Designing and Optimizing Thermoelectric Materials	*260*
9.1.3	Thin-Film Thermoelectric Devices	*262*
9.2	ALD Processes for Thermoelectrics	*263*
9.2.1	Thermoelectric Oxide Thin Films	*263*
9.2.2	Thermoelectric Selenide and Telluride Thin Films	*266*
9.3	Superlattices for Enhanced Thermoelectric Performance	*266*
9.4	Prospects and Future Challenges	*271*
	References	*272*

Index *275*

Preface

Julien Bachmann

Friedrich-Alexander University of Erlangen-Nürnberg, Department of Chemistry and Pharmacy,
Egerlandstrasse 1, 91058 Erlangen, Germany

The Past of Energy Conversion

The ability to harvest energy from the environment and utilize it is a defining characteristic of life. It also represents an activity of central importance to mankind. In fact, several major (pre)historic events are intimately related to the control of novel energy forms. Mastering fire and subsequently the labor of domestic animals are two most crucial achievements of the Paleolithic and the Neolithic, respectively. Much later, craftsmanship became powered by water or wind. Subsequently, the industrial revolution originated from the invention of the heat engine, with coal and steam featured as the most prominent energy carriers (Figure 1). In the twentieth century, they were replaced with petroleum derivatives in the internal combustion engine, used for decentralized energy conversion. Simultaneously, electrical power was established as the most versatile energy form. However, even to date, its generation still mostly relies on coal and steam in a highly centralized infrastructure based on very large energy-converting units (power plants).

The Future of Energy Conversion

This hybrid situation, in which electrical power is generated from fossil fuels and complemented by them, especially for mobile applications, might represent an intermediary stage toward a society in which electricity has become the universal energy carrier and fossil fuels are eventually outdated. One can envision that in a not-too-distant future, mankind will harvest solar power, be it directly or via wind, water, and biomass (all of which originate from it) into electrical current without the intermediacy of any heat engine. Given that mankind consumes approximately 18 TW (18×10^{12} J s^{-1}, or roughly the equivalent of 18 000 large "traditional" power plants) whereas sunlight provides 120 000 TW

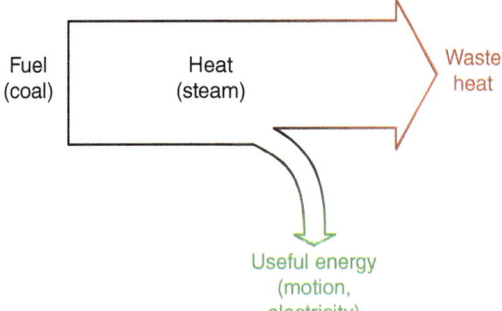

Figure 1 Diagram illustrating the principle of energy harvesting from fossil fuels and highlighting the massive losses inherent to the heat engine.

(120×10^{15} J s^{-1}), capturing only a small fraction of the incoming solar energy would suffice [1].

Of course, this transformation of our energy economy can only be gradual, and it is unlikely to be driven significantly by a shortage of fossil fuels and the corresponding increase of their costs. Transitioning toward renewable energy sources, however, would present a number of opportunities (Table 1). First and foremost,

Table 1 Contrasting features of economies based on fossil fuels and renewable energy sources.

	Fossil	Renewable
	Nineteenth and twentieth centuries	Twenty-first century?
Primary energy sources	Coal, oil, natural gas (and uranium)[a]	Sunlight (possibly via wind, water, and biomass)
Available amounts of primary sources	Limited[b]	Practically unlimited[b]
Secondary form of energy	Heat (as steam or combustion gases)	Electricity
Storable form of energy	Fuels as refined petroleum derivatives	Fuels in or from (photo)electrochemical devices
Emissions	Greenhouse gas CO_2 [a]	No net CO_2
Major type of energy-converting devices	Mechanical: heat engine	Solid-state, no moving parts
Most typical scaling behavior of power conversion efficiency	Increases with unit size	Decreases slightly when unit size increases
Organization of energy-harvesting and distribution infrastructure	Centralized: large units owned and operated by few large companies	Distributed: small units owned and operated by companies and individuals

a) Nuclear fission power relies on a fossil, nonrenewable source (fissile material such as ^{235}U) but does not contribute to greenhouse gas emissions. Nuclear fusion, if or when technically realized, would represent a practically unlimited energy source.
b) Fossil fuel reserves may last for some decades, centuries, or perhaps millennia. Solar power will be available for billions of years.

it would halt anthropogenic emissions of greenhouse gases (or, at least, cut down on them very significantly) and thereby avoid dramatic climate changes with all their social and geopolitical consequences. Second, it would eliminate the heat engine, another relic of the nineteenth century, the efficiency of which is limited by thermodynamics independently of all engineering feats. Third, it would put energy harvesting at the (financial and technical) reach of individual private persons and small corporations. This possible decentralization of the whole energy infrastructure would reverse the trend followed in the past centuries and would represent a formidable empowerment of the individual citizens [2].

Technical Ingredients Needed

The vision of a fully renewable, decentralized energy economy can only be realized if technologies are available for the inexpensive and efficient conversion of energy between its various forms (Figure 2). Nowadays, electricity can be harvested from sunlight in photovoltaic devices (solar cells) with efficiencies of ~20% in mass-market products and up to 46% in the current record-holding laboratory device [3]; thus, photovoltaics have long beaten the efficiency numbers theoretically possible with heat engines. Electricity, however, is much more difficult to store and transport compared to fuels. Therefore, the intermittency of energy harvesting inherent to renewable sources (which are tributary of weather conditions) also renders efficient schemes for the reversible interconversion of electrical and chemical energy forms (electrical current and fuels) just as crucial. This is the role of electrochemical devices such as rechargeable batteries, electrolyzers, and fuel cells. Optionally, the direct conversion of sunlight into fuels in photoelectrochemical cells (it can also be called artificial photosynthesis) represents an

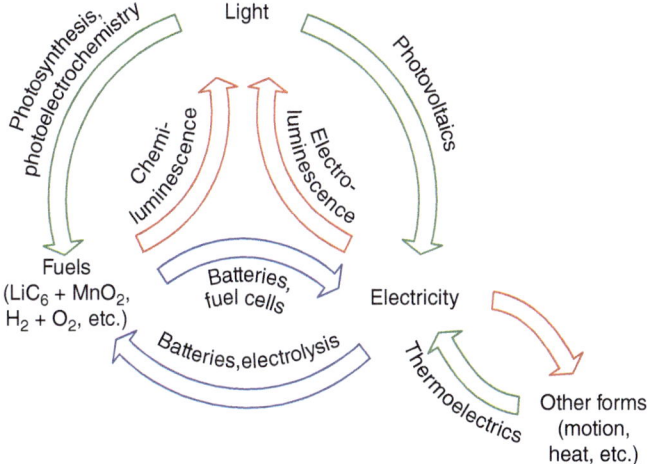

Figure 2 Diagram illustrating energy harvesting (green), storage (blue), and exploitation (red) in a renewable energy economy. The forms of energy harvesting and storage displayed are treated in this book.

integrated solution of simultaneous harvesting and storage. Finally, the direct conversion of heat into electrical power in thermoelectric devices may contribute to improving the overall energy efficiency of industrial processes by recovering a fraction of the waste heat that is otherwise lost and by enabling the operation of low-power devices off-grid.

No single technology can possibly provide a general solution. Instead, each specific application defines specific requirements and calls for an individual solution best suited to it. For example, different types of electrochemical storage chemistries may be optimized with respect to maximizing energy density (for sustained, regular delivery over an extended period) or power density (for occasional but intense usage). Some are of particular interest in terms of volumetric density (compact), others in terms of gravimetric density (lightweight). The technical performance of energy-converting devices cannot be defined by a single parameter and must be considered in the context of other parameters of economic or practical nature. For example, the relative relevance of purchasing cost, maintenance costs, reliability, and longevity may be very different, depending on whether a certain storage device is used in consumer products, mobility, health care, or space applications. Correspondingly, the future of renewable energy will be diverse.

Scope of This Book

Accordingly, this book presents selected types of devices serving to interconvert the solar, electrical, chemical, and heat forms of energy in nonmechanical devices. Thus, wind, hydroelectric power, geothermal power, solar thermal conversion, heat pumps, and other approaches based on moving parts and "classical" engineering will not be treated. Instead, we focus on approaches based on solids and their interfaces, in which atomic layer deposition, a thin-film coating technique, is of relevance. We also exclude the generation of light from electricity (a scientifically challenging topic of industrial importance in lighting and displays but not directly involved in renewables), of heat from electricity (since resistive heating presents no fundamental scientific hurdle), and of heat from fuels (given that combustion is well suited to it and well established).

All other types of energy conversion rely on the transport of charge carriers (mostly electrons, but also ions) inside condensed phases and their transfer across interfaces (which may be associated with rearrangements in chemical bonding). The throughput of energy conversion (the power density) is often determined by the rate at which charge carriers are exchanged at the interface between two solids or between a solid and a liquid phase. In that case, increasing the geometric area of the interface while maintaining short transport distances toward each point of the interface is possible via a nanostructured interface (Figure 3). Nanostructuring presents the advantage that it allows for a systematic approach to performance engineering by adjusting the geometric parameters of the structures to the bulk charge transport and interface charge transfer kinetics of the materials involved. It usually implies the additional difficulty, however, that the very thin (<2 µm,

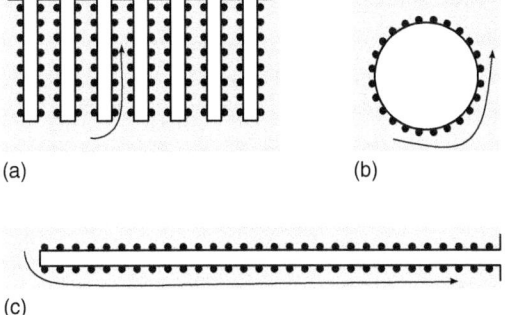

Figure 3 Diagram illustrating the inherent relationship between nanostructuring and the optimization of exchange phenomena as they occur in energy conversion. (a) A nanostructured interface between two phases allows one to engineer a large volumetric density of exchange sites (black dots) at the interface while maintaining short transport distances within each phase (arrow). (b) Bulk phases exhibit lower exchange site densities. (c) The nanostructures cannot be elongated beyond a certain length, because transport distances increase correspondingly and may become limiting.

often <20 nm, sometimes <2 nm) solid layers mostly involved at the interface(s) must be deposited in a homogeneous manner despite the complex geometry. The ability to perform this is a unique feature of atomic layer deposition (ALD, Chapter 1), which is one reason why the method is intimately related to modern energy conversion. Another advantageous feature of ALD in photovoltaic, electrochemical, photoelectrochemical, and thermoelectric devices is its outstanding thickness control in the subnanometer range, which is often required for engineering the interfaces of semiconductors.

Photovoltaics: Strategies, Length Scales, and ALD

The various types of solar cells are often categorized in "generations" (Figure 4). The first utilizes homojunctions between thick (>100 μm), planar crystalline silicon layers. Second-generation cells are also often referred to as the "thin layer cells" and base on inorganic direct-band-gap semiconductors combined as thin (<1 μm) films in a planar manner. A number of distinct approaches are gathered under the denomination "third generation" (dye-sensitized, extremely thin absorber, excitonic), which share a common feature: the necessity to exploit nanostructured interfaces.

Crystalline silicon photovoltaics make up the bulk of commercial solar cells and base on an indirect-band-gap semiconductor, which must be thick (>100 μm) in order to absorb enough light. Minimizing transport losses in those thick layers, therefore, necessitates a material of extreme purity and crystallinity. This is rendered possible by the outstanding level of technological control gained on silicon in the information technology industry. In this type of cells, in which the fundamental semiconductor junction is generated inside a bulk piece of solid, ALD emerges as a most promising method for the passivation of surface traps (Chapter 2). If III–V semiconductors are utilized instead of Si, the level of control

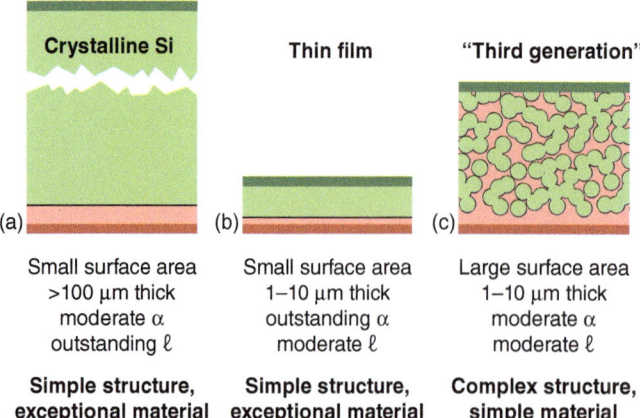

Figure 4 Diagram illustrating the morphology (in cross section) of the main three "generations" of solar cells and the corresponding requirements on material properties (light absorption coefficient α and carrier diffusion length ℓ) and interface geometry. The requirements on the materials are relaxed in the third-generation approaches by the large specific surface area.

possible by band-gap engineering makes it possible to assemble multijunction solar cells of very large efficiencies for performance applications [4].

Working with amorphous silicon relaxes the constraints associated with the indirect band gap of crystalline silicon. The layers can therefore be made thinner (on the order of 1 µm), which renders plasma-enhanced chemical vapor deposition (CVD) amenable to the generation of the active p–n junction layer structure. The pattern of impurities and thereby the electronic properties of the material can be adjusted by the deposition conditions [5, 6]. However, the very nature of the doping agents, prominently hydrogen, limits the stability of the devices and has represented a significant hurdle toward commercial applications. Alternatively, a heterojunction can be generated between a p-type absorber (a couple of micrometer thick) and a very thin (<100 nm) n-doped so-called buffer layer. The latter consists mostly of CdS, and the former of $Cu(In,Ga)(S,Se)_2$. The whole stack is usually combined using thermal evaporation, sputter coating, and complementary treatments with reactive gases. This technology represents a significant market share but cannot be scaled indefinitely due to the presence of elements which may be associated with availability and safety issues. Although ALD has not integrated the thin-film photovoltaics mainstream (yet), recent results have suggested that it may become relevant to the use of alternative materials such as p-type $Cu(Zn,Sn)S_2$ and n-doped ZnO [7, 8].

A novel paradigm was introduced when the possibility was demonstrated that molecules be used as the light-absorbing material in so-called dye-sensitized solar cells [9]. A molecular monolayer must be adsorbed on a solid wide-band-gap semiconductor surface (originally and still mostly $n\text{-}TiO_2$) in a way that ensures efficient charge transfer from the excited state. Because of the limited absorption cross section of such a single monolayer, the semiconductor surface must be nanostructured, typically in the form of a dried colloidal suspension, in order to maximize light absorption. The p side of the junction would initially consist of an electrolytic solution such as I^-/I_3^-, but it can also be realized with a polymer. Of course, maximizing light absorption and charge separation while minimizing recombination losses can be performed in part by adjusting the geometric parameters of the cell, but ALD furnishes the ideal way of introducing ultrathin (<1 nm) tunnel barriers to achieve that goal as well (Chapter 4).

In an all-organic variant (in which ALD is not relevant) of the third-generation cells, the large exciton binding energy forces one to follow a "bulk heterojunction" strategy in which both immiscible organic phases must be intertwined with a characteristic lateral length on the order of 10 nm [10]. Alternatively, an all-solid, all-inorganic type of nanostructured cell also exists, the extremely thin absorber (ETA) cell, where the intrinsic absorber of a p–i–n junction can be made thin, thanks to the enhanced geometric interfacial area in the nonplanar geometry [11–13]. Thus, absorbers with relatively poor transport properties can be considered. If the geometry is perfectly interdigitated, the ETA cell enables the experimentalist to adjust the length and width of the elongated structures to the light absorption coefficient and carrier transport lengths. In this system, ALD is ideally suited to depositing the thin (~10 nm) light-absorbing layer (Chapter 3). The most dramatic recent development in this field is the discovery that ammonium plumbate perovskites such as the prototypical $H_3CNH_3PbI_3$ (which are not accessible by ALD to date) combine excellent light absorption and charge transport properties [14, 15]. Currently, their stability and unclear photophysics are their most stringent limitations.

The current state of the art in the various cell types, as updated by the National Renewable Energy Laboratory, is summarized in Figure 5.

Electrochemical Energy Storage: Principles, Chemistries, and ALD

Electrical energy can be stored by driving an endergonic chemical reaction, which can later be reversed for release. In the prototypical example, water is electrolyzed to dihydrogen and dioxygen, and the reverse reaction drives a fuel cell (the device capable of running the reaction in both directions would be called a regenerative fuel cell). However, the disproportionation of lead(II) to lead(0) and lead(IV) serves the same purpose in an old-fashioned lead battery. Thus, similar principles underline very different kinds of chemical reactivity in batteries and fuel cells. The performance parameters are usually power density and energy density, both volumetric and gravimetric (in addition to practical constraints such as cost, safety, and longevity). In this respect, electrolyzers and fuel cells, which generate and consume a chemical reducing agent that can be isolated in pure form (the fuel), whereas the oxidant is usually dioxygen in the air, have two major design advantages. Firstly, they gain in energy and power densities by circumventing the local storage (and transport) of the oxidant. Secondly, they allow the engineer to define separately energy density (set by the fuel tank size) and power density (which depends on the electrode area). Such flexibility is not offered by batteries, in which electrodes consist of the oxidants and reductants themselves in a

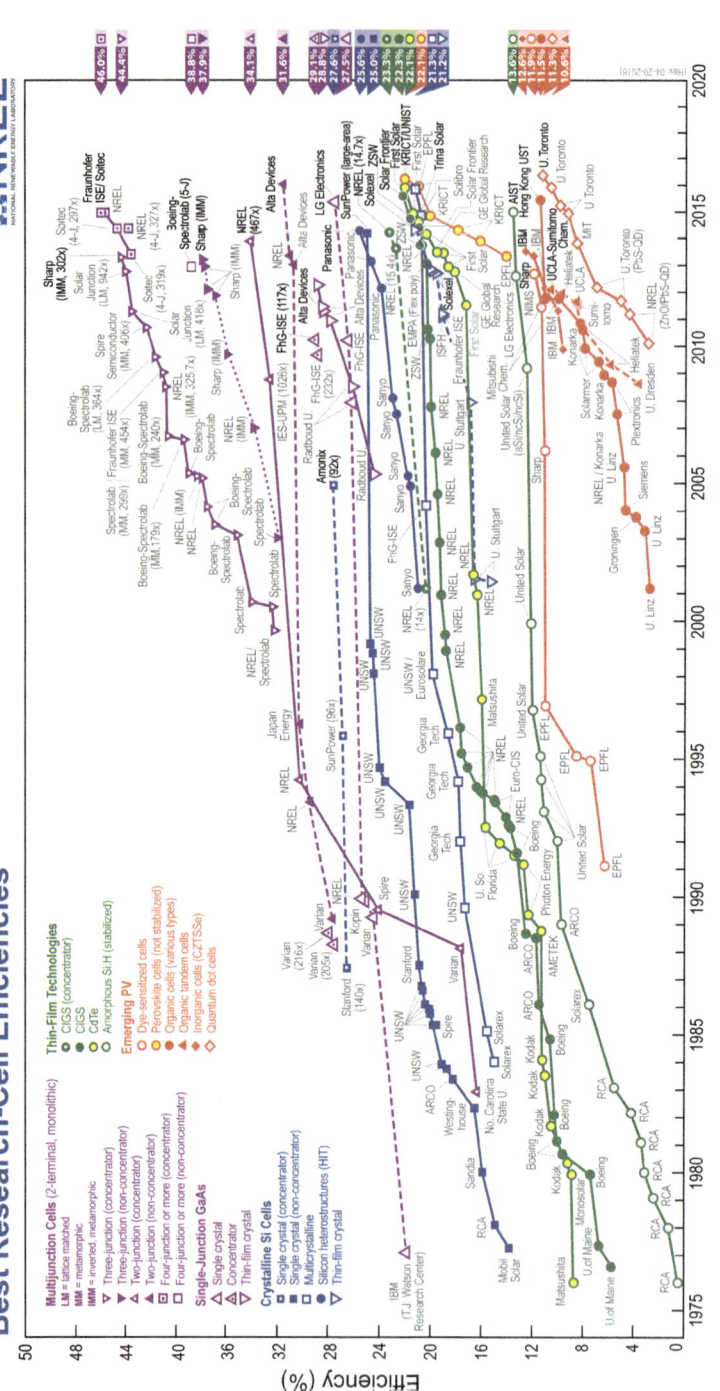

Figure 5 Diagram of the historical development of solar cell efficiency records in the different categories of cells, as compiled by the NREL [3]. Copyright 2016, The National Renewable Energy Laboratory, Golden, CO.

completely closed system. Batteries do, however, simplify operations in that they need not be catalyzed: electron transfers occur as limited only by the transport of ions in the electrolyte and the electrodes. Since the formation of molecular fuels implies the rearrangement of chemical bonds and typically involves several electrons and several ions and/or molecules, the surface chemistry of electrocatalysts is of primordial importance.

The various types of batteries differ in the nature of ions being exchanged between the electrodes. In the most contemporary family of rechargeable batteries, one of the lightest elements, lithium, inserts into either a transition metal oxide or graphite during discharge and charge, respectively (Chapter 6). The miniaturization of devices necessary for mitigating transport losses and optimizing power density renders ALD attractive in battery research as well (albeit not in commercial applications to date). In fact, materials can be deposited as thin layers by ALD not only for the negative and positive electrodes but also for some (solid) ion-exchange separators. One challenge to the application of ALD in lithium ion batteries has been the volatility and reactivity of molecular metal–organic lithium precursors.

In the electrolyzer and fuel cell realm, significant effort is dedicated to the deposition of catalytically active particles based on noble metals by ALD (Chapter 5) for alkaline fuel cells (AFC), polymer electrolyte fuel cells (PEFC) and methanol fuel cells (MFC). Given the high costs of the elements (Ru, Ir, Pd, Pt) that best catalyze the electrochemical transformations of dioxygen/water, dihydrogen/water, methanol/carbon dioxide, as well as some other related reactions, their most efficient usage by minimized loadings would represent the best possible drive of cost reduction, achievable thanks to the outstanding control possible in ALD. An additional possible application of the technique is found in solid oxide fuel cells (SOFC, Chapter 7), namely, the ion-conducting layer. Indeed, these large, stationary cells operated at elevated temperatures circumvent the need of noble metal catalysts, replaced by ceramics. The electrolyte separating them must be able to exchange and transport oxide ions and consists of a ternary oxidic ceramic. Electrode and electrolyte materials can all be deposited as films by ALD, with quite drastic requirements on stoichiometry and crystallinity, which define their chemical and physical properties (ionic conductivity, electronic conductivity, thermal stability).

Other Energy Conversion Strategies Based on Interfaces

As mentioned earlier, energy conversion (from sunlight) and storage (in chemical form) can be combined into a single, photoelectrochemical, device (Chapter 8). Of course, the challenge stems from the combination of requirements from the photovoltaic and electrochemical fields, quite stringent individually, which must all be met simultaneously for efficient photoelectrochemistry to be mastered. Photon energies, band-edge positions of the semiconductors, thermodynamics of the fuels, and performance of the catalysts are all intertwined. However, these challenges render the ALD technique, applicable to a variety of functional materials and to nonplanar substrates, all the more crucial. The success of the dye-sensitized solar cell, which in essence uses half of a photoelectrochemical device (albeit without generating a practically storable fuel), also proves that the principle is tenable.

Our closing Chapter 9 focuses on an approach to renewable energy harvesting that runs in parallel to all those mentioned above and quite independently of them: thermoelectrics. The ability of using heat to generate electrical energy presents an extreme appeal in at least two distinct respects. Firstly, it can power distributed devices off-grid and without moving parts, as has been done in a number of (rather small-volume) applications for decades. Secondly, it can exploit a form of energy that is often considered as waste, albeit with low efficiency. The efficiency can be improved if the temperature difference between warm and cold reservoirs is large, but many typical thermoelectric materials have limited thermal stability. Thus, future advances in thermoelectric energy harvesting might be related to the development of novel nanostructures based on materials which give rise to limited thermoelectric effects individually but which are particularly suitable in combination. The length scale of the structures considered is thereby of prime importance, since they must scatter phonons efficiently without introducing any hurdles to electron transport [16, 17]. Here again, ALD has emerged as a method of choice for generating such structures, although the corresponding chemistry has proven challenging.

References

1 U.S. Energy Information Administration (2016) *International Energy Outlook 2016*: U.S. Department of Energy, www.eia.gov/forecasts/ieo (accessed 23 November 2016).
2 Armaroli, N. and Balzani, V. (2007) *Angew. Chem. Int. Ed.*, **46**, 52–66.
3 NREL *Data courtesy of the National Renewable Energy Laboratory*, Golden, CO: U.S. Department of Energy 2016, www.nrel.gov/ncpv/images/efficiency_chart.jpg (accessed 23 November 2016).
4 King, R.R., Law, D.C., Edmondson, K.M., Fetzer, C.M., Kinsey, G.S., Yoon, H., Sherif, R.A., and Karam, N.H. (2007) *Appl. Phys. Lett.*, **90**, 183516.
5 Rech, B. and Wagner, H. (1999) *Appl. Phys. A*, **69**, 155–167.
6 Terakawa, A. (2013) *Sol. Energy Mater. Sol. Cells*, **119**, 204–208.
7 Mughal, M.A., Engelken, R., and Sharma, R. (2015) *Sol. Energy*, **120**, 131–146.
8 Naghavi, N., Abou-Ras, D., Allsop, N., Barreau, N., Bucheler, S., Ennaoui, A., Fischer, C.-H., Guillen, C., Hariskos, D., Herrero, J., Klenk, R., Kushiya, K., Lincot, D., Menner, R., Nakada, T., Platzer-Bjorkman, C., Spiering, S., Tiwari, A.N., and Torndahl, T. (2010) *Progr. Photovoltaics*, **18**, 411–433.
9 Grätzel, M. (2004) *J. Photochem. Photobiol., A*, **164**, 3–14.
10 Benten, H., Mori, D., Ohkita, H., and Ito, S. (2016) *J. Mater. Chem. A*, **4**, 5340–5365.
11 Tennakone, K., Kumara, G.R.R.A., Kottegoda, I.R.M., Perera, V.P.S., and Aponsu, G.M.L.P. (1998) *J. Phys. D: Appl. Phys.*, **31**, 2326–2330.
12 Kaiser, I., Ernst, K., Fischer, C.-H., Könenkamp, R., Rost, C., Sieber, I., and Lux-Steiner, M.C. (2001) *Sol. Energy Mater. Sol. Cells*, **67**, 89–96.

13 Hodes, G. and Cahen, D. (2012) *Acc. Chem. Res.*, **45**, 705–713.
14 Sum, T.C. and Mathews, N. (2014) *Energy Environ. Sci.*, **7**, 2518–2534.
15 Boix, P.P., Nonomura, K., Mathews, N., and Mhaisalkar, S.G. (2014) *Mater. Today*, **17**, 16–23.
16 Nielsch, K., Bachmann, J., Kimling, J., and Böttner, H. (2011) *Adv. Energy Mater.*, **1**, 713–731.
17 Chen, Z.-G., Han, G., Yang, L., Cheng, L., and Zou, J. (2012) *Progr. Nat. Sci. Mater. Int.*, **22**, 535–554.

List of Contributors

Michel Cassir
PSL Research University
Chimie ParisTech – CNRS
Institut de Recherche de Chimie Paris
Paris Cedex 05
75005 Paris
France

Marie-Hélène Chavanne
PSL Research University
Chimie ParisTech – CNRS
Institut de Recherche de Chimie Paris
Paris Cedex 05
75005 Paris
France

Dorra Dallel
PSL Research University
Chimie ParisTech – CNRS
Institut de Recherche de Chimie Paris
Paris Cedex 05
75005 Paris
France

Jolien Dendooven
Ghent University
CoCooN Group, Department of
Solid-state Sciences
Krijgslaan 281
9000 Ghent
Belgium

Christophe Detavernier
Ghent University
CoCooN Group
Department of Solid-State Sciences
Krijgslaan 281
9000 Ghent
Belgium

Helmer Fjellvåg
University of Oslo
Department of Chemistry
Centre for Materials Science and
Nanotechnology (SMN)
P.O. Box 1033, Blindern
0315 Oslo
Norway

Knut B. Gandrud
University of Oslo
Department of Chemistry
Centre for Materials Science and
Nanotechnology (SMN)
P.O. Box 1033, Blindern
0315 Oslo
Norway

Carlos Guerra-Nuñez
EMPA, Swiss Federal Laboratories for
Materials Science and Technology
Laboratory for Mechanics of
Materials and Nanostructures
Feuerwerkerstrasse 39
CH-3602 Thun
Switzerland

List of Contributors

Maarit Karppinen
Aalto University
Department of Chemistry and
Materials Science
Kemistintie 1
02150 Espoo
Finland

Antti J. Karttunen
Aalto University
Department of Chemistry and
Materials Science
Kemistintie 1
02150 Espoo
Finland

W. M. M. Kessels
Eindhoven University of Technology
Department of Applied Physics
De Rondom 1, Building TNO
5612 AP Eindhoven
The Netherlands

Lifeng Liu
International Iberian Nanotechnology
Laboratory (INL)
Avenida Mestre José Veiga
4715-330 Braga
Portugal

B. Macco
Eindhoven University of Technology
Department of Applied Physics
De Rondom 1, Building TNO
5612 AP Eindhoven
The Netherlands

Alex Martinson
Argonne National Laboratory
9700 South Cass Avenue B109
Lemont, IL 60439
USA

Arturo Meléndez-Ceballos
PSL Research University
Chimie ParisTech – CNRS
Institut de Recherche de Chimie Paris
Paris Cedex 05
75005 Paris
France

Ola Nilsen
University of Oslo
Department of Chemistry
Centre for Materials Science and
Nanotechnology (SMN)
P.O. Box 1033, Blindern
0315 Oslo
Norway

Hyung Gyu Park
ETH Zürich
Nanoscience for Energy Technology
and Sustainability
Department of Mechanical and
Process Engineering
Tannenstrasse 3
CH-8092 Zürich
Switzerland

Armelle Ringuedé
PSL Research University
Chimie ParisTech – CNRS
Institut de Recherche de Chimie Paris
Paris Cedex 05
75005 Paris
France

Amund Ruud
University of Oslo
Department of Chemistry
Centre for Materials Science and
Nanotechnology (SMN)
P.O. Box 1033, Blindern
0315 Oslo
Norway

Lionel Santinacci
Aix Marseille Univ, CNRS
Center for Interdisciplinary
Nanoscience of Marseille
(CINaM, UMR 7325)
Campus de Luminy – Case 913
13288 Marseille
France

Ivo Utke
EMPA, Swiss Federal Laboratories for
Materials Science and Technology
Laboratory for Mechanics of
Materials and Nanostructures
Feuerwerkerstrasse 39
CH-3602 Thun
Switzerland

B. W. H. van de Loo
Eindhoven University of Technology
Department of Applied Physics
De Rondom 1, Building TNO
5612 AP Eindhoven
The Netherlands

Part I

Introduction to Atomic Layer Deposition

1

Basics of Atomic Layer Deposition: Growth Characteristics and Conformality

Jolien Dendooven and Christophe Detavernier

Ghent University, Department of Solid-State Sciences, CoCooN Group, Krijgslaan 281, 9000 Ghent, Belgium

This chapter introduces some fundamentals and key advantages of the atomic layer deposition (ALD) technique. Following the standard example of a typical TMA/H_2O ALD cycle, the essential characteristics of true layer-by-layer growth (linearity, growth per cycle, saturation, and temperature window) are discussed. It is explained how the surface-controlled nature of the reactions ensures atomic-level thickness control and excellent conformality on 3D substrates, and the concept of plasma-enhanced ALD is introduced. In the second part of this chapter, the focus moves to *in situ* characterization methodologies that are often used in ALD research. The advantages of using *in situ* techniques to determine ALD growth characteristics are discussed, and examples of quartz crystal microbalance, quadrupole mass spectroscopy, spectroscopic ellipsometry, and optical emission spectroscopy (OES) are provided. In the third and final part, the conformality of ALD is reviewed. The use of macroscopic and microscopic lateral test structures for quantifying conformality will be addressed, as well as approaches to characterize conformal ALD in nanoporous materials.

1.1 Atomic Layer Deposition

1.1.1 Principle of ALD

ALD is a self-limited film growth method that is characterized by alternating exposure of the growing film to chemical precursors, resulting in the sequential deposition of (sub)monolayers [1, 2]. ALD was invented in the 1970s and further developed in the 1980s for the fabrication of luminescent ZnS and Al_2O_3 insulator films for electroluminescent flat-panel displays. It was only in the 1990s with the decreasing device dimensions and the resulting need for high-k oxides in microelectronics that the ALD technique has become a commercial success. Since then, a wide range of materials have been deposited by ALD including several oxides, nitrides, chalcogenides, and metals [3, 4]. As an example, we discuss the growth of an Al_2O_3 film using ALD. The basic process is illustrated in Figure 1.1. The initial situation is shown in Figure 1.1a, where a SiO_2 substrate,

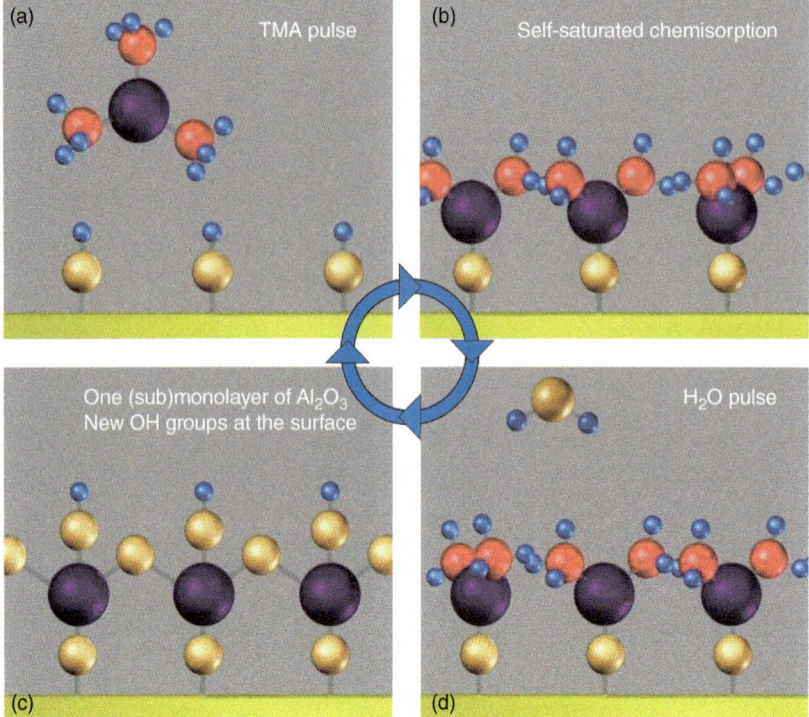

Figure 1.1 Principle of ALD, illustrated by the process for deposition of Al_2O_3 using TMA and H_2O. (Detavernier et al. 2011 [5]. Reproduced with permission of Royal Society of Chemistry.)

which is terminated by OH groups, is exposed to a pulse of trimethylaluminum vapor (TMA, typical exposure time of seconds). The TMA molecules adsorb on all exposed surfaces in the chamber and within pores, holes, and so on, in the sample. This deposition is self-limited, since the TMA molecules are chemisorbed through reaction with OH groups on the surface. Once all accessible OH groups have been consumed, no more TMA will adsorb (Figure 1.1b). The TMA pulse is followed by an evacuation of the reaction chamber through purging or pumping, where after a pulse of the reactant, for example, water vapor is introduced (Figure 1.1d). The water vapor reacts with the adsorbed TMA and hydrolyzes the residual methyl groups. This surface reaction results in the formation of a (sub)monolayer of alumina. In Figure 1.1c, one notices the presence of OH groups terminating the first alumina layer. Therefore, the ALD process can be repeated over and over again to deposit films, one (usually fractional) atomic layer at a time.

When compared to other film deposition techniques such as chemical vapor deposition (CVD), physical vapor deposition (PVD, i.e., evaporation or sputter deposition), or electrochemical deposition, ALD offers several advantages. The key advantage is the ability to deposit conformally into high aspect ratio (AR) structures. Other advantages include (i) control of the layer thickness at the Angstrom level (the limited deposition rate that is regarded as a disadvantage of ALD is nowadays considered a unique advantage because it allows for the deposition of ultrathin films (e.g., <10 nm)); (ii) chemical selectivity, enabling area-selective ALD where deposition only occurs in those regions of the surface where reactive surface sites are present [6–13]; and (iii) industrial scalability. Indeed, because ALD is based on the exposure of a surface to a precursor vapor and, therefore, is nondirectional, one can design batch reactors, where many substrates can be coated simultaneously [14, 15]. In addition, large area processing based on spatially separated reagent flows is being actively explored today, especially for photovoltaic and flexible electronic applications [16].

Over the past decades, hundreds of ALD chemistries have been found for depositing a variety of materials. A review paper by Puruunen [3] from 2005 and the more recent update by Miikkulainen *et al.* [4] provide an excellent review of the available process chemistries for deposition of oxides, nitrides, chalcogenides, and metals using ALD (Figure 1.2).

ALD currently is mainly used in the microelectronics industry, for example, for growing high-k gate oxides. Its ability for conformal deposition into high AR features offers the potential for breakthroughs in fields where ultrathin coatings are required on nanostructured, nanoporous, or fibrous substrates. Since the early 2000s, researchers have been exploring ALD as a generic coating technique for a variety of nanostructures, as illustrated in a number of review papers [5, 17–21]. ALD deposition has, for example, been reported for coating anodic alumina [22–24], aerogel [25–28], nano-sized powder [29–34], nanowires [35–38], and fibrous materials [39–42]. Potential application fields include catalysis, gas separation, sensors, batteries, capacitors, fuel cells, photovoltaics, and photonics.

1.1.2 ALD Growth Characteristics – Linearity, Saturation, and ALD Window

The ideal ALD process is characterized by a linear increase of the amount of deposited material as a function of the number of ALD cycles. One usually defines the "growth per cycle" or GPC as the amount of material deposited (or the equivalent thickness increase) per ALD cycle. While GPC is a practical concept that is often used in everyday communication in the laboratory as well as in the literature, it is important to realize that the "apparent" GPC value does not reflect in any way the chemical reaction kinetics during the deposition process but is determined by the number of chemisorption sites on the growth surface, which will depend on the reactivity and number of accessible surface sites and even on surface morphology. As illustrated in Figure 1.3, one often observes that the apparent GPC is substrate dependent at the start of the ALD process and that it takes a certain number of cycles before a steady-state GPC value is obtained. This is caused by the fact that the chemical sites on the original substrate can have a different reactivity compared to the chemical sites on the surface of the as-grown material. While substrate inhibition and the resulting delay in film nucleation/growth may be considered a disadvantage at first sight, this effect can in fact be very beneficial when one is targeting area selective ALD. The data for Pt ALD shown in Figure 1.3 illustrates that the "apparent" GPC can strongly depend on the surface morphology. Indeed, ALD growth on a rough surface will have an apparently higher GPC, as more material can be deposited per cycle in view of the larger effective surface area that is available for growth.

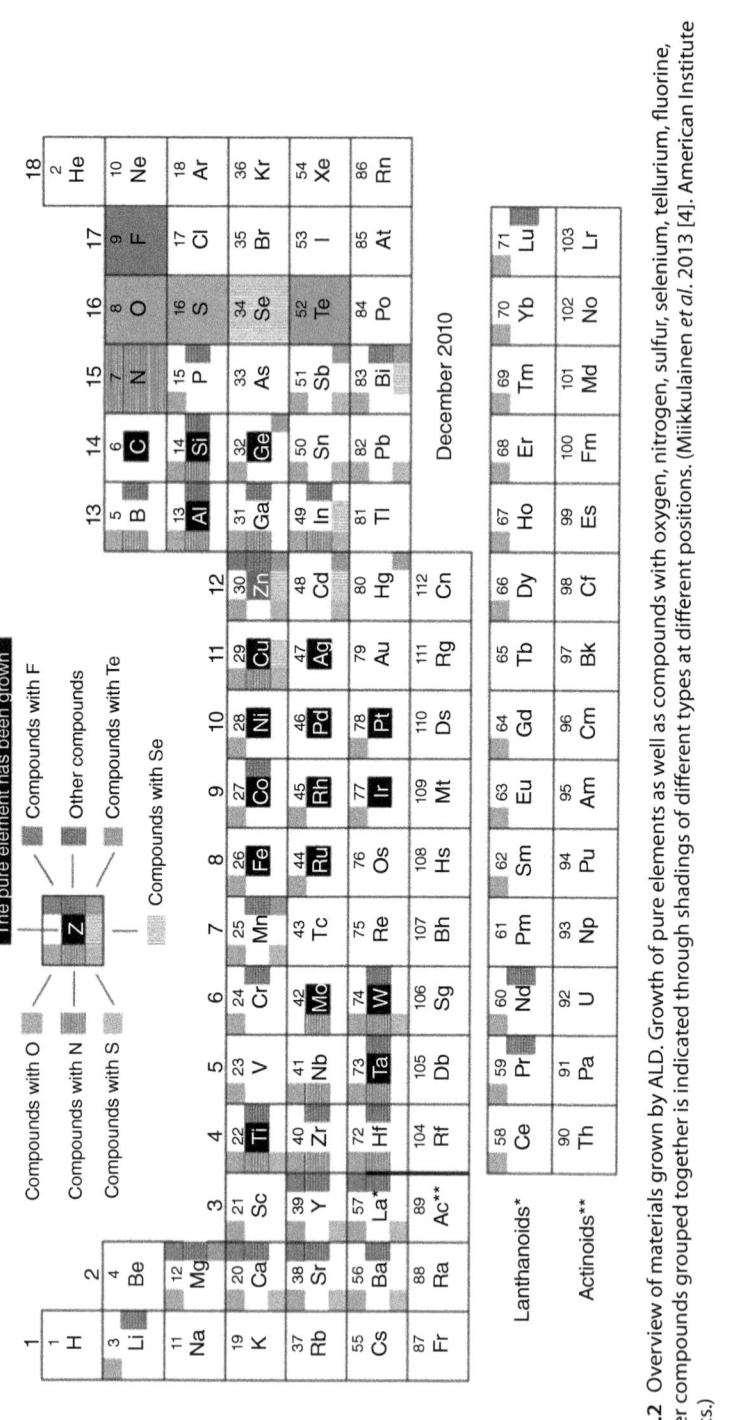

Figure 1.2 Overview of materials grown by ALD. Growth of pure elements as well as compounds with oxygen, nitrogen, sulfur, selenium, tellurium, fluorine, and other compounds grouped together is indicated through shadings of different types at different positions. (Miikkulainen et al. 2013 [4]. American Institute of Physics.)

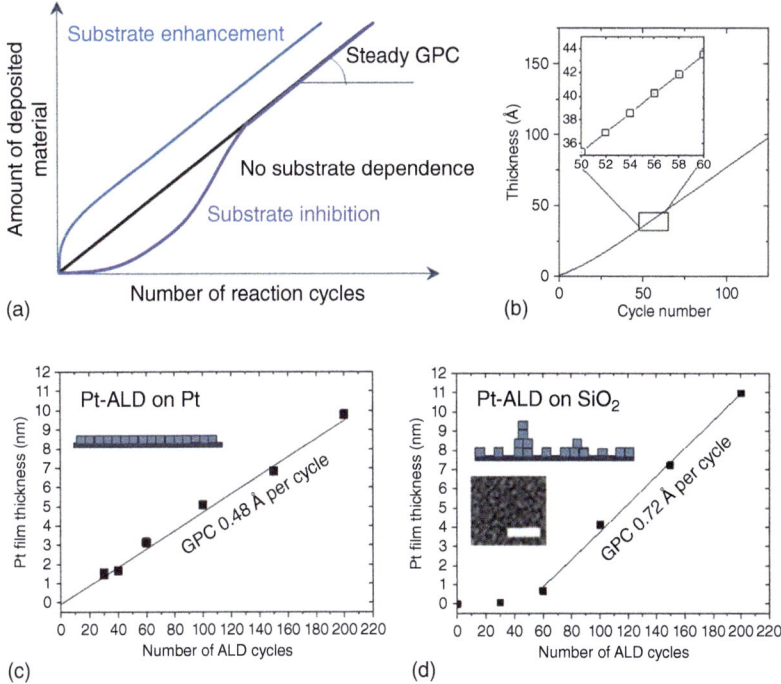

Figure 1.3 Linear increase of the amount of deposited material as a function of the number of ALD cycles: principle (a: [43]), *in situ* ellipsometry data for Al_2O_3 ALD (b), and film thickness data for a $MeCpPtMe_3/O_3$ process at 150 °C with 2D growth (on a sputtered Pt surface (c)) versus island-type growth on a SiO_2 surface (d) [44], illustrating that GPC values should be interpreted as "apparent" values that depend on the surface conditions (the scale bar on the electron microscopy image is 100 nm). (Reproduced with kind permission of Annelies Delabie.)

The self-saturating nature of the surface reactions occurring during both half cycles can be considered as the defining characteristic of ALD. When developing novel ALD processes, demonstrating "saturation" is a key goal. Saturation curves are typically achieved by running several deposition experiments at the same substrate temperature, while varying the exposure dose during one of the half cycles. The GPC is then plotted as a function of the exposure dose (most often as a function of exposure time), as illustrated in Figure 1.4 for the TMA/H_2O process.

Finally, the "ALD window" is defined as the temperature range in which saturated growth conditions prevail (Fig. 1.5a). When the temperature is too low, one often observes that growth is no longer possible, because the thermal energy is insufficient to drive the surface chemistry. In some cases, an apparently faster growth is observed, which is often related to the occurrence of physisorption instead of the self-saturating chemisorption that is desirable. At high temperatures, higher growth rates are often observed, which are caused by thermal decomposition of the precursor on the hot surface (essentially resulting in CVD-type growth). In some cases, the growth rate is observed to decrease at high temperature because of thermal desorption of chemisorbed species that

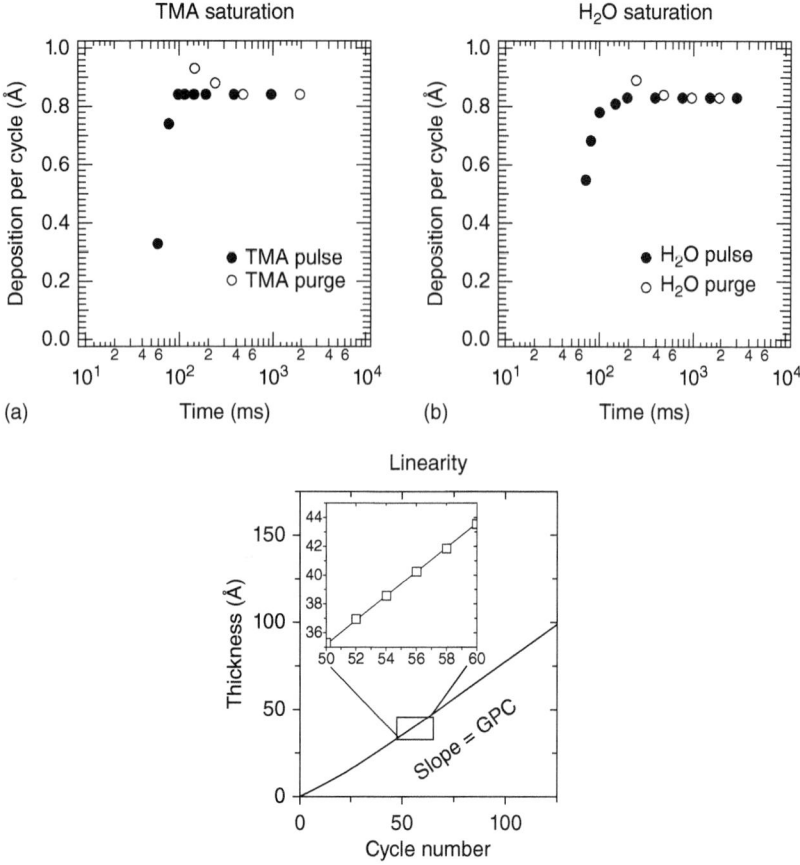

Figure 1.4 Saturation during the TMA and H_2O exposures for ALD of Al_2O_3. (Puurunen 2005 [3]. Reproduced with permission of American Institute of Physics.)

are required for growth. It is important to note that the GPC is not necessarily constant within the ALD window. Indeed, even for the TMA/H_2O process, which exhibits saturated growth from 180 to 380 °C [45], the GPC decreases for increasing temperature, as illustrated in Figure 1.5b. Although self-saturating chemisorption is achieved across the entire ALD window, the GPC decreases because of dehydroxylation of the growth surface at higher temperatures.

1.1.3 Plasma-Enhanced ALD

In ALD processes such as the described TMA/H_2O process, the activation energy required for the surface reactions is solely provided by heating the sample. Therefore, these processes are called thermal ALD processes. Alternatively, the reactant can be preactivated by means of a plasma source (plasma-enhanced ALD or PE-ALD) [46]. A plasma is a gaseous mixture of neutral and charged particles that is macroscopically neutral. Most materials processing plasmas are generated

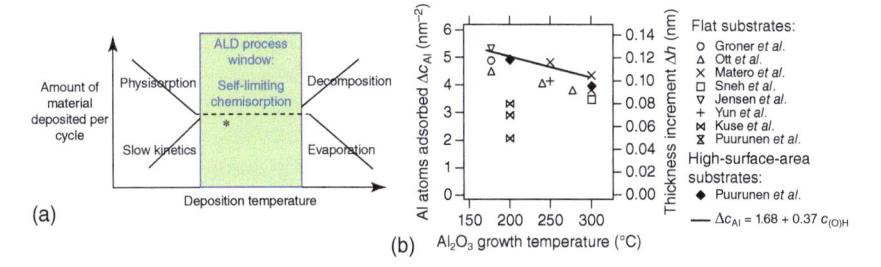

Figure 1.5 Effect of substrate temperature on ALD: (a) schematic [43] and (b) a compilation of data for the TMA/H$_2$O process by Puruunen [3], illustrating that the GPC may vary as a function of temperature within the "ALD window." (Puruunen 2005 [3]. Reproduced with permission of American Institute of Physics.)

by a strong electric field at low gas pressures (<10 Torr). Any electrons present in the gas are accelerated to high kinetic energies. When these electrons collide with the atoms and/or molecules of the background gas, they are able to ionize, excite, or dissociate these gas species, thus yielding electrons, ions, reactive atomic and/or molecular neutrals (radicals), and photons. The created ions and electrons are in turn accelerated by the applied field. Because of the large difference in mass, the electrons will gain more kinetic energy compared to the ions, thus leading to "hot" electrons with an average temperature of several 10^4 K (several electron volts), while the other gas species remain approximately at reactor temperature (300–500 K). Consequently, low-pressure plasmas are not in thermal equilibrium. The degree of ionization, that is, the fraction of ionized particles in the plasma, is typically in the range 10^{-6} to 10^{-3}.

1.1.3.1 Plasma Configurations for Plasma-Enhanced ALD

Plasma-enhanced ALD (PE-ALD) uses the species generated in a plasma as reactants. Mainly O$_2$, N$_2$, NH$_3$, and H$_2$ plasmas (or mixtures thereof) have been used for the growth of oxides, nitrides, and metal films [46]. As mentioned before, plasmas produce energetic ions and reactive radicals, for example, O$_2$ can dissociate in two O radicals. Because the degree of ionization is rather low in typical PE-ALD plasmas, the radicals are expected to play the key role in the surface reactions. Nevertheless, ions that arrive at the sample surface can provide additional energy to the surface, leading to physical changes in the sample, such as smoothening or densification. However, they can also be implanted in the growing film or substrate, which is often unwanted or induce defect creation. Because radicals have a longer lifetime compared to charged particles, they are less confined to the plasma discharge region. Therefore, the level of ion bombardment can be controlled by the plasma configuration.

The different types of plasma configurations used for PE-ALD are schematically depicted in Figure 1.6. In the first configuration, the plasma is formed in a cavity, which is separated from the deposition chamber. The ions and electrons recombine during transport to the ALD reaction zone, and only the longer lived radicals can reach the substrate. Therefore, this configuration is called "radical-enhanced" or "radical-assisted" ALD. Its main advantage is that ion-induced damage of the film or substrate is completely avoided. However, also the flux of reactive radicals will be reduced compared to plasma configurations where the plasma is in contact with the sample. This is clearly the case in a "direct plasma" configuration. Here, a capacitively coupled radio-frequency (RF) generator powers one electrode, while the substrate is placed on the second, generally grounded, electrode. The sample is in close contact with the plasma and is thus exposed to high fluxes of radicals and ions. Consequently, uniform coatings can be achieved with short plasma exposures, but, depending on the processing conditions, severe ion bombardment can be an issue. In the "remote plasma" configuration, the plasma source is placed upstream of the substrate. Very often, an inductively coupled plasma is generated in a glass or quartz tube surrounded by a radio-frequency (RF) coil. For a remote O_2 plasma, ion fluxes in the range 10^{12}–10^{14}/cm^{2}s have been measured at the sample stage, with ion energies below 35 eV. These values are considered low enough to not create substantial film or substrate damage. On the other hand, however, it was shown that the UV photons (9.5 eV) created in the plasma can induce electrical defects [47]. The fourth plasma configuration can be considered as a direct plasma operating in a remote configuration by placing a grid in between the top electrode and the sample [48]. Because this grid acts as the bottom electrode, the sample stage is no longer involved in the plasma generation. As such, significant ion bombardment of the substrate is avoided.

1.1.3.2 Reactions in Plasma-Enhanced ALD

Most reaction mechanism studies of PE-ALD in the literature concern O_2 plasma-based processes, mostly resulting in the growth of oxides [49]. Heil *et al.* studied the reaction products formed during the Al_2O_3 process from trimethylaluminum (TMA) and O_2 plasma by means of *in situ* mass spectroscopy and OES [50]. These measurements revealed the formation of CO, CO_2, and H_2O during the plasma step, which was attributed to the combustion of the methyl ligands (of the adsorbed TMA molecules) on the surface by the O radicals. Using *in situ* infrared spectroscopy, Rai *et al.* showed that these combustion-like reactions produce OH groups and carbonates on the surface during the O_2 plasma step in the PE-ALD of Al_2O_3 [51] and TiO_2 [52]. They furthermore demonstrated that a prolonged exposure of the carbonates to the O_2 plasma decomposes them in CO_2 and CO, meaning that the OH groups are the dominant chemisorption sites in the subsequent precursor step, as in thermal ALD. Pure metallic films are often grown with H_2 plasma as the reducing reactant. Kim *et al.* studied the reaction mechanism underlying the PE-ALD of Ti using $TiCl_4$ and H_2 plasma. They proposed that the reaction proceeds via an Eley–Rideal mechanism: the H radicals react with the adsorbed Cl species from the gas phase to form HCl that then is desorbed from the surface [53].

1.1.3.3 Advantages and Challenges of Plasma-Enhanced ALD

Key advantages of PE-ALD include a higher film density, lower impurity content, better stoichiometry, and improved electronic properties. From a process perspective, the use of radicals enables deposition at lower substrate temperatures, which can be a key advantage when coating polymers. Moreover, PE-ALD

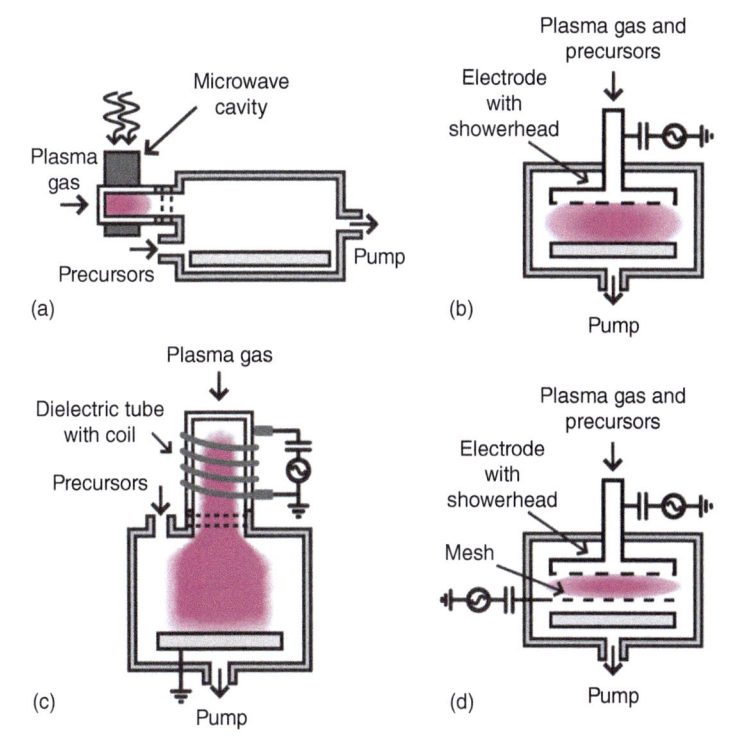

Figure 1.6 Plasma configurations used for PE-ALD. (a) Radical-enhanced ALD. (b) Direct plasma. (c) Remote plasma. (d) Direct plasma in remote configuration. (Profijt et al. 2011 [46]. Reproduced with permission of American Institute of Physics.)

typically enables a slightly higher growth rate, shortening the overall deposition time. More importantly, the use of radicals from a plasma increases the choice of precursors for deposition of a specific coating, for example, the beta-diketonate precursors show low reactivity to water vapor but react readily with oxygen radicals. Some of these benefits of PE-ALD are illustrated in Figure 1.7.

There are, unfortunately, also specific disadvantages to using plasma during an ALD process. Firstly, recombination of the radicals on the sidewalls limits the conformality in high AR structures. Secondly, the ions and UV photons from the plasma may generate specific defects in the growing layer. Thirdly, from a process point of view, the use of a plasma requires more complicated and therefore expensive reactor designs.

1.2 *In Situ* Characterization for Studying ALD Processes

When exploring ALD processes, *in situ* characterization techniques offer the advantage that the ALD process no longer occurs in a "black box" but that the surface chemistry and the properties of the growing film can be monitored in real time. Since ALD is, in essence, surface chemistry, one would ideally like to use, for example, the electron- and ion-spectroscopy techniques that have been

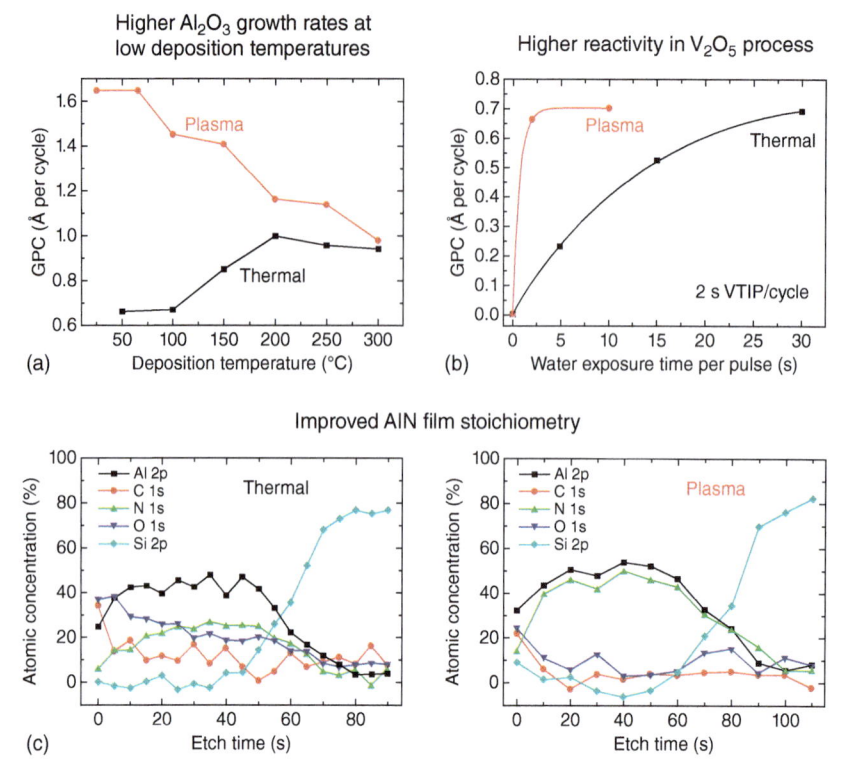

Figure 1.7 Illustration of some of the advantages of (remote) PE-ALD as compared to thermal ALD. (a) Growth per cycle (GPC) as a function of deposition temperature for the growth of Al_2O_3 using 2 s TMA and 5 s H_2O or H_2O plasma. (b) GPC of V_2O_5 as a function of H_2O exposure time for thermal and PE-ALD using vanadyl triisopropoxide (VTIP) as V precursor (data taken from [54]). (c) X-ray photoelectron spectroscopy profiles for AlN films grown by thermal and PE-ALD using 2 s TMA and 5 s NH_3 or NH_3 plasma at 250 °C. The plasma power was in all cases 300 W.

developed by surface scientists. Unfortunately, it is very difficult to implement these techniques truly "*in situ*" on an ALD reactor, since the ultrahigh vacuum that is typically required by these techniques is not compatible with ALD process conditions.[1] Over the past decade, a variety of *in situ* techniques have been developed, and several have become available on commercial ALD reactors.

1.2.1 Quartz Crystal Microbalance

Quartz crystal microbalance (QCM) is a well-known technique to monitor thin-film deposition [55, 56]. As material is deposited onto an oscillating

---

1 X-ray photoelectron spectroscopy, low-energy ion spectroscopy, and scanning tunneling microscopy have been implemented as "*in vacuo*" techniques, where ALD is performed in a dedicated reactor and the sample is moved from the ALD reactor under vacuum conditions (i.e., without exposure to ambient air) into a ultrahigh vacuum (UHV) chamber for characterization. Evidently, this can provide a wealth of useful data on the surface reactions during ALD. However, because of the need for moving the sample, these techniques are not discussed here, as we have limited the discussion to true "*in situ*" techniques.

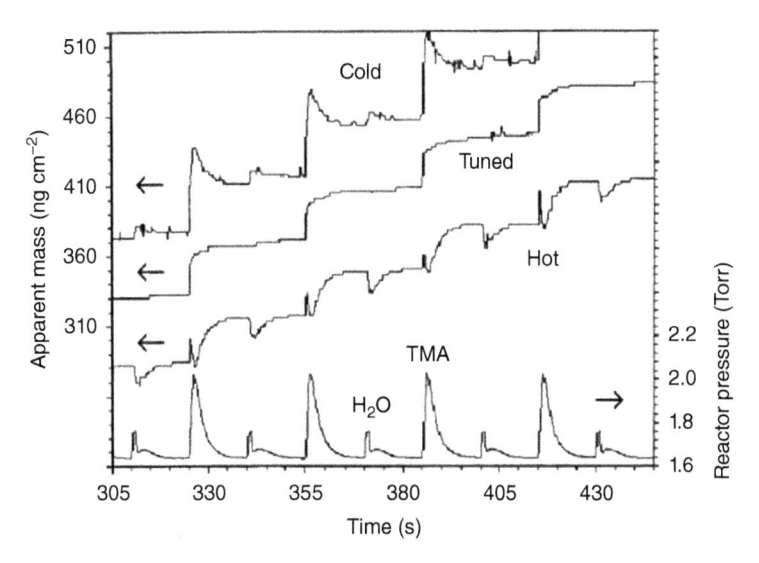

Figure 1.8 QCM measurement during TMA/H$_2$O ALD, illustrating the impact of temperature effects on the data. Only the "TUNED" data reflects the actual deposition onto the crystal. For the "COLD" and "HOT" traces, apparent mass changes were recorded due to temperature fluctuations during gas pulsing. (Rocklein and George 2003 [58]. Reproduced with permission of American Chemical Society.)

piezoelectric crystal, the resonant frequency of the crystal decreases. By measuring this shift in resonant frequency, the added mass can be determined with a resolution below $1\,\text{ng}\,\text{cm}^{-2}$. The key challenge to implement QCM during ALD concerns (i) preventing backside deposition [57] and (ii) dealing with temperature effects [58]. Indeed, the resonant frequency is also strongly dependent on the temperature of the QCM crystal. For AT-cut quartz crystals, the specified temperature range is −45 to 90 °C. At higher temperatures, small temperature changes can lead to significant changes in resonant frequency, and hence in apparent mass changes, even when pulsing inert probe gases onto the crystal. Lower temperatures prior to the QCM lead to positive mass transients. More elevated temperatures prior to the QCM lead to negative mass transients. Unreactive probe gases can be employed to optimize the temperature profile to minimize the temperature-induced apparent mass changes. QCM can be a simple and powerful technique, provided that sufficient care is taken during the measurement and analysis, as apparent mass transients and apparent mass drifting can lead to misinterpretation of ALD surface chemistry and produce error in measured ALD growth rates, as illustrated in Figure 1.8.

1.2.2 Quadrupole Mass Spectrometry (QMS)

QMS can be used during ALD to identify and monitor the gaseous species that are present in the reactor and/or the exhaust line [59, 60]. Commercial systems provide an inlet orifice for gas sampling. Electron impact from a filament then results in ionization and molecular fragmentation of the gas molecules. The ionized species are then sent through a quadrupole filter, which transmits a specific mass/charge ratio. As the quadrupole is scanned, for example, a Faraday cup is used to measure the ion current.

Ritala *et al.* reported a dedicated system with a sample point near the hot zone of the ALD reactor to avoid precursor condensation [61, 62]. While this can be beneficial for easily condensable vapors, many gases can also be detected using a standard commercial QMS, for example, attached to the exhaust line of the reactor [63]. To enhance the QMS signal, a very useful trick can be to increase the total surface area in the ALD reactor. As $\sim 10^{14}$ reaction products will be released per square centimeter of sample during ALD, increasing the surface area of the sample (e.g., by including a large number of glass slides or powder particles in the reactor) will increase the amount of gaseous reaction products, facilitating detection.

1.2.3 Spectroscopic Ellipsometry

Spectroscopic ellipsometry (SE) is a powerful optical technique for the determination of the optical constants and thickness of thin (multi)layers deposited on a substrate. SE does not directly measure these film properties; instead, it measures the change in polarization of a light beam upon reflection from the sample. Therefore, from an instrumentation point of view, ellipsometry requires (i) a light source and a polarizer to define the polarization of the incoming beam and (ii) a polarization analyzer and a detector unit to determine the polarization of the reflected light.

An ellipsometric measurement is commonly described in terms of Ψ and Δ with

$$\tan(\Psi)\exp(i\Delta) = R_p/R_s$$

where R_p and R_s are the complex Fresnel coefficients of the sample for p- (in the plane of incidence) and s- (perpendicular to the plane of incidence) polarized light, respectively (see Figure 1.9). In SE, the Ψ and Δ parameters are measured for a range of photon wavelengths. In order to relate the SE data to the actual properties of the sample, a model of the sample must be constructed from which the modeled ellipsometric parameters Ψ_{mod} and Δ_{mod} can be calculated using the Fresnel equations. In general, a multilayer model is built, where each layer is characterized by a thickness and a certain dispersion relationship of the optical

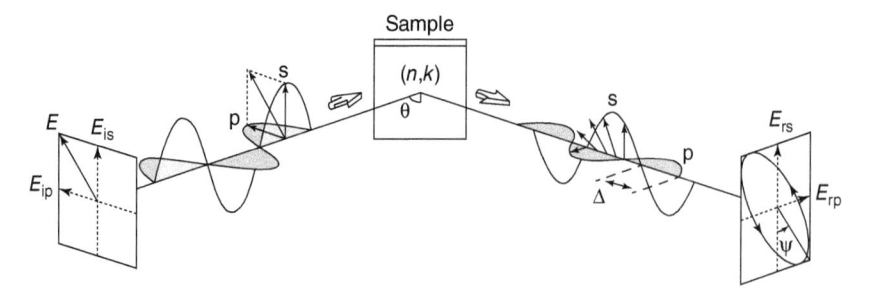

Figure 1.9 Schematic representation of a basic ellipsometer system. (Fujiware 2007 [64]. Reproduced with permission of John Wiley and Sons.)

constants. Finally, the adjustable model parameters are varied to find the best fit of the calculated Ψ_{mod} and Δ_{mod} values to the measured SE data [65]. The data in Figure 1.3b illustrates the Angstrom-level thickness sensitivity of the SE technique during ALD of Al_2O_3.

1.2.4 Fourier Transform Infrared Spectroscopy

Fourier transform infrared spectroscopy (FTIR) detects vibration modes of molecular bonds in the sample [66]. The key challenge in using FTIR during ALD concerns the fact that one is trying to measure the IR absorption caused by a (sub)monolayer on the surface of the sample. This requires an excellent signal/noise ratio and a very "clean" measurement, as traces of gaseous H_2O and CO_2 along the beam path may create spurious signals in the measurement [67]. Several dedicated setups have been designed for IR surface spectroscopy, including transmission geometries through a double polished Si wafer [68], transmission geometries using ZrO_2 powder pressed into metal grid to enlarge the effective surface area and hence enhance the signal [69], attenuated total reflection measurements [70], and, more recently, also IR reflection absorption spectroscopy [71]. Figure 1.10 shows typical *in situ* FTIR difference spectra before/after specific steps during a TMA/H_2O ALD process.

1.2.5 Optical Emission Spectroscopy

Optical emission spectroscopy (OES) can be used to measure the intensity of light emitted by a plasma as a function of wavelength and, as such, is a very convenient technique for *in situ* monitoring of PE-ALD processes. The spectral lines in

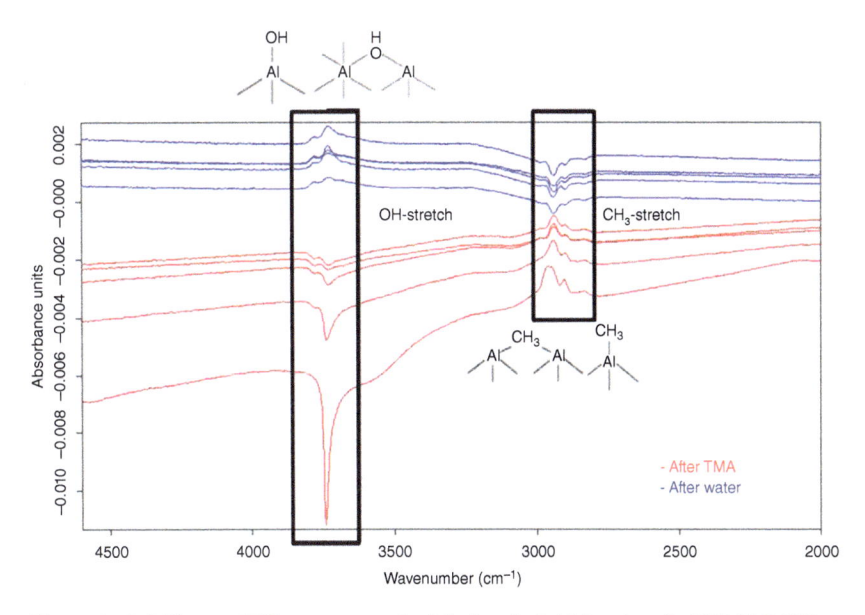

Figure 1.10 Difference FTIR spectra acquired during the initial cycles of a TMA/H_2O ALD process.

the emission spectrum originate from the radiative decay of electronically excited states of the plasma species. Therefore, OES spectra can reveal information on the presence of particular excited ions, atoms, and molecules in the plasma, and, because these species can be both reactant and reaction products, also about the reactions that take place in the gas phase, at the chamber walls, or at the surface of the growing ALD film [72].

1.2.6 Other *In Situ* Techniques

Other *in situ* techniques that have been implemented on ALD reactors include reflection high-energy electron diffraction [73], a thermopile for measuring reaction enthalpy [74], sheet resistance measurements for monitoring coalescence of metal layers [75], gravimetry [76], optical reflectometry [77], gas conductance measurements [78] for monitoring pore size reduction, and *in situ* ellipsometric porosimetry (EP) [79] for monitoring ALD in nanopores.

Optical techniques have proven to be the easiest to use in practice, since no measuring equipment needs to be incorporated within the ALD reactor. They can remotely probe the sample surface and only require entry and exit windows to pass light into and out of the reactor. Recently, several groups have explored expanding the spectral range from UV/Vis (used for SE, with quartz windows) and IR (used for FTIR, with KBr windows) toward X-rays (with Be windows), thus opening up an even wider variety of characterization possibilities [80]. Although standard X-ray-based analysis techniques such as X-ray reflectivity and diffraction using lab-based X-ray sources have proven valuable for *ex situ* characterization of ALD deposited thin films, most *in situ* experiments during ALD require synchrotron-based X-rays. To enable *in situ* studies during ALD, it is important to limit the impact of the prolonged purge or evacuation times that are introduced between subsequent ALD (half)cycles to perform the measurements. The high photon flux at a synchrotron facility is beneficial in this respect, because it allows for shorter acquisition times compared to lab-based X-ray sources. The high intensity X-ray flux also lowers the detection limit, enabling the study of layer growth from the very first ALD cycle onward. A second main advantage of synchrotron sources is their unique ability to tune the photon energy to a specific experiment and material system. In recent years, the use of synchrotron-based X-rays has broadened the available tool box of *in situ* methods to X-ray fluorescence (XRF, measuring film composition) [81, 82], X-ray absorption spectroscopy (XAS/EXAFS, measuring the local atomic environment) [83, 84], X-ray photoelectron spectroscopy (XPS, measuring surface composition and oxidation state), X-ray diffraction (XRD, measuring film crystallinity, grain size), X-ray reflectivity (XRR, measuring film thickness, roughness, and density), and grazing-incidence small-angle X-ray scattering (GISAXS, measuring surface morphology) [85, 86].

1.3 Conformality of ALD Processes

The deposition of uniform coatings into deep structures such as holes, trenches, and (nano)pores is becoming increasingly important in the rapidly growing field of nanotechnology. Among all the thin-film growth techniques, ALD can achieve the highest conformality, that is, the most uniform thickness over micro- and nanoscale 3D features. The excellent conformality of ALD is a direct consequence of the self-saturated surface reaction control, as opposed to flux-controlled deposition in, for example, PVD and CVD. In CVD, the growing film is usually exposed to simultaneous flows of precursor vapor and reactant gas (instead of the sequential pulses in ALD), and the growth rate of the film generally depends on the local gas flux. During coating of deep features in which the gas transport is diffusion-limited, the surface region near the entrance of these features will receive reactant fluxes that can be several orders of magnitude larger than the fluxes arriving at the surfaces deeper in the structures. Therefore, deposition techniques where the growth rate is flux-controlled are less convenient for coating high AR structures or porous materials, because the entrance region of features such as holes, trenches, and pores tends to get clogged during the early stages of deposition. In addition, in ALD, the regions near the entrance of the holes, trenches, or pores in a material will experience a larger flux of precursor vapor and, therefore, will become saturated much sooner than the interior surfaces. Once saturated, however, no further reaction will occur near the openings, and one can "simply" expose the entire sample for a sufficient amount of time until the precursor molecules have diffused into the deep features and have saturated the available chemisorption sites throughout the interior surface of the material.

ALD has proven to be an effective technique for the deposition of conformal coatings in features with a high AR [87–89]. Nevertheless, achieving a good conformality along the entire depth of high AR structures requires careful optimization of the ALD process parameters.

1.3.1 Quantifying the Conformality of ALD Processes

The most common way to demonstrate the conformality of an ALD process is by deposition into deep microscopic trenches (typical width <1 μm) and subsequent characterization by cross-sectional scanning electron microscopy (SEM) or transmission electron microscopy (TEM), as illustrated in Figure 1.11. This method yields a stop/go-type result in the sense that it provides a means to verify whether full coverage could be achieved or not. Obtaining a quantitative film thickness profile along the length of the trench is, however, difficult.

Several groups successfully obtained thickness profiles for ALD films deposited in anodic aluminum oxide (AAO) [23, 91]. AAO can be prepared by a two-step electrochemical anodization of aluminum films [92–94]. It consists of well-defined parallel cylindrical pores (typical diameters 30–500 nm) and is therefore an attractive material for model studies of ALD in high AR nanostructures. After ZnO ALD in an AAO membrane, Elam *et al.* obtained Zn coverage profiles as a function of depth in the membrane using electron probe microanalysis (EPMA) line scans along the axis of the pores [22]. Figure 1.12 shows the diffusion-limited behavior of ZnO ALD in AAO nanopores with a diameter of 65 nm. Indeed, increasing the Zn-precursor exposure time allowed for a deeper penetration and deposition of the precursor molecules in the hole.

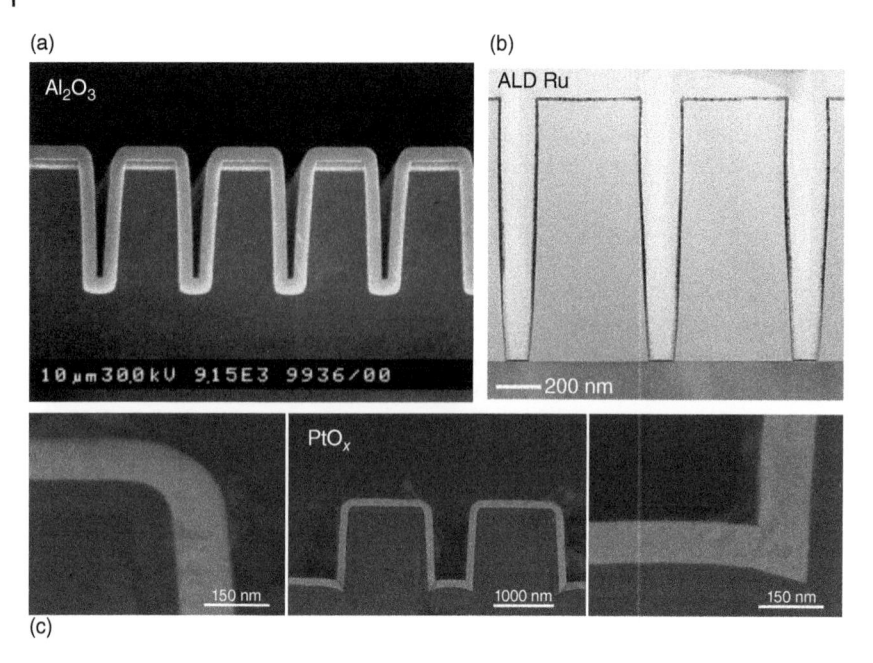

Figure 1.11 (a) Cross-sectional SEM image of a 300 nm-thick Al_2O_3 film deposited on a Si wafer with trench structures (Ritala *et al.* 1999 [89]. Reproduced with permission of Wiley.). (b) Cross-sectional TEM of a conformal Ru ALD coating on a trench-patterned substrate (Kim *et al.* 2009 [18]. Reproduced with permission of Wiley.). (c) Cross-sectional Field Emission Scanning Electron Microscopy (FESEM) images of a conformal PtO_x film deposited on a Si substrate with trench structures. (Hämäläinen *et al.* 2008 [90]. Reproduced with permission for Elsevier.)

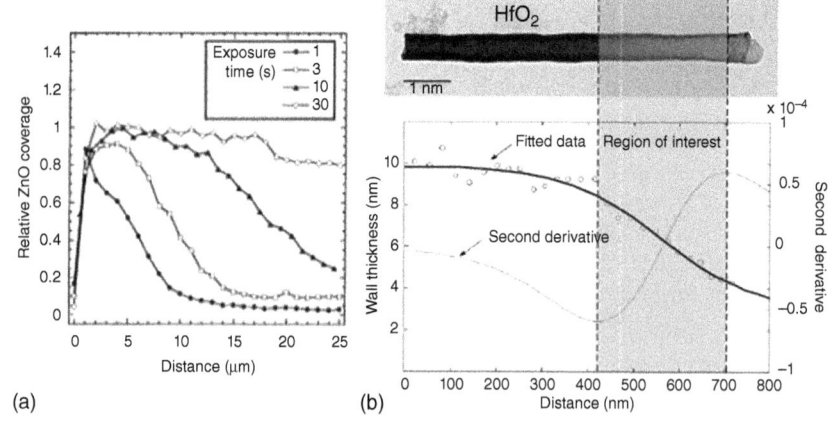

Figure 1.12 Film thickness/coverage profiles obtained for ALD in AAO nanopores. (a) Increasing the Zn-precursor exposure time results in an improved ZnO coverage in AAO pores with a diameter of 65 nm (Elam *et al.* 2003 [22]. Reproduced with permission of American Chemical Society.). (b) Wall thickness measured along the length of a HfO_2 nanotube obtained by ALD of HfO_2 into an AAO template followed by dissolution of the template. (Perez *et al.* 2008 [23]. Reproduced with permission of Wiley.)

Perez et al. investigated the conformality of ALD processes in AAO by means of TEM [23]. After depositing a HfO$_2$ ALD layer into AAO structures and selective dissolution of the AAO, nanotubes could be obtained by replicating the AAO pores (Figure 1.12b). Using TEM, the conformality of the ALD process could be quantified by locally measuring the thickness of the replicated nanotube as a function of depth in the original AAO nanopore. Note that the diffusion-limited behavior of ALD in AAO pores can be exploited to achieve patterned ALD on the interior surface of an AAO membrane [95]. Adsorption of a first ALD precursor to a certain depth in the pore can block the active surface sites and prevent the subsequent adsorption of a second ALD precursor on that part of the pore walls, allowing for depth-controlled deposition. Using the passivation effect of TMA, ALD of ZnO, TiO$_2$, V$_2$O$_5$ has been achieved at controlled depths in AAO pores.

Under standard ALD conditions, the precursor vapor penetrates into microscopic holes or trenches through molecular flow (or Knudsen diffusion), because the mean free path of the precursor molecules is much larger than the diameter of the hole or the width of the trench. For instance, at 200 °C and a pressure of 1 Torr, the mean free path of the TMA molecules is about 35 µm. At sufficiently low pressure, molecular flow can also be achieved in macroscopic holes (Figure 1.15). Therefore, when ALD is performed at sufficiently low pressure, relatively simple test structures can be used to determine the depth of infiltration of ALD material into a structure with a given AR. Becker et al. introduced the use of fused-silica capillary tubes with a diameter of 20 µm to measure the conformality of ALD layers [96]. The tubes were exposed to an ALD process, heated to burn off the ALD coating on the outside, and filled with a fluid having a refractive index matching that of fused silica. By placing the tubes in an optical microscope, the depth of infiltration could be visually determined. The authors demonstrated successful ALD of WN in a capillary with an AR of about 200 : 1 (Figure 1.13a). Dendooven et al. introduced another macroscopic approach for quantifying the conformality of ALD [97, 98]. A macroscopic test structure was created by cutting a rectangular-shaped structure from a sheet of aluminum foil and clamping the resulting foil in between two silicon wafers (Figure 1.13b). This simple structure allowed for a direct and straightforward test of the ability of an ALD process to coat "around the corner." After deposition, the structure can be disassembled to inspect the deposition of material on the interior surface of the rectangular hole. A successful conformal coating with TiN is illustrated in Figure 1.13c, while a nonconformal Ru coating is shown in Figure 1.13d. The method was later extended by Musschoot et al. to investigate the penetration of thermal and plasma-enhanced ALD into fibrous materials [99].

The macroscopic test structures allow for quantitative analysis of the film thickness profile via spectroscopic ellipsometry (SE) measurements. Figure 1.14 shows the Al$_2$O$_3$ film thickness as a function of depth in a hole with an AR of about 200 : 1 for two TMA exposure times. As for the ZnO process shown in Figure 1.12a, the conformality could be improved by increasing the TMA exposure time. The experimental results depicted in Figure 1.12a,b show similar trends. The relative coverage or thickness is (nearly) at its maximum near the entrance of the hole/pore. At a certain depth, depending on the (unsaturated)

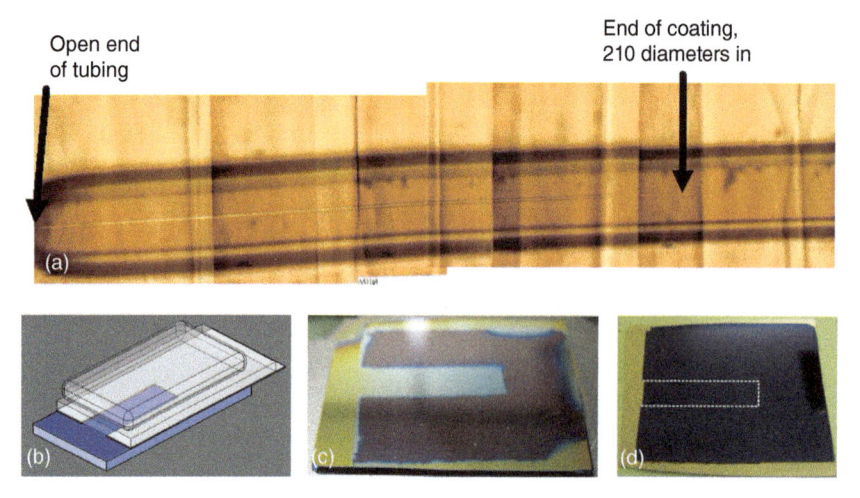

Figure 1.13 Macroscopic test structures that allow for visual inspection of the penetration depth of the ALD material. (a) WN coating inside a fused-silica capillary tube having an inner diameter of 20 μm. (Becker *et al*. 2003 [96]. Reproduced with permission of American Chemical Society.) (b) Schematic representation of a macroscopic hole with rectangular cross section (blue area) that can be disassembled into planar pieces of SiO_2 wafer after ALD. (c) The conformal deposition of TiN on the interior surface of a hole with an AR of 200 : 1 (defined as depth:width). In (d), the yellow coating outside of the cover is due to Ru deposition, while there is no coating inside the marked hole region, illustrating the lack of conformality for this Ru ALD process.

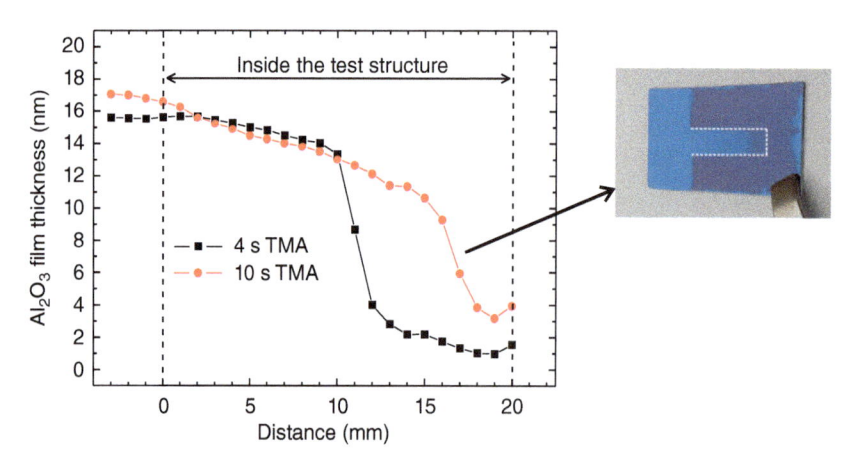

Figure 1.14 Film thickness profiles obtained for Al_2O_3 ALD in a macroscopic rectangular hole with an AR of 200 : 1. Increasing the TMA exposure time causes the Al_2O_3 coating to penetrate deeper into the test structure.

exposure time, the coverage or thickness gradually decreases as a function of depth in the hole/pore. The resulting data on film thickness as a function of penetration depth can be compared to simulations based on models for the diffusion of ALD precursors in deep holes or trenches (see the following discussion).

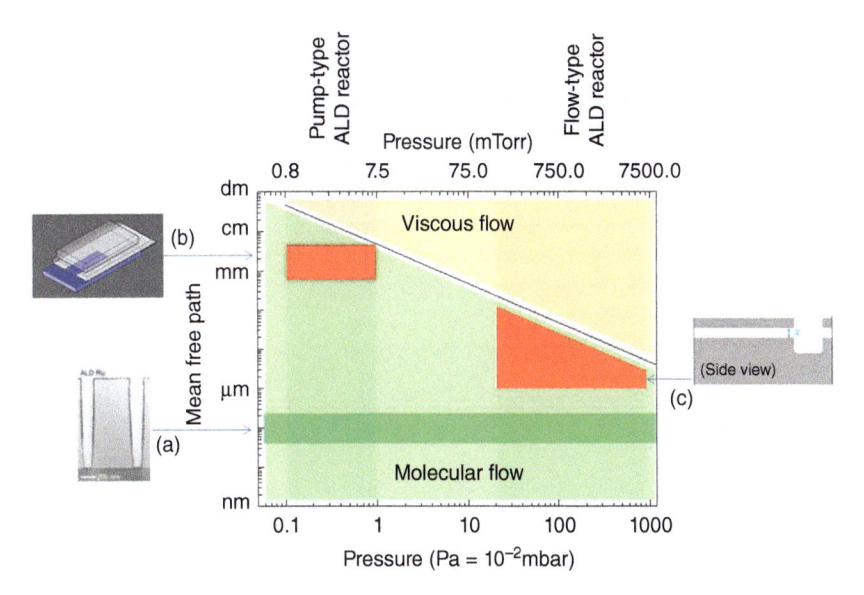

Figure 1.15 Mean free path of TMA molecules at 200 °C as a function of pressure in the ALD reactor. At all relevant pressures and for all relevant nano- and microstructures, precursor transport is governed by molecular flow. Different approaches have been developed for quantifying the conformality of ALD processes in different pressure regimes: (a) deposition into nano-sized features, followed by cross-sectional electron microscopy, (b) dedicated macroscopic lateral trench structures, as proposed by Dendooven et al. for low-pressure processes, and (c) dedicated micrometer-sized lateral trench structures as recently proposed by Puruunen et al. for high-pressure processes.

As illustrated in Figure 1.15, the approach using millimeter-sized macroscopic structures as proposed by Dendooven *et al.* is limited to low-pressure ALD processes as typically encountered in the pump-type ALD reactors that are used for PE-ALD research (where the chamber is evacuated by pumping in between half cycles). The more traditional flow-type ALD reactors (where the chamber is evacuated by purging in between half cycles) typically operate at significantly higher pressures, and at these pressures, the flow in millimeter-sized structures is determined by viscous flow conditions. Puurunen *et al.* recently proposed an approach using MEMS fabrication techniques to fabricate horizontal trenches with ARs up to 25 000:1 under thin membranes. After ALD, the thin membrane can be easily peeled off, and the deposited film can be investigated as a function of "depth" (in fact, lateral position) on the interior surface of the "trench" [100].

1.3.2 Modeling the Conformality of ALD

Gordon *et al.* [101] proposed an analytical model to predict the required exposure (defined as the product of precursor partial pressure present at the opening of the hole, P, and the precursor pulse duration, t) to conformally coat a cylindrical hole with a certain aspect ratio $AR = L/D$, with L being the depth and D the diameter

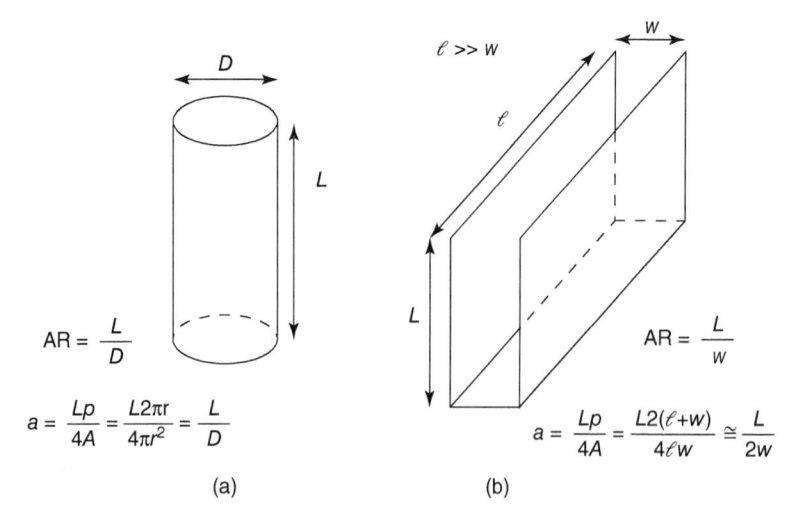

Figure 1.16 Aspect ratio (AR) and generalized aspect ratio (a) for a cylindrical hole (a) and a trench (b).

of the cylindrical hole (see also Figure 1.16):

$$P \cdot t = K_{max} \sqrt{2\pi m k T} \cdot \left(1 + \frac{19}{4}a + \frac{3}{2}a^2\right) \quad (1.1)$$

In this equation, K_{max} is the saturated coverage per surface area (molecules per square meter), m is the mass of the precursor molecules, k is the Boltzmann constant, and T is the temperature. For large ARs, the required exposure increases approximately quadratically with a. The idea behind the model is schematically shown in Figure 1.17a. Saturation of the pore walls starts at the top part of the hole and then propagates as a "front" through the hole. It is assumed that the precursor molecules stick to the wall upon their first collision with an uncoated part of the wall, that is, the sticking probability is unity. For unsaturated exposures, the coverage profile consists of a fully saturated part that abruptly stops at the "front" position. Mathematically, the model can be derived using gas conductance equations: the fully covered part of the hole acts as a "tube" leading to the reactive uncoated part of the hole, which can be considered as a "vacuum pump." The authors showed good agreement between the predicted and the experimentally derived saturation dose for ALD of HfO_2, Ta_2O_5, WN, and V_2O_5 in cylindrical holes with a known AR [102].

Although the formulae used to develop Equation 1.1 were derived for cylindrical holes [103], Equation 1.1 can also (approximately) be applied for trenches and holes that have a noncircular cross section by introducing a generalized expression for a,

$$a = \frac{Lp}{4A}$$

where L is the depth of the hole, p its perimeter, and A its cross-sectional area. For cylindrical holes, a then still equals depth/diameter. For a trench of width w, the expression simplifies to $L/(2w)$, which is only half of the AR that is conventionally

Figure 1.17 (a) Schematic representation of the idea behind the model proposed by Gordon et al. The resulting coverage profile is characterized by a step. (b) Schematic representation of the model proposed by Dendooven et al. The resulting coverage profile is characterized by a slope of decreasing thickness. (c) TMA coverage as a function of depth for a cylindrical hole with an AR of 100 : 1 for simulations with an initial sticking probability $s_0 = 1, 0.1$, and 0.001 ($P = 0.3$ Pa, $t = 5$ s, $K_{max} = 4.7 \times 10^{18}$ m^{-2}).

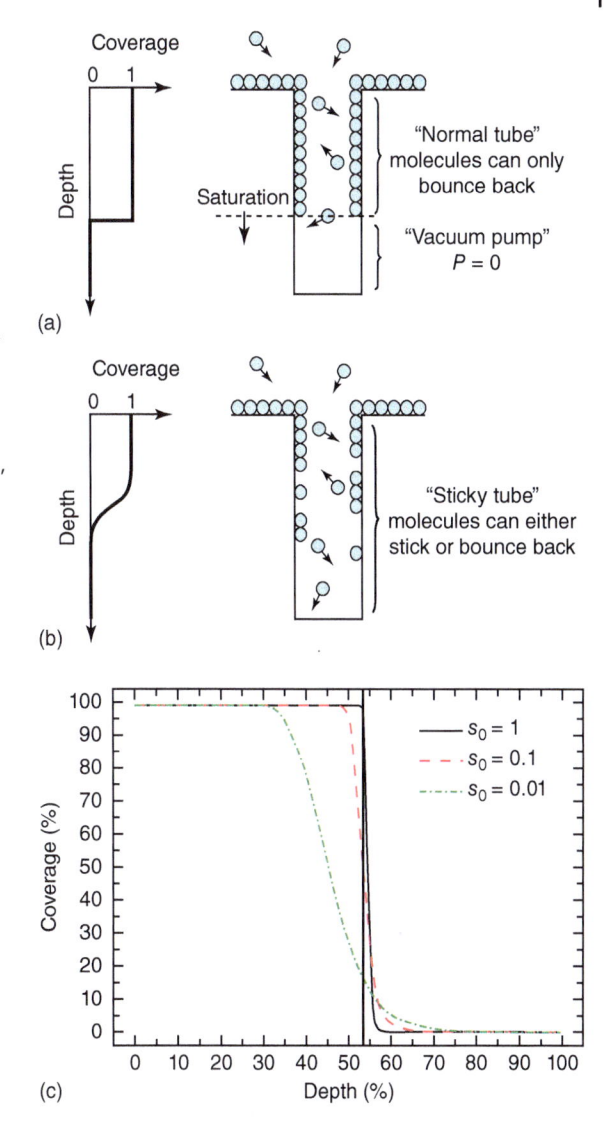

used for trenches, that is, L/w. Given that $(P \cdot t) \sim a^2$ for large ARs, the model thus predicts that a trench requires four times less exposure compared to a hole with the same AR (depth-to-width ratio).

During ALD in a nanoscopic hole, such as an AAO pore, the coating deposited during each ALD cycle decreases the pore diameter. Consequently, for a fixed unsaturated exposure, the Gordon model predicts a decrease in the penetration depth with each ALD cycle deposited. This gives rise to a slope of decreasing film thickness along the depth of the pore in the final film thickness profile, as also observed in the experimentally obtained profiles. Perez et al. showed good agreement between the slope obtained in the depth profiles for HfO$_2$ ALD in AAO pores (Figure 1.12b) and the slope predicted by iteratively applying

Gordon's model to a pore that is gradually getting clogged as the ALD process progresses. However, the slope observed in the thickness profiles for Al_2O_3 ALD in macroscopic holes (Figure 1.14) cannot be explained by an increasing AR during the deposition, because the deposited film thickness (typically a few nm) is negligible compared to the width of the hole (~100 µm). In this case, the observed slope is related to a sticking probability, which is less than 1 (as can be expected for any real ALD process).

Dendooven et al. extended the kinetic model of Gordon et al. for sticking probabilities of less than unity [97]. The extended model is based on conductance formulae that were derived for a "sticky" tube [104]. In this case, the precursor molecules can either stick on an uncoated part of the wall or bounce back from it (Figure 1.17b). Moreover, the sticking probability on a reactive part of the wall is assumed to decrease linearly with increasing surface coverage according to Langmuir's law. Figure 1.17c shows the effect of the initial sticking probability, s_0, on the simulated coverage profile for a cylindrical hole with an AR of 100:1. For $s_0 = 100\%$, a step-like profile is obtained, in agreement with the model by Gordon et al. For lower values of s_0, a slope of decreasing thickness is predicted. Using an initial sticking probability of 10% for TMA, good agreement was achieved between the simulated and the experimentally obtained Al_2O_3 thickness profiles.

Elam et al. developed a one-dimensional (1D) Monte Carlo (MC) model for simulating experimental Zn coverage profiles obtained by cross-sectional analysis of ZnO layers deposited by ALD in AAO membranes [22] (Figure 1.12a). Based on their 1D MC model, Elam et al. predicted a reaction-limited behavior rather than a diffusion-limited behavior for sufficiently low reaction probabilities and reasonable ARs. This classification between diffusion- and reaction-limited regimes was later confirmed by Dendooven et al. [105] and Knoops et al. [106] Figure 1.18 shows the pressure evolution and coverage evolution in a rectangular hole with an AR of 66:1 for an initial sticking probability of 10%. In this case, the deposition is clearly diffusion-limited. With progressing time, the precursor molecules penetrate deeper into the hole, leading to a moving deposition "front" toward the bottom of the hole. On the other hand, if an initial sticking probability of 0.1% is used in the simulation, the deposition becomes reaction-limited. From the very start of the exposure, a large fraction of the precursor molecules can reach the bottom of the hole, but the reaction is slow as observed in the coverage profile evolution.

1.3.3 The Conformality of Plasma-Enhanced ALD

As mentioned earlier, achieving a good conformality in high AR structures is considered to be more challenging for PE-ALD than for thermal ALD. This is because radicals can recombine upon collision with a surface. For instance, an oxygen radical can recombine with an oxygen atom that resides at the surface to form molecular O_2, which is often nonreactive to the adsorbed metal precursor on the surface. When coating deep holes or trenches with PE-ALD, the reactive species have to undergo multiple wall collisions during which they may be lost through surface recombination before they can reach the surfaces deeper in the hole. It is therefore inevitable that the elimination of radicals through recombination on the sidewalls of high AR structures will limit the conformality of PE-ALD.

The recombination probability r describes the probability that a reactive atom (radical) will recombine upon collision with a surface and is typically determined

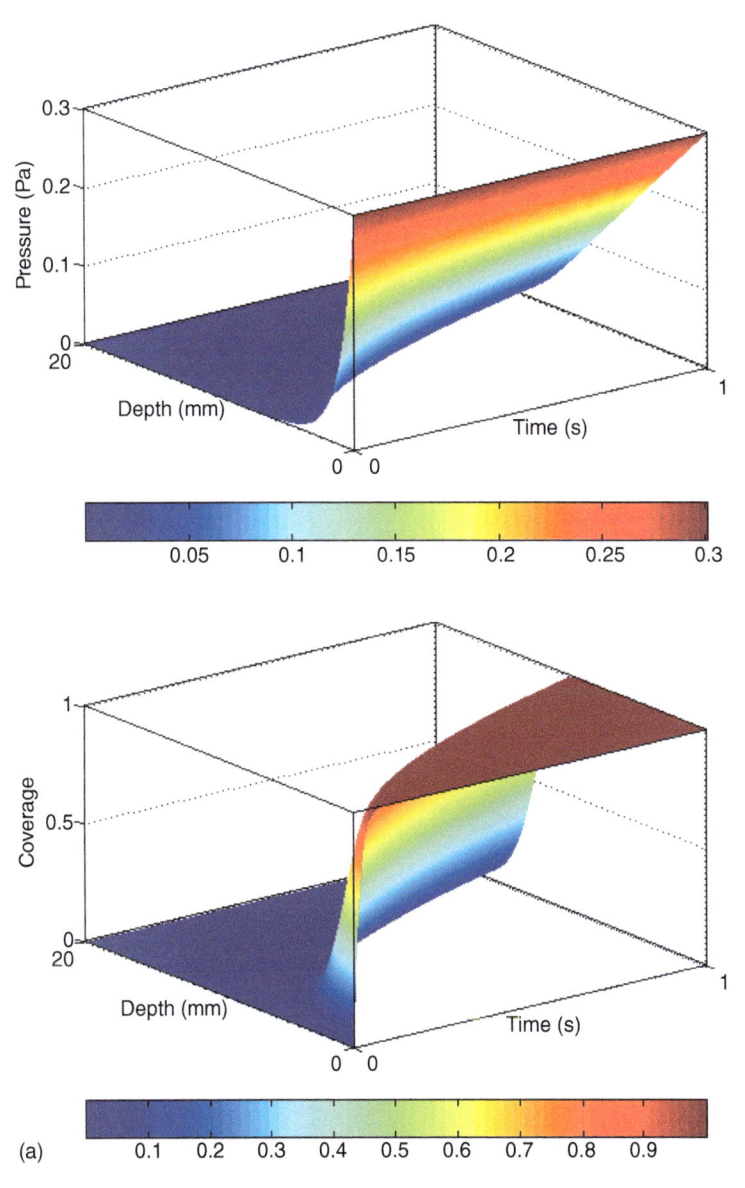

Figure 1.18 Simulated pressure (a) and coverage (b) profiles calculated for a TMA pulse in a rectangular hole with an AR of 66 : 1 ($P = 0.3$ Pa, $t = 1$ s, $K_{max} = 4.7 \times 10^{18}$ m^{-2}). For panel (a), the initial sticking probability was 10%, resulting in a diffusion-limited deposition behavior. For panel (b), the initial sticking probability was 0.1%, resulting in a reaction-limited deposition behavior.

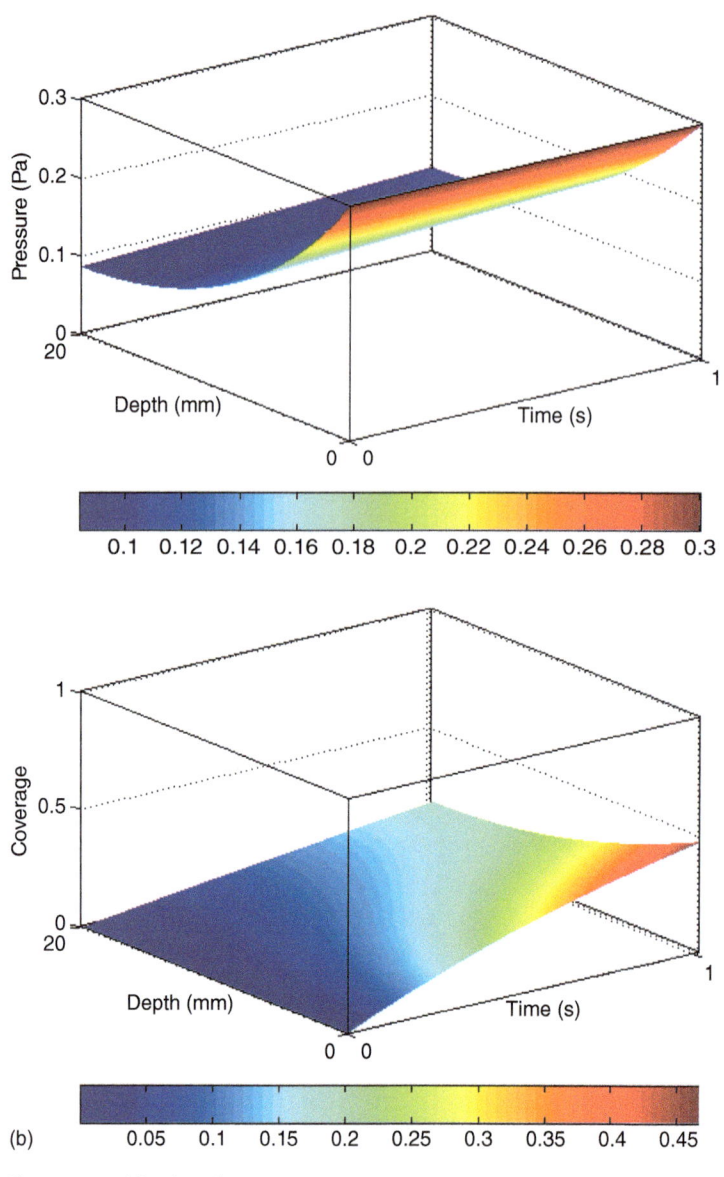

Figure 1.18 (Continued).

for a certain atom type on a certain surface material. Reported values for recombination probabilities of O, N, and H atoms on various surfaces span a large range, from 0.000094 for the recombination of O atoms on Pyrex to 0.8 for the recombination of H atoms on silicon [106]. There is quite a discrepancy in the r values published in the literature. This is likely related to the large variety of (often indirect) techniques that have been used to monitor atom recombination (including actinometric OES, two-photon laser-induced fluorescence, use of dual thermocouples, or catalytic probes, and mass spectroscopy) [107, 108] and to the fact that the atom recombination coefficient depends not only on the surface material but also on the surface conditions (e.g., presence of impurities, adsorbed gas species, certain surface pretreatments), the gas pressure and temperature, the surface temperature, and the plasma configuration [109–112]. This latter issue has, among others, been addressed by Cartry *et al.*, who measured a value of 0.0004 for O atom recombination on a silica surface positioned in the afterglow region of a microwave plasma, while a two orders of magnitude higher value of 0.03 was obtained on a silica surface directly submitted to the plasma [113]. A possible explanation for this difference is that ions (which are predominantly present in the core of the plasma) can create active sites for surface recombination. As mentioned earlier, the gas pressure can also affect the recombination probability. Adams *et al.* reported a decreasing N atom recombination with increasing pressure for silicon, aluminum, and stainless steel surfaces. For instance, on silicon, they measured a recombination probability of 0.0026 at 1 Torr, 0.0016 at 3 Torr, and 0.0005 at 5 Torr. Gomez *et al.* observed a similar trend for O atom recombination [111]. Moreover, significant dependencies on the surface temperature can exist. For instance, Guyon *et al.* measured an increase in the O atom recombination probability on alumina from 0.0097 at room temperature to 0.061 at 500 °C [114]. On the other hand, Wood and Wise reported quasi-constant H atom recombination on several metals over a large range of temperatures [115]. It is learned from the given overview that the conformality of a PE-ALD process will depend, via the recombination probability, on the type of radicals used as reactant in the process, on the material that is deposited, and on the process parameters such as the gas pressure, deposition temperature, and plasma configuration. Note that the gas pressure and the plasma configuration not only affect the recombination probability but also the radical density at the sample surface and the entrance of high AR structures, which in turn will have an effect on the conformality. Furthermore, the recombination coefficient will likely vary over the duration of a plasma exposure, because the sample surface changes (due to reaction with the radicals) from a surface that is covered with metal precursor ligands to the oxide, nitride, or metal that is being deposited.

There exist only a few systematic studies on the conformality of PE-ALD in the literature. Dendooven *et al.* used macroscopic test structures to study the influence of the gas pressure, the RF power, the plasma exposure time, and the directionality of the plasma plume on the conformality of the remote PE-ALD of Al_2O_3 from TMA and O_2 plasma [98]. To investigate the effect of the plasma type on the conformality, they compared the conformality of Al_2O_3 to the conformality of AlN deposited from TMA and NH_3 plasma. In addition, a Monte Carlo (MC) model was used to evaluate the effect of radical recombination. For the Al_2O_3 process using O_2 plasma, conformal coatings in holes with an AR of 40 : 1 (defined as depth/width) were achievable by optimizing the process parameters. The conformality of the AlN process was more limited, and an AR of 20 : 1 already seemed impractical. This suggests that the radicals generated in the NH_3 plasma suffer from faster recombination compared to the O radicals. The conformality of the Al_2O_3 PE-ALD process could be improved by increasing the radical density via the gas pressure or the RF power or by prolonging the plasma exposure time. It should be noted that the H_2O formed as reaction product during the combustion reaction that takes place during the O_2 plasma step contributed to the apparent conformality of PE-ALD processes via a secondary thermal ALD reaction.

Knoops et al. used an MC model to obtain insights into the effect of the recombination probability r on the conformality of PE-ALD processes in high AR trenches. The required saturation dose increases considerably with increasing r values, especially for high ARs. Therefore, besides the diffusion-limited and reaction-limited regimes that are also observed in thermal ALD, they distinguished a recombination-limited regime for PE-ALD processes with high r values (or lower r values in combination with high ARs). It was further speculated that conformal coating in trenches with an AR of 30:1 should be achievable for PE-ALD processes with low r values. On the other hand, for high surface recombination probabilities, as observed on many metallic surfaces, impractically large exposures seem to be required to coat trenches with ARs larger than 10:1.

Kariniemi et al. verified the conformality of various PE-ALD processes by deposition into deep microscopic trenches and subsequent characterization by cross-sectional SEM [48]. They showed good conformality of metal oxide coatings deposited in trenches with ARs considerably larger than what had been achieved earlier (up to 60:1). The key difference with other remote PE-ALD studies [116–118] are the reactor design and, related to that, the two to three orders of magnitude higher pressures used during the O_2 plasma step. Compared to the inductively coupled RF plasma sources, the remote capacitively coupled RF plasma configuration used by Kariniemi et al. is expected to result in higher radical densities at the sample surface and the entrance of the trench and thus also in higher radical fluxes deeper in the trench and improved conformality. In addition, it is possible that a higher pressure during the plasma exposure causes the O radicals to recombine less efficiently at the trench walls. In the case of Ag PE-ALD using H_2 plasma, Kariniemi et al. observed Ag growth near the bottom of a 60:1 trench, but the coating was far from conformal. This can be related to the rather high recombination probability typically observed for H radicals on metals and/or to the absence of a secondary thermal ALD reaction contributing to the conformality.

Increasing the radical exposure, via the radical density available at the entrance of the hole or the exposure time, is found to be the key factor in improving the conformality of PE-ALD processes. The influence of the radical density was demonstrated in the work of Dendooven et al. by increasing the gas pressure and the RF power and in the work of Kariniemi et al. by modifying the plasma configuration. This latter method appears to be more effective.

The experimental results also show that reasonable ARs can be achieved for O_2-based PE-ALD processes. The MC model of Knoops et al. solely attributes this to the relatively low recombination probability of O radicals on oxide surfaces. However, the experimental film thickness profiles obtained for PE-ALD of Al_2O_3 by Dendooven et al. and Musschoot et al. could only be "reproduced" by MC simulations if a superposition of two reactions was assumed, that is, (i) combustion reactions of O radicals with adsorbed TMA molecules at the entrance of the hole and (ii) a secondary thermal ALD reaction of H_2O molecules that are being generated during the combustion reactions with adsorbed TMA molecules deeper in the hole. On the other hand, Kariniemi et al. concluded that the secondary H_2O effect plays a minor role in their depositions as good conformality was also achieved for the SiO_2 process, while the Si precursor reacts only slowly with H_2O. This might be explained by the difference in radical fluxes inside the high AR structures. Indeed, for sufficiently large O radical fluxes, as is also the case for planar substrates, the effect of the secondary H_2O reaction will be minor as it has to compete with the combustion-like O radical reactions, which are likely to occur faster. If the radical flux is low, secondary reactions with the H_2O will have a relatively larger impact.

When NH_3, N_2, or H_2 plasmas are used, H_2O is usually not formed as a reaction product in the plasma step, and secondary thermal ALD reactions are not expected. The results by Dendooven et al. demonstrated that the growth of nitrides in high AR structures using NH_3 plasma is not trivial, pointing to a high recombination probability of the radicals generated in the plasma. The chemistry of a NH_3 plasma is, however, very complex, and an understanding of the surface recombination of its radicals remains elusive [119]. Even for the pure N_2 plasma, there are only a few studies concerning the surface recombination of N radicals. Although not stated explicitly in the literature, the lack of papers reporting conformal growth of nitrides with PE-ALD using NH_3 or N_2 plasma could indicate that it is indeed challenging to achieve good conformality for those processes.

1.3.4 Conformal Coating of Nanoporous Materials

ALD in combination with nanoporous materials and membranes allows for the creation of nanomaterials with improved compositional and structural properties through either replication or coating of the porous network. These materials have applications in photonics, catalysis, gas separation, green energy conversion, sensing, and so on. The unique potential of ALD to conformally coat mesopores was demonstrated in 1996 for SnO_2 ALD in porous silicon [87]. Since then, many authors have used ALD in porous films and membranes to fabricate nanostructures, to functionalize the pore walls, or to tune the pore size.

The first freestanding nanotubes realized by ALD were fabricated by the infiltration of a polycarbonate filter (200 nm diameter pores) with TiO_2 or ZrO_2 followed by the dissolution of the filter [120]. Later, several groups have used AAO (typically 30–500-nm pores) as a dissoluble template for the fabrication of nanotubes (Figure 1.19) [123–126]. For instance, Daub et al. synthesized ferromagnetic Ni nanotubes by ALD of Ni oxide in the pores of an AAO template followed by reduction of the metal oxide in hydrogen atmosphere [121]. Bae et al. demonstrated the possibility of producing coaxial nested TiO_2 nanotubes by introducing dissoluble Al_2O_3 spacer layers between two or more TiO_2 coatings in an AAO template [127]. Gu et al. applied the same trick to synthesize multiwalled HfO_2 nanotubes [122]. Furthermore, several authors have investigated the photocatalytic activity of TiO_2 nanotube arrays created by TiO_2 ALD in AAO templates [128–130].

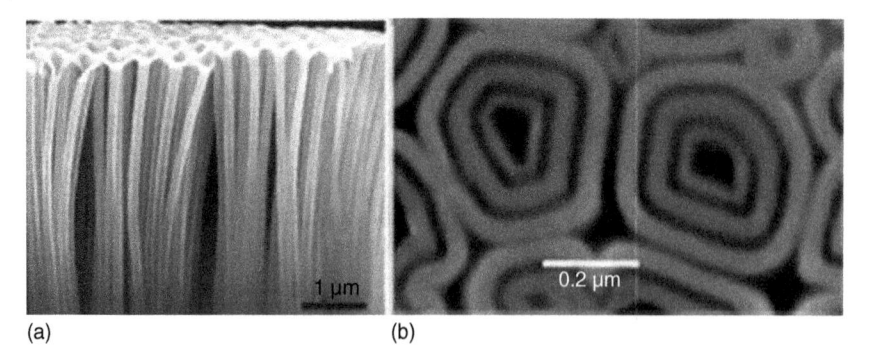

Figure 1.19 (a) SEM image of freestanding $TiO_2/Ni/TiO_2$ nanotubes obtained via ALD in an AAO template. (Daub et al. 2007 [121]. Reproduced with permission of American Institute of Physics.) (b) Top view SEM image of triple coaxial HfO_2 nanotubes obtained by using AAO and Al_2O_3 ALD layers as template and spacer layers, respectively. (Gu 2010 [122]. Reproduced with permission of American Chemical Society.)

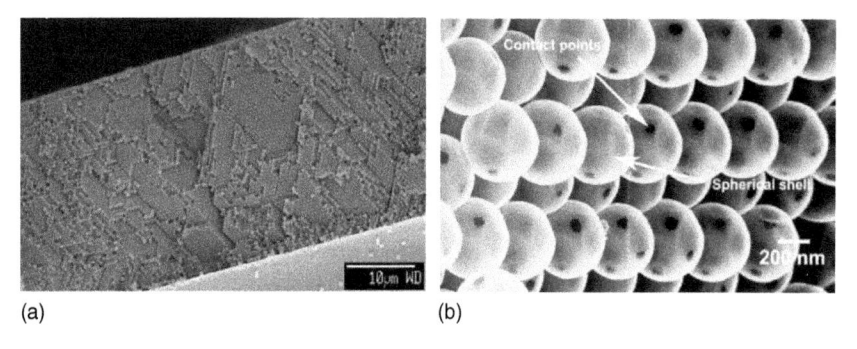

Figure 1.20 (a) SEM image of an opal film. (Karuturi et al. 2010 [138]. Reproduced with permission of American Chemical Society.) (b) SEM image of a TiO_2 inverse opal synthesized by ALD. (King et al. 2005 [132]. Reproduced with permission of Wiley.)

ALD has also been used for infiltration and replication of opal structures made of close-packed colloidal silica or polystyrene spheres [131–138]. Opal replicas, or inverse opals, thus consist of a regular arrangement of submicrometer air voids embedded in a solid matrix material. Because of their periodically modulated dielectric constant, these structures can exhibit a photonic band gap (photon wavelengths for which wave propagation is not possible inside the material) and are therefore promising candidates as 3D photonic crystals. Figure 1.20 shows an opal structure consisting of 510-nm polystyrene particles. It is clear that only a highly conformal deposition method such as ALD can achieve uniform infiltration of the opal film. Successful replication of an opal structure by means of TiO_2 ALD and subsequent etching of the silica spheres could be demonstrated.

Most research has focused on ALD coatings in materials with pore sizes >30 nm, for example, using the aforementioned Si-based trench structures, AAO and opal structures. Fewer studies have focused on ALD coatings in sub-10 nm pores. George and coworkers investigated ALD of Al_2O_3, TiO_2, and SiO_2 in 5-nm tubular alumina membranes [78, 139]. In each ALD half cycle, the pore diameter was derived from *in situ* N_2 conductance measurements (assuming Knudsen flow in the pores). The pore size was smaller after a precursor exposure than after the subsequent H_2O exposure, in accordance with the replacement of the bulky precursor ligands on the pore walls by the smaller OH groups during the H_2O step. The pore diameter was successfully reduced to molecular diameters (estimated in the range 3–10 Å), demonstrating the potential of ALD to tailor nanoporous membranes for specific gas separation purposes. Lin and coworkers used ALD of Al_2O_3 to modify sol–gel prepared alumina membranes with 4 nm pores [140, 141]. ALD resulted in an improved separation of water vapor over O_2 gas via the capillary condensation mechanism. McCool and DeSisto studied the pore size reduction of a mesoporous silica membrane via catalyzed ALD of SiO_2 [142, 143]. For a sufficient number of ALD cycles, the temperature dependence of the N_2 permeance through the membrane revealed a shift from Knudsen to configurational diffusion, suggesting pore sizes in the microporous regime [144]. Separation experiments with H_2 and CH_4 also showed deviation from the Knudsen diffusion mechanism in the direction of molecular sieving. Velleman *et al.* combined pore size tuning of AAO membranes by ALD with wet chemical functionalization of the coated membrane [145]. The surface modification with highly hydrophobic silane species was employed to improve the selectivity of the membrane for hydrophobic molecules. Due to hydrophobic–hydrophilic repulsions, the chemically modified membrane showed enhanced sensitivity to the transport of hydrophobic molecules over hydrophilic molecules. Chen *et al.* performed ALD of TiO_2 to reduce the pore size of kinked silica nanopores from 2.6 to 2 nm [146]. The authors demonstrated great potential of this structure for DNA sequencing.

In a series of recent papers, Dendooven *et al.* explored the limit of ALD for coating the interior surface of nano-sized pores. They used meso- and microporous SiO_2 and TiO_2 films that were deposited onto silicon substrates. Because of the well-defined sample structure, ALD into the nanoporous layers could be monitored *in situ* using XRF, GISAXS [80], and EP [79].

A first series of experiments focused on ALD into mesoporous silica thin films consisting of an unordered 3D network of silica nanoslabs, with controllable average pore diameters in the range of 6–20 nm. The films exhibited high porosity (70–80%) and excellent 3D pore accessibility. Figure 1.21 summarizes the results obtained for ALD of TiO_2 from tetrakis(dimethylamino) titanium (TDMAT) and H_2O in the 3D mesoporous network of nanoslab-based silica thin films [79, 147]. The ALD conditions for reaching saturation in the mesoporous films were investigated via determination of the chemical composition using *in situ* XRF. Films were successively exposed to 1 s TDMAT pulses, each of them followed by 20 s XRF data collection. The Ti XRF intensity (Ti Kα peak area), which is proportional to the amount of Ti atoms deposited in the mesoporous thin film, is plotted against the TDMAT exposure time in Figure 1.21b. About 4 s of exposure was needed to reach saturation. Prolongation of the exposure up to 20 s did not significantly enhance the uptake. It is concluded that the penetration of the ALD precursor proceeds readily in the 3D network of interconnected channels. Exposure of a mesoporous thin film to repeated cycles of the TDMAT/H_2O ALD process is expected to result in the conformal

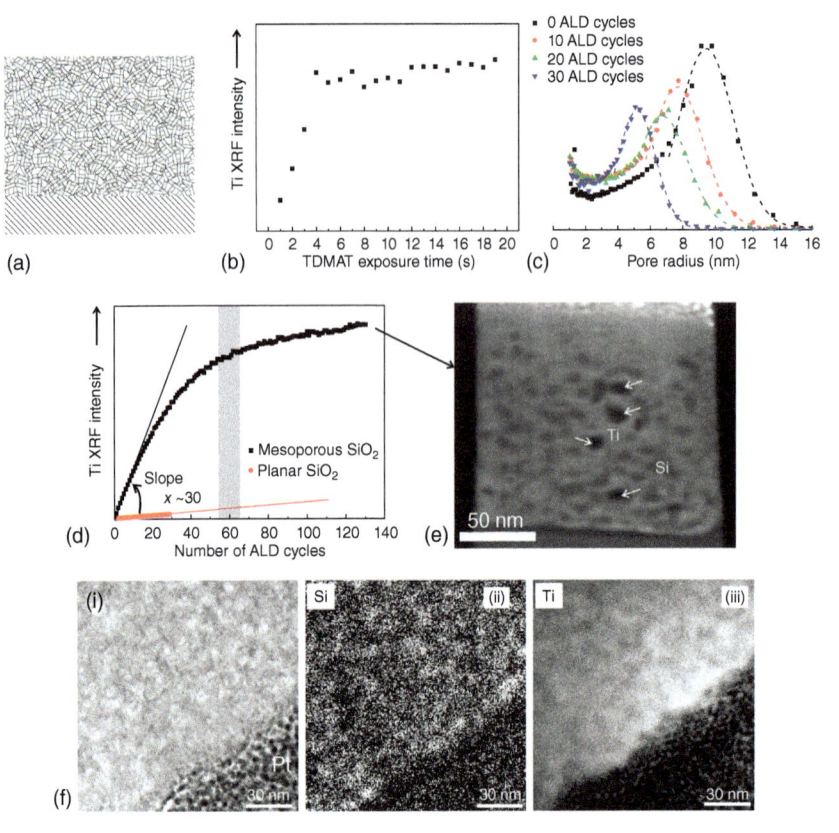

Figure 1.21 TiO_2 ALD in mesoporous silica thin films. (a) Schematic representation of the nanoslab-based mesoporous films. (b) Ti XRF intensity as a function of the TDMAT exposure time on a 115-nm-thick film with about 6.5 nm pores and about 75% porosity. (c) Pore radius distribution calculated from *in situ* EP data measured every 10 ALD cycles on a 150-nm thick film with about 18 nm pores and about 80% porosity. (d) Ti XRF intensity against the number of ALD cycles on a 120-nm film with about 7.5 nm pores and about 75% porosity and on a planar SiO_2 substrate. (e) Electron tomography study of the TiO_2-coated film in (d): out-of-plane orthoslice through the 3D reconstruction of a micropillar sample. Dark gray, silica; light gray, TiO_2; and black (arrows), voids. (f) TEM image of a cross-sectional sample of the TiO_2-coated silica film (i), Si energy-filtered TEM map of this region (ii), and Ti energy-filtered TEM map of this region (iii).

deposition of a TiO_2 film on its pore walls and thus in a reduction of its pore size. Figure 1.21c confirms a gradual decrease in pore radius with each 10 ALD cycles deposited in channel-like mesopores with an initial average pore radius of about 9 nm. It should be noted that the pore radius reduction was not only caused by the ALD coating but also influenced by shrinkage of the porous network during ALD. Figure 1.21d shows the Ti uptake against the number of ALD cycles in a film with an initial average pore diameter of about 7.5 nm. In an ALD process performed on a planar reference substrate, the XRF intensity increased linearly with the number of ALD cycles, as expected. XRR revealed a growth rate of 0.5 Å per cycle on the planar substrate. During ALD on the mesoporous film, the slope of the XRF intensity curve was initially much larger, proving that TiO_2 got deposited onto the interior surface of the channel-like mesopores. The slope decreased gradually with the number of ALD cycles because the TiO_2 coating on the pore walls caused a gradual decrease in the pore diameter and, related to that, a decline in the interior surface area of the porous network. Evidently, the amount of Ti atoms deposited per ALD cycle is directly linked to the available surface area. Finally, the slope of the Ti XRF intensity curve became constant, suggesting that the pores were no longer accessible for the TDMAT molecules and that deposition continued on top of the filled mesoporous film. Shrinking the pore diameter below the estimated kinetic diameter of the TDMAT molecule, 0.7 nm [148], took about 60 ALD cycles, indicating a diameter decrease by 0.11 nm per cycle. The TiO_2 growth rate in the pores was thus about 0.55 Å per cycle, which is in reasonable agreement with the value found for deposition on the planar substrate (0.5 Å per cycle). The TiO_2-filled mesoporous silica film was further investigated by electron tomography. Figure 1.21e shows an out-of-plane orthoslice through the 3D reconstruction. The result confirms the deposition of TiO_2 throughout the whole film. It furthermore revealed the presence of larger pores (diameter >7.5 nm) that were not completely filled with TiO_2 ALD. These pores most likely became inaccessible for the ALD precursors due to filling of the smaller pores. Elemental distribution maps from energy-filtered TEM confirmed the presence of TiO_2 throughout the mesoporous silica film (Figure 1.21f). Similar results were obtained for ALD of TiO_2 and HfO_2 into mesoporous TiO_2 with ink-bottle-shaped mesopores [86, 149].

Dendooven *et al.* also studied the conformal deposition within microporous silica films with an average pore size of about 1 nm [81]. TDMAT molecules have a molecular diameter of about 0.7 nm and therefore managed to penetrate into the micropores of this film (Figure 1.22a). After application of one ALD cycle, about 18 times more Ti atoms were deposited in the microporous film than on a planar reference substrate (Fig. 1.22(b)). The increment of Ti loading became lower during the subsequent ALD cycles, as the micropores became too narrow after 1–3 cycles of TiO_2 deposition and were no longer accessible for the TDMAT molecules.

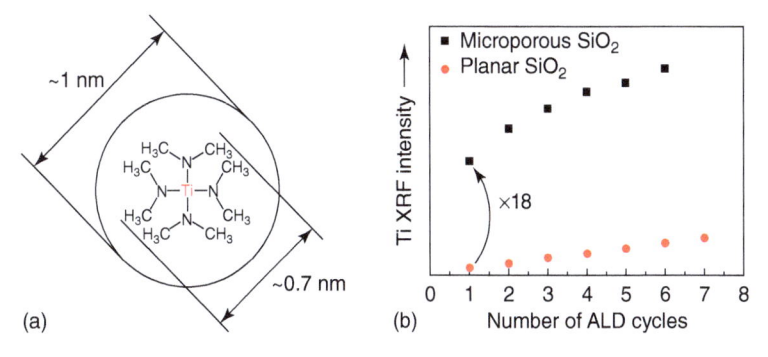

Figure 1.22 TiO_2 ALD in a microporous silica thin film with a porosity of about 40% and a thickness of about 80 nm. (a) Size of the TDMAT molecule as compared to the average pore size. (b) Ti XRF intensity against the number of ALD cycles on the microporous film and a planar SiO_2 substrate.

This systematic study clearly demonstrates the ability of ALD to deposit conformal coatings on the pore walls of both channel-like and ink-bottle-shaped mesopores with diameters in the low mesoporous and even microporous regime, indicating that ALD is ideally suited for conformal deposition into porous materials and for atomic level tuning of the pore size down to near molecular dimensions.

References

1 George, S.M. (2010) *Chem. Rev.*, **110**, 111–131.
2 Leskelä, M., Ritala, M., and Nilsen, O. (2011) *MRS Bull.*, **36**, 877–884.
3 Puurunen, R.L. (2005) *J. Appl. Phys.*, **97**, 121301.
4 Miikkulainen, V., Leskelä, M., Ritala, M., and Puurunen, R.L. (2013) *J. Appl. Phys.*, **113**, 021301.
5 Detavernier, C., Dendooven, J., Sree, S.P., Ludwig, K.F., and Martens, J.A. (2011) *Chem. Soc. Rev.*, **40**, 5242–5253.
6 Park, M.H., Jang, Y.J., Sung-Suh, H.M., and Sung, M.M. (2004) *Langmuir*, **20**, 2257–2260.
7 Chen, R., Kim, H., McIntyre, P.C., Porter, D.W., and Bent, S.F. (2005) *Appl. Phys. Lett.*, **86**, 191910.
8 Chen, R. and Bent, S.F. (2006) *Adv. Mater.*, **18**, 1086.
9 Sinha, A., Hess, D.W., and Henderson, C.L. (2006) *J. Electrochem. Soc.*, **153**, G465–G469.
10 Farm, E., Kemell, M., Ritala, M., and Leskelä, M. (2008) *J. Phys. Chem. C*, **112**, 15791–15795.
11 Mackus, A.J.M., Mulders, J.J.L., van de Sanden, M.C.M., and Kessels, W.M.M. (2010) *J. Appl. Phys.*, **107**, 116102.
12 Lee, W., Dasgupta, N.P., Trejo, O., Lee, J.-R., Hwang, J., Usui, T., and Prinz, F.B. (2010) *Langmuir*, **26**, 6845–6852.
13 Kim, W.-H., Lee, H.-B.-R., Heo, K., Lee, Y.K., Chung, T.-M., Kim, C.G., Hong, S., Heo, J., and Kim, H. (2011) *J. Electrochem. Soc.*, **158**, D1–D5.
14 Okuyama, Y., Barelli, C., Tousseau, C., Park, S., and Senzaki, Y. (2005) *J. Vac. Sci. Technol., A*, **23**, L1–L3.
15 Granneman, E., Fischer, P., Pierreux, D., Terhorst, H., and Zagwijn, P. (2007) *Surf. Coat. Technol.*, **201**, 8899–8907.
16 Poodt, P., Cameron, D.C., Dickey, E., George, S.M., Kuznetsov, V., Parsons, G.N., Roozeboom, F., Sundaram, G., and Vermeer, A. (2012) *J. Vac. Sci. Technol., A*, **30**, 010802.
17 Knez, M., Nielsch, K., and Niinistö, L. (2007) *Adv. Mater.*, **19**, 3425–3438.
18 Kim, H., Lee, H.-B.-R., and Maeng, W.-J. (2009) *Thin Solid Films*, **517**, 2563–2580.
19 Bae, C., Shin, H., and Nielsch, K. (2011) *MRS Bull.*, **36**, 877–884.
20 Elam, J.W., Dasgupta, N.P., and Prinz, F.B. (2011) *MRS Bull.*, **36**, 899–906.
21 Marichy, C., Bechelany, M., and Pinna, N. (2012) *Adv. Mater.*, **24**, 1017–1032.

22 Elam, J.W., Routkevitch, D., Mardilovich, P.P., and George, S.M. (2003) *Chem. Mater.*, **15**, 3507–3517.
23 Perez, I., Robertson, E., Banerjee, P., Henn-Lecordier, L., Son, S.J., Lee, S.B., and Rubloff, G.W. (2008) *Small*, **4**, 1223–1232.
24 Banerjee, P., Perez, I., Henn-Lecordier, L., Lee, S.B., and Rubloff, G.W. (2009) *Nat. Nanotechnol.*, **4**, 292–296.
25 Kucheyev, S.O., Biener, J., Wang, Y.M., Baumann, T.F., Wu, K.J., van Buuren, T., Hamza, A.V., Satcher, J.H., Elam, J.W., and Pellin, M.J. (2005) *Appl. Phys. Lett.*, **86**, 083108.
26 Elam, J.W., Libera, J.A., Pellin, M.J., Zinovev, A.V., Greene, J.P., and Nolen, J.A. (2006) *Appl. Phys. Lett.*, **89**, 053124.
27 Biener, J., Baumann, T.F., Wang, Y., Nelson, E.J., Kucheyev, S.O., Hamza, A.V., Kemell, M., Ritala, M., and Leskelä, M. (2007) *Nanotechnology*, **18**, 055303.
28 Ghosal, S., Baumann, T.F., King, J.S., Kucheyev, S.O., Wang, Y., Worsley, M.A., Biener, J., Bent, S.F., and Hamza, A.V. (2009) *Chem. Mater.*, **21**, 1989–1992.
29 Hakim, L.F., George, S.M., and Weimer, A.W. (2005) *Nanotechnology*, **16**, S375–S381.
30 King, D.M., Spencer, J.A. II, Liang, X., Hakim, L.F., and Weimer, A.W. (2007) *Surf. Coat. Technol.*, **201**, 9163–9171.
31 Hakim, L.F., Vaughn, C.L., Dunsheath, H.J., Carney, C.S., Liang, X., Li, P., and Weimer, A.W. (2007) *Nanotechnology*, **18**, 345603.
32 King, D.M., Liang, X., Zhou, Y., Carney, C.S., Hakim, L.F., Li, P., and Weimer, A.W. (2008) *Powder Technol.*, **183**, 356–363.
33 Zhou, Y., King, D.M., Li, J., Barrett, K.S., Goldfarb, R.B., and Weimer, A.W. (2010) *Ind. Eng. Chem. Res.*, **49**, 6964–6971.
34 Longrie, D., Deduytsche, D., and Detavernier, C. (2014) *J. Vac. Sci. Technol., A*, **32**, 010802.
35 Min, B., Lee, J., Hwang, J., Keem, K., Kang, M., Cho, K., Sung, M., Kim, S., Lee, M.-S., Park, S., and Moon, J. (2003) *J. Cryst. Growth*, **252**, 565–569.
36 Kang, M., Lee, J.-S., Sim, S.-K., Min, B., Cho, K., Kim, H., Sung, M.-Y., Kim, S., Song, S.A., and Lee, M.-S. (2004) *Thin Solid Films*, **466**, 265–271.
37 Fan, H.J., Knez, M., Scholz, R., Nielsch, K., Pippel, E., Hesse, D., Zacharias, M., and Gösele, U. (2006) *Nat. Mater.*, **5**, 627–631.
38 Law, M., Greene, L.E., Radenovic, A., Kuykendall, T., Liphardt, J., and Yang, P. (2006) *J. Phys. Chem. B*, **110**, 22652–22663.
39 Peng, Q., Sun, X.-Y., Spagnola, J.C., Hyde, G.K., Spontak, R.J., and Parsons, G.N. (2007) *Nano Lett.*, **7**, 719–722.
40 Hyde, G.K., Park, K.J., Stewart, S.M., Hinestroza, J.P., and Parsons, G.N. (2007) *Langmuir*, **23**, 9844–9849.
41 Lee, S.-M., Pippel, E., Gösele, U., Dresbach, C., Qin, Y., Chandran, C.V., Brauniger, T., Hause, G., and Knez, M. (2009) *Science*, **324**, 488–492.
42 Roth, K.M., Roberts, K.G., and Hyde, G.K. (2010) *Text. Res. J.*, **80**, 1970–1981.
43 Figure courtesy of A. Delabie, private communication.

44 Dendooven, J., Ramachandran, R.K., Devloo-Casier, K., Rampelberg, G., Filez, M., Poelman, H., Marin, G.B., Fonda, E., and Detavernier, C. (2013) *J. Phys. Chem. C*, **117**, 20557–20561.
45 George, S.M., Ott, A.W., and Klaus, J.W. (1996) *J. Phys. Chem.*, **100**, 13121.
46 Profijt, H.B., Potts, S.E., van de Sanden, M.C.M., and Kessels, W.M.M. (2011) *J. Vac. Sci. Technol., A*, **29**, 050801.
47 Profijt, H.B., Kudlacek, P., van de Sanden, M.C.M., and Kessels, W.M.M. (2011) *J. Electrochem. Soc.*, **158**, G88–G91.
48 Kariniemi, M., Niinistö, J., Vehkamäki, M., Kemell, M., Ritala, M., Leskelä, M., and Putkonen, M. (2012) *J. Vac. Sci. Technol., A*, **30**, 01A115.
49 Potts, S.E., Keuning, W., Langereis, E., Dingemans, G., van de Sanden, M.C.M., and Kessels, W.M.M. (2010) *J. Electrochem. Soc.*, **157**, P66–P74.
50 Heil, S.B.S., van Hemmen, J.L., van de Sanden, M.C.M., and Kessels, W.M.M. (2008) *J. Appl. Phys.*, **103**, 103302.
51 Rai, V.R., Vandalon, V., and Agarwal, S. (2010) *Langmuir*, **26**, 13732–13735.
52 Rai, V.R. and Agarwal, S. (2009) *J. Phys. Chem. C*, **113**, 12962–12965.
53 Kim, H. and Rossnagel, S.M. (2002) *J. Vac. Sci. Technol., A*, **20**, 802–808.
54 Musschoot, J., Deduytsche, D., Poelman, H., Haemers, J., Van Meirhaeghe, R.L., Van den Berghe, S., and Detavernier, C. (2009) *J. Electrochem. Soc.*, **156**, P122–P126.
55 Sauerbrey, G. (1959) *Z. Angew. Phys.*, **155**, 206–222.
56 Ballantine, D.S. (1997) *Acoustic Wave Sensors*, Academic Press.
57 Elam, J.W., Groner, M.D., and George, S.M. (2002) *Rev. Sci. Instrum.*, **73**, 2981.
58 Rocklein, M.N. and George, S.M. (2003) *Anal. Chem.*, **75**, 4975–4982.
59 Paul, W. and Steinwedel, H. (1953) *Z. Naturforsch.*, **80**, 448.
60 Dawson, P. (1976) *Quadrupole Mass Spectrometry*, Elsevier, Amsterdam.
61 Ritala, M., Juppo, M., Kukli, K., Rahtu, A., and Leskelä, M. (1999) *J. Phys. IV*, **9**, 8.
62 Juppo, M., Rahtu, A., Ritala, M., and Leskelä, M. (2000) *Langmuir*, **16**, 4034–4039.
63 Henn-Lecordier, L., Lei, W., Anderle, M., and Rubloff, G.W. (2007) *J. Vac. Sci. Technol., B*, **25**, 130–139.
64 Fujiware, H. (2007) *Spectroscopic Ellipsometry – Principles and Applications*, Wiley.
65 Langereis, E., Heil, S.B.S., Knoops, H.C.M., Keuning, W., van de Sanden, M.C.M., and Kessels, W.M.M. (2009) *J. Phys. D: Appl. Phys.*, **42**, 073001.
66 Chabal, Y.J. (1988) *Surf. Sci. Rep.*, **8**, 211–357.
67 Dillon, A.C., Ott, A.W., Way, J.D., and George, S.M. (1995) *Surf. Sci.*, **322**, 230.
68 Kwon, J., Dai, M., Halls, M.D., Langereis, E., Chabal, Y.J., and Gordon, R.G. (2009) *J. Phys. Chem. C*, **113**, 654.
69 Goldstein, D.N., Mccormick, J.A., and George, S.M. (2008) *J. Phys. Chem.*, **112**, 19530–19539.
70 Rai, V.R. and Agarwal, S. (2012) *J. Vac. Sci. Technol., A*, **30**, 01A158.
71 Sperling, B., Kimes, W., and Maslar, J.E. (2010) *Appl. Surf. Sci.*, **256**, 5035–5041.

72 Mackus, A.J.M., Heil, S.B.S., Langereis, E., Knoops, H.C.M., van de Sanden, M.C.M., and Kessels, W.M.M. (2010) *J. Vac. Sci. Technol., A*, **28**, 77–87.
73 Bankras, R., Holleman, J., Schmitz, J., Sturm, M., Zinine, A., Wormeester, H., and Poelsema, B. (2006) *Chem. Vap. Deposition*, **12**, 275–279.
74 Nilsen, O. and Fjellväg, H. (2011) *J. Therm. Anal. Calorim.*, **105**, 33–37.
75 Schuisky, M., Elam, J.W., and George, S.M. (2002) *Appl. Phys. Lett.*, **81**, 180–182.
76 Koukitu, A., Kumagai, T., Taki, T., and Seki, H. (1999) *Jpn. J. Appl. Phys.*, **38**, 4980–4982.
77 Rosental, A., Adamson, P., Gerst, A., and Niilisk, A. (1996) *Appl. Surf. Sci.*, **107**, 178–183.
78 Berland, B.S., Gartland, I.P., Ott, A.W., and George, S.M. (1998) *Chem. Mater.*, **10**, 3941–3950.
79 Dendooven, J., Devloo-Casier, K., Levrau, E., Van Hove, R., Sree, S.P., Baklanov, M.R., Martens, J.A., and Detavernier, C. (2012) *Langmuir*, **28**, 3852.
80 Devloo-Casier, K., Ludwig, K.F., Detavernier, C., and Dendooven, J. (2014) *J. Vac. Sci. Technol., A*, **32**, 010801.
81 Dendooven, J., Pulinthanathu Sree, S., De Keyser, K., Deduytsche, D., Martens, J.A., Ludwig, K.F., and Detavernier, C. (2011) *J. Phys. Chem. C*, **115**, 6605.
82 Fong, D.D., Eastman, J.A., Kim, S.K., Fister, T.T., Highland, M.J., Baldo, P.M., and Fuoss, P.H. (2010) *Appl. Phys. Lett.*, **97**, 191904.
83 Setthapun, W., Williams, W.D., Kim, S.M., Feng, H., Elam, J.W., Rabuffetti, F.A., Poeppelmeier, K.R., Stair, P.C., Stach, E.A., Ribeiro, F.H., Miller, J.T., and Marshall, C.L. (2010) *J. Phys. Chem. C*, **114**, 9758–9771.
84 Filez, M., Poelman, H., Ramachandran, R.K., Dendooven, J., Devloo-Casier, K., Fonda, E., Detavernier, C., and Marin, G.B. (2014) *Catal. Today*, **229**, 2–13.
85 Devloo-Casier, K., Dendooven, J., Ludwig, K.F., Lekens, G., D'Haen, J., and Detavernier, C. (2011) *Appl. Phys. Lett.*, **98**, 231905.
86 Dendooven, J., Devloo-Casier, K., Ide, M., Grandfield, K., Kurttepeli, M., Ludwig, K.F., Bals, S., Van Der Voort, P., and Detaverniera, C. (2014) *Nanoscale*, **6**, 14991–14998.
87 Dücsö, C., Khanh, N.Q., Horvath, Z., Barsony, I., Utriainen, M., Lehto, S., Nieminen, M., and Niinistö, L. (1996) *J. Electrochem. Soc.*, **143**, 683–687.
88 Ott, A.W., Klaus, J.W., Johnson, J.M., George, S.M., McCarley, K.C., and Way, J.D. (1997) *Chem. Mater.*, **9**, 707–714.
89 Ritala, M., Leskelä, M., Dekker, J.-P., Mutsaers, C., Soininen, P.J., and Skarp, J. (1999) *Chem. Vap. Deposition*, **5**, 7–9.
90 Hämäläinen, J., Munnik, F., Ritala, M., and Leskelä, M. (2008) *Chem. Mater.*, **20**, 6840–6846.
91 Diskus, M., Nilsen, O., and Fjellväg, H. (2011) *Chem. Vap. Deposition*, **17**, 135–140.
92 Masuda, H. and Satoh, M. (1996) *Jpn. J. Appl. Phys., Part 2*, **35**, L126–L129.
93 Thompson, G.E. (1997) *Thin Solid Films*, **297**, 192–201.

94 Jessensky, O., Muller, F., and Gösele, U. (1998) *Appl. Phys. Lett.*, **72**, 1173–1175.

95 Elam, J.W., Libera, J.A., Pellin, M.J., and Stair, P.C. (2007) *Appl. Phys. Lett.*, **91**, 243105.

96 Becker, J.S., Suh, S., Wang, S.L., and Gordon, R.G. (2003) *Chem. Mater.*, **15**, 2969–2976.

97 Dendooven, J., Deduytsche, D., Musschoot, J., Vanmeirhaeghe, R.L., and Detavernier, C. (2009) *J. Electrochem. Soc.*, **156**, P63–P67.

98 Dendooven, J., Deduytsche, D., Musschoot, J., Vanmeirhaeghe, R.L., and Detavernier, C. (2010) *J. Electrochem. Soc.*, **157**, G111–G116.

99 Musschoot, J., Dendooven, J., Deduytsche, D., Haemers, J., Buyle, G., and Detavernier, C. (2012) *Surf. Coat. Technol.*, **206**, 4511–4517.

100 Gao, F., Arpiainen, S., and Puurunen, R.L. (2015) *J. Vac. Sci. Technol., A*, **33**, 010601.

101 Gordon, R.G., Hausmann, D., Kim, E., and Shepard, J. (2003) *Chem. Vap. Deposition*, **9**, 73–78.

102 Gordon, R. G. (2008) *Step Coverage by ALD Films: Theory and Examples of Ideal and Non-ideal Reactions*. Presented at the AVS Topical Conference on ALD, Bruges, Belgium, June 29–July 2.

103 Knudsen, M. (1909) *Ann. Phys.*, **333**, 75–130.

104 In, S.R. (1998) *J. Vac. Sci. Technol., A*, **16**, 3495–3501.

105 Dendooven, J., Musschoot, J., Deduytsche, D., Vanmeirhaeghe, R., and Detavernier, C. (2008) *Conformality of Thermal and Plasma Enhanced ALD*. Presented at the AVS Topical Conference on ALD, Bruges, Belgium, June 29–July 2.

106 Knoops, H.C.M., Langereis, E., van de Sanden, M.C.M., and Kessels, W.M.M. (2010) *J. Electrochem. Soc.*, **157**, G241–G249.

107 Macko, P., Veis, P., and Cernogora, G. (2004) *Plasma Sources Sci. Technol.*, **13**, 251–262.

108 Adams, S.F. and Miller, T.A. (2000) *Plasma Sources Sci. Technol.*, **9**, 248–255.

109 Cvelbar, U., Mozetic, M., and Ricard, A. (2005) *IEEE Trans. Plasma Sci.*, **33**, 834–837.

110 Tserepi, A.D. and Miller, T.A. (1994) *J. Appl. Phys.*, **75**, 7231–7236.

111 Gomez, S., Steen, P.G., and Graham, W.G. (2002) *Appl. Phys. Lett.*, **81**, 19–21.

112 Kim, Y.C. and Boudart, M. (1991) *Langmuir*, **7**, 2999–3005.

113 Cartry, G., Duten, X., and Rousseau, A. (2006) *Plasma Sources Sci. Technol.*, **15**, 479–488.

114 Guyon, C., Cavadias, S., Mabille, I., Moscosa-Santillan, M., and Amouroux, J. (2004) *Catal. Today*, **89**, 159–167.

115 Wood, B.J. and Wise, H. (1961) *J. Phys. Chem.*, **65**, 1976.

116 van Hemmen, J.L., Heil, S.B.S., Klootwijk, J.H., Roozeboom, F., Hodson, C.J., van de Sanden, M.C.M., and Kessels, W.M.M. (2007) *J. Electrochem. Soc.*, **154**, G165–G169.

117 Dingemans, G., van Helvoirt, C.A.A., Pierreux, D., Keuning, W., and Kessels, W.M.M. (2012) *J. Electrochem. Soc.*, **159**, H277–H285.

118 Kubala, N.G., Rowlette, P.C., and Wolden, C.A. (2009) *J. Phys. Chem. C*, **113**, 16307–16310.
119 van den Oever, P.J., van Helden, J.H., Lamers, C.C.H., Engeln, R., Schram, D.C., van de Sanden, M.C.M., and Kessels, W.M.M. (2005) *J. Appl. Phys.*, **98**, 093301.
120 Shin, H.J., Jeong, D.K., Lee, J.G., Sung, M.M., and Kim, J.Y. (2004) *Adv. Mater.*, **16**, 1197.
121 Daub, M., Knez, M., Gösele, U., and Nielsch, K. (2007) *J. Appl. Phys.*, **101**, 09J111.
122 Gu, D., Baumgart, H., Abdel-Fattah, T.M., and Namkoong, G. (2010) *ACS Nano*, **4**, 753–758.
123 Sander, M.S., Cote, M.J., Gu, W., Kile, B.M., and Tripp, C.P. (2004) *Adv. Mater.*, **16**, 2052.
124 Yang, C.-J., Wang, S.-M., Liang, S.-W., Chang, Y.-H., Chen, C., and Shieh, J.-M. (2007) *Appl. Phys. Lett.*, **90**, 033104.
125 Kim, W.-H., Park, S.-J., Son, J.-Y., and Kim, H. (2008) *Nanotechnology*, **19**, 045302.
126 Chong, Y.T., Goerlitz, D., Martens, S., Yau, M.Y.E., Allende, S., Bachmann, J., and Nielsch, K. (2010) *Adv. Mater.*, **22**, 2435.
127 Bae, C., Yoon, Y., Yoo, H., Han, D., Cho, J., Lee, B.H., Sung, M.M., Lee, M., Kim, J., and Shin, H. (2009) *Chem. Mater.*, **21**, 2574–2576.
128 Kemell, M., Pore, V., Tupala, J., Ritala, M., and Leskelä, M. (2007) *Chem. Mater.*, **19**, 1816–1820.
129 Ng, C.J.W., Gao, H., and Tan, T.T.Y. (2008) *Nanotechnology*, **19**, 445604.
130 Liang, Y.-C., Wang, C.-C., Kei, C.-C., Hsueh, Y.-C., Cho, W.-H., and Perng, T.-P. (2011) *J. Phys. Chem. C*, **115**, 9498–9502.
131 Rugge, A., Becker, J.S., Gordon, R.G., and Tolbert, S.H. (2003) *Nano Lett.*, **3**, 1293–1297.
132 King, J.S., Graugnard, E., and Summers, C.J. (2005) *Adv. Mater.*, **17**, 1010.
133 King, J.S., Heineman, D., Graugnard, E., and Summers, C.J. (2005) *Appl. Surf. Sci.*, **244**, 511–516.
134 Graugnard, E., Chawla, V., Lorang, D., and Summers, C.J. (2006) *Appl. Phys. Lett.*, **89**, 211102.
135 Sechrist, Z.A., Schwartz, B.T., Lee, J.H., McCormick, J.A., Piestun, R., Park, W., and George, S.M. (2006) *Chem. Mater.*, **18**, 3562–3570.
136 Povey, I.M., Bardosova, M., Chalvet, F., Pemble, M.E., and Yates, H.M. (2007) *Surf. Coat. Technol.*, **201**, 9345–9348.
137 Hwang, D.-K., Noh, H., Cao, H., and Chang, R.P.H. (2009) *Appl. Phys. Lett.*, **95**, 091101.
138 Karuturi, S.K., Liu, L., Su, L.T., Zhao, Y., Fan, H.J., Ge, X., He, S., and Yoong, A.T.I. (2010) *J. Phys. Chem. C*, **114**, 14843–14848.
139 Cameron, M.A., Gartland, I.P., Smith, J.A., Diaz, S.F., and George, S.M. (2000) *Langmuir*, **16**, 7435–7444.
140 Pan, M., Cooper, C., Lin, Y., and Meng, G. (1999) *J. Membr. Sci.*, **158**, 235–241.
141 Cooper, C.A. and Lin, Y.S. (2002) *J. Membr. Sci.*, **195**, 35–50.
142 McCool, B.A. and DeSisto, W.J. (2004) *Chem. Vap. Deposition*, **10**, 190.

143 McCool, B.A. and DeSisto, W.J. (2004) *Ind. Eng. Chem. Res.*, **43**, 2478–2484.
144 Schuring, D. (2002) Diffusion in zeolites: Towards a microscopic understanding, PhD thesis. Eindhoven University of Technology.
145 Velleman, L., Triani, G., Evans, P.J., Shapter, J.G., and Losic, D. (2009) *Microporous Mesoporous Mater.*, **126**, 87–94.
146 Chen, Z., Jiang, Y., Dunphy, D.R., Adams, D.P., Hodges, C., Liu, N., Zhang, N., Xomeritakis, G., Jin, X., Aluru, N.R., Gaik, S.J., Hillhouse, H.W., and Brinker, C.J. (2010) *Nat. Mater.*, **9**, 667–675.
147 Dendooven, J., Pulinthanathu Sree, S., De Keyser, K., Deduytsche, D., Martens, J.A., Ludwig, K.F., and Detavernier, C. (2011) *J. Phys. Chem. C*, **115**, 6605–6610.
148 Davie, M.E., Foerster, T., Parsons, S., Pulham, C., Rankin, D.W.H., and Smart, B.A. (2006) *Polyhedron*, **25**, 923–929.
149 Dendooven, J., Goris, B., Devloo-Casier, K., Levrau, E., Biermans, E., Baklanov, M.R., Ludwig, K.F., Van Der Voort, P., Bals, S., and Detavernier, C. (2012) *Chem. Mater.*, **24**, 1992–1994.

Part II

Atomic Layer Deposition in Photovoltaic Devices

2

Atomic Layer Deposition for High-Efficiency Crystalline Silicon Solar Cells

Bart Macco, Bas W. H. van de Loo, and Wilhelmus M. M. Kessels

Eindhoven University of Technology, Department of Applied Physics, De Rondom 1, Building TNO, 5612 AP Eindhoven, The Netherlands

2.1 Introduction to High-Efficiency Crystalline Silicon Solar Cells

At present, crystalline silicon (Si) solar cells are being mass-produced at a typical rate of 43 GWp per year, and they dominate the photovoltaic (PV) market with a share of >90% [1]. The success of Si solar cells over other solar cell technologies lies in their ever-improving cost-effectiveness. In fact, the Si solar cells themselves are not even the main contributor to the total cost of solar electricity anymore [1]. Because of this, higher conversion efficiencies have become the main driver for a further cost reduction of solar electricity. As is the case for most semiconductor devices, also for Si solar cells a precise engineering of interfaces has proven to be key in achieving optimal device performance and consequently higher conversion efficiencies. For instance, by reducing the interface defect density of Si by a thin passivation layer of Al_2O_3, the efficiency of Si solar cells could significantly be enhanced. It was for this particular application that the potential of atomic layer deposition (ALD) for Si solar cells was for the first time revealed in 2004 [2]. Due to the development of novel high-throughput ALD reactors, ALD of Al_2O_3 has since then even found its way to the high-volume manufacturing (HVM) of Si solar cells. Today, Al_2O_3 nanolayers prepared by ALD can account for a ~1% absolute increase in conversion efficiency of commercial solar cells, and they are incorporated in solar cells with record efficiencies of >25% on a lab scale [3]. In a broader context, the success of ALD Al_2O_3 also triggered the exploration of ALD for the preparation of other functional layers for a variety of commercial solar cells. For example, ALD is used for the deposition of transparent conductive oxides (TCOs). TCOs should be simultaneously transparent and conductive and therefore should have a high electron mobility. It turns out that ALD is well suited for the deposition of such films, as is, for instance, evident from ALD TCOs with record-high electron mobilities of 138 $cm^2 V^{-1} s^{-1}$ [4].

Macco and van de Loo contributed equally to this work.

Atomic Layer Deposition in Energy Conversion Applications, First Edition. Edited by Julien Bachmann.
© 2017 Wiley-VCH Verlag GmbH & Co. KGaA. Published 2017 by Wiley-VCH Verlag GmbH & Co. KGaA.

This chapter illustrates that ALD is in fact an enabler of novel high-efficiency Si solar cells, owing to its (unique) merits such as a high material quality, precise thickness control, and the ability to prepare film stacks in a well-controlled way. In the remainder of this section, a brief overview of the field of Si solar cells is given, where for each concept, the present and potential role of ALD is discussed.

In short, a solar cell can deliver power when excess electrons and holes, created by the absorption of light in the semiconductor, are extracted separately by electrodes. To enable this, the two types of electrodes should be made selective for the extraction of only electrons or holes. Commonly, Si solar cells are classified by the type of junction that is formed in the creation of such *carrier selective contacts*. Based on such a classification, in Figure 2.1, a concise overview of the most common Si solar cell concepts is given. Note that each individual concept is discussed in more detail as follows. The first class of solar cells (which is industrially by far the most used) is based on *homojunctions*, which are formed by doping separate regions of the Si, making them strongly n- or p-type (also referred to as n^+ or p^+ Si). In this way, the metal electrode that contacts the n^+ or p^+ Si is made selective for the extraction of either electrons or holes, respectively. *Heterojunctions*, which are junctions between Si and other materials, are also used to achieve such charge carrier selectivity. Interestingly, for these types of cells, that is, whether involving homojunctions, heterojunctions, or a combination of both, record conversion efficiencies over 25% have been demonstrated on a lab scale [3, 5–8]. Although such efficiencies are already close to the predicted fundamental limit of 29.4% for Si solar cells [9], research efforts are still ongoing to attain more cost-effective approaches for the HVM of such cells in the industry.

2.1.1 ALD for Si Homojunction Solar Cells

The current workhorse of the PV industry is the aluminum back-surface field (Al-BSF) solar cell, which is based on p-type Si (Figure 2.1). It typically yields moderate conversion efficiencies of 19–20% and can be manufactured very cost-effectively. On the front side, an n^+ Si region is present (conventionally termed an "emitter"), which makes the front contacts selective for the extraction of electrons. Moreover, the n^+ Si also serves as lateral conduction pathway for electrons to the front contact grid. In between the metal contacts, hydrogenated amorphous silicon nitride, a-SiN$_x$:H (also referred to as SiN$_x$), is typically present as antireflection coating (ARC). Additionally, the SiN$_x$ reduces electron–hole recombination at the Si surface, which is referred to as *surface passivation*. The latter is vital to achieve a high open-circuit voltage, V_{oc}, and hence a high solar cell efficiency. At the full back side of the Al-BSF solar cell, Al is screen-printed and subsequently (partially) alloyed with the Si by a high-temperature "firing" step at 800 °C. In this way, Al contacts Al-doped (p^+-type) Si, and a hole-selective contact is formed. In the Al-BSF concept, generally no layers are prepared by ALD.

A more advanced concept, which is still industrial but allows for higher conversion efficiencies, is the so-called passivated emitter rear contact (PERC) solar cell. Instead of using Al over the full area (as is the case for the Al-BSF cell), it uses *local* Al contacts at the rear side. In this way, the Si surface in between the contacts can

Homojunction-based c-Si solar cells

Al-BSF cell
- Mass-produced
- No ALD
- Cell efficiency ~16–20%

High-efficiency homojunctions cells:
- ALD for surface passivation (Al$_2$O$_3$, SiO$_2$)

PERC/PERL
23.9%

Interdigitated back contact (IBC)
25%

Heterojunction-based c-Si solar cells

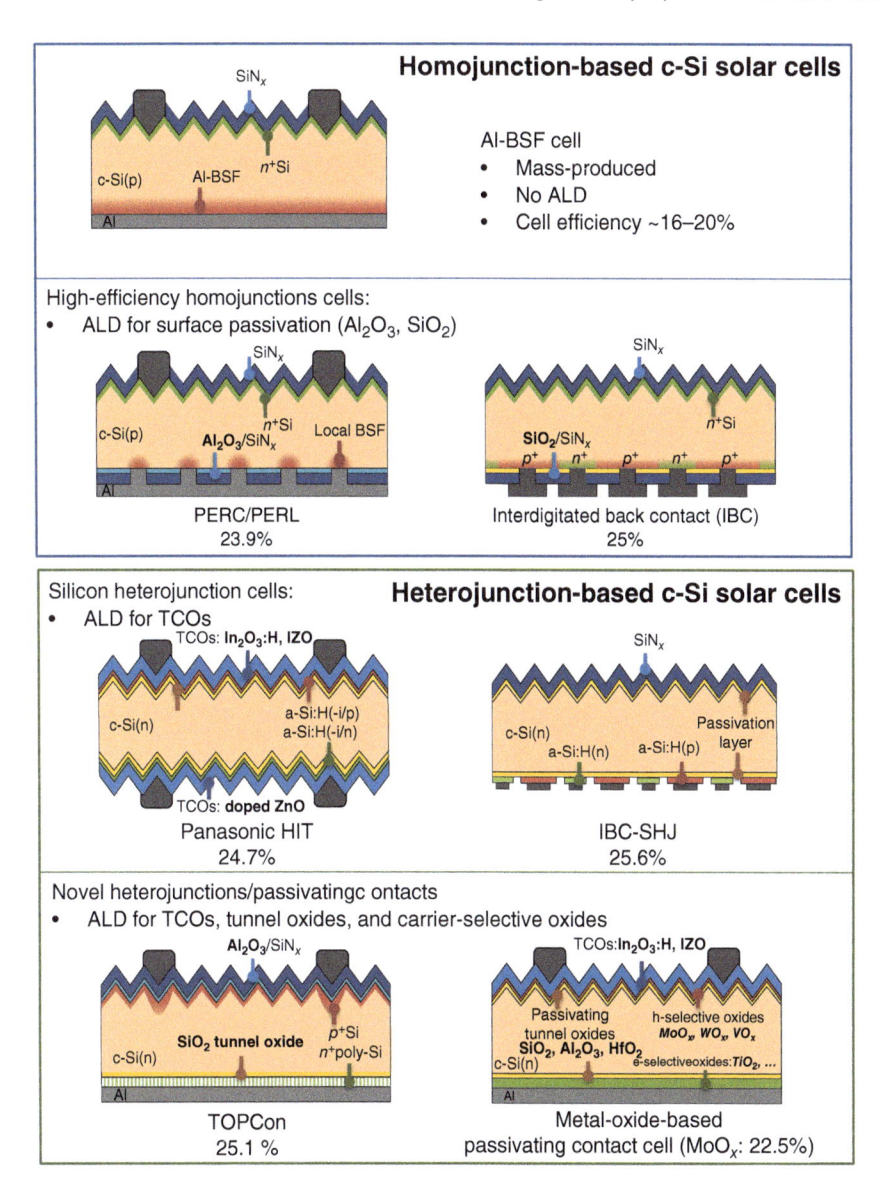

Silicon heterojunction cells:
- ALD for TCOs

Panasonic HIT
24.7%

IBC-SHJ
25.6%

Novel heterojunctions/passivatingc ontacts
- ALD for TCOs, tunnel oxides, and carrier-selective oxides

TOPCon
25.1 %

Metal-oxide-based
passivating contact cell (MoO$_x$: 22.5%)

Figure 2.1 Various silicon solar cell concepts and their current-record efficiencies (January 2016). Note that these efficiencies might be several percent absolutely lower when produced industrially. The first class of cells is based on homojunctions, the second class on heterojunctions. The functional thin films, which could be prepared by ALD, are indicated in bold.

be passivated by a thin film, which reduces charge carrier recombination and in this way enhances the conversion efficiency. In the past decade, ALD of Al$_2$O$_3$ has been proven to be very successful for the passivation of this lowly doped p-type Si back surface, and this application even paved the way for ALD for implementation in solar cell manufacturing.

Besides reducing charge carrier recombination at the Si back surface, the recombination via defects in the Si bulk should also be minimized to attain high conversion efficiencies. For instance, rather than the low-cost multicrystalline Si, which has defects at its grain boundaries, more expensive monocrystalline Si wafers can be used. In addition, *n*-type Si wafers can be used instead of the conventionally used *p*-type Si wafers, as they generally have a higher bulk material quality due to, for example, the absence of boron–oxygen defects. Therefore, *n*-type cell concepts compete with (*p*-type) PERC concepts for a spot in the high-efficiency segment. Most solar cells based on *n*-type Si require *p*-type doping of the full Si front surface (such a cell is not shown in Figure 2.1). For a long time, this approach was considered to be challenging, in part due to a lack of a suitable passivation scheme for p^+ Si. In addition, for the passivation of these p^+ Si surfaces, Al_2O_3 prepared by ALD resulted in a breakthrough, yielding solar cell conversion efficiencies of up to 23.2% at that time [10].

Finally, even higher efficiencies can in principle be achieved by using interdigitated back-contact (IBC) cell concepts [6], albeit at the price of requiring more processing steps. In IBC concepts, the electron- and the hole-selective contacts are localized at the back side to avoid parasitic absorption and reflection by front metallization. In addition, for IBC solar cells, ALD layers are being investigated, for instance, to passivate defects at the n^+ or p^+ Si surfaces at the back side of the solar cell simultaneously [11].

2.1.2 ALD for Si Heterojunction Solar Cells

A radically different solar cell design, which has successfully been developed by Panasonic (formerly Sanyo), is the heterojunction with intrinsic thin layer (HIT)[12], generally referred to as the silicon heterojunction (SHJ) concept. Here, intrinsic and doped hydrogenated amorphous Si (a-Si:H) layers are used for the passivation of the Si surfaces and for achieving selectivity for the extraction of charge carriers, respectively. For conventional Si homojunction solar cells, the lateral conduction of excess charge carriers toward the metal grid takes place via the highly doped regions. However, in SHJ cells, such regions are absent, which is why TCOs are used on both sides of the SHJ cell. In fact, TCOs prepared by ALD can fulfill the stringent requirements set by the SHJ cells in terms of processing (e.g., a low deposition temperature <200 °C and *soft* deposition) as well as in material quality (e.g., high conductivity and transparency and a suitable work function). Despite its high potential, ALD is not yet implemented in HVM for this application. Nonetheless, the deposition of TCOs in an industrial ALD reactor has already been achieved [13].

The discrepancy between the maximal efficiency for conventional SHJ cells (25.1% [14]) and the fundamental limit of Si solar cells (i.e., 29.4% [9]) is partially attributed to the parasitic absorption or reflection by the front a-Si:H layers and the TCO. This can be overcome in several ways. For instance, by using an IBC-SHJ concept, no front TCO is required and the a-Si:H layers are only used at the back side. In this way, the current world-record efficiency for silicon solar cells of 25.6% has been achieved by Panasonic [7]. It is not disclosed at this moment whether ALD is used in this concept.

2.1.3 Novel Passivating Contacts and ALD

Due to the limit in processing temperature, which can be a challenge for HVM, and due to parasitic absorption of a-Si:H-based heterojunction cells, research efforts are also targeting other types of materials that make the electrodes carrier-selective. These are generally referred to as *passivating contacts*. For example, a novel passivating contact approach is the tunnel oxide passivated contact (TOPCon). It has excellent thermal stability and so far has reached a conversion efficiency of 25.1% [3]. TOPCon is in fact a *hybrid* cell design, as on the front, a classical homojunction is formed, which is passivated by Al_2O_3, whereas at the full back side, a highly doped (partially) crystalline Si passivating contact forms a heterojunction. The contact is made selective to extract either electrons or holes by the choice of dopant. Moreover, a very thin, passivating tunnel oxide at the interface between Si and the partially crystalline Si is key in achieving the high conversion efficiencies [15]. Conceivably, the precise thickness control offered by ALD can play a key role in studying and optimizing the often encountered trade-off between surface passivation and contact resistance in passivating contacts.

Interestingly, there is also increasing interest in novel heterojunctions based on metal oxides as passivating contacts. Preferably, they are fully transparent, in contrast to partially crystalline Si and a-Si:H. If such a passivating contact can be used on the full area of the solar cell, they could yield a significant process simplification. Metal oxides of interest include, for example, MoO_x [16], WO_x [17], NiO_x [18], and TiO_x [19], for which ALD processes readily exist and are also being explored [20]. In addition, well-defined stacks of metal oxides or other materials can be used, which provide surface passivation, carrier selectivity as well as lateral conduction [21, 22]. Potentially, such stacks could find their way into novel silicon solar cell concepts, allowing for very high conversion efficiencies in combination with cost-effective manufacturing.

2.1.4 Outline of this Chapter

The chapter is organized as follows. In Section 2.2, the role of ALD in preparing passivation layers for homojunction Si solar cells is discussed. Special attention is given to the physics of surface passivation, the surface passivation by ALD Al_2O_3, ALD as a high-throughput deposition technique in the PV industry, and recent developments in the field of passivation layers prepared by ALD. Section 2.3 focuses on TCOs prepared by ALD for use in heterojunction Si solar cells, such as doped ZnO and In_2O_3 films. Therefore, the physics of TCOs and the specific requirements of TCOs for SHJ cells are briefly discussed first. Finally, in Section 2.4, novel passivating contacts based on metal oxides are considered. Various unique aspects of ALD, such as the ability to prepare precisely tailored stacks play an important role here.

2.2 Nanolayers for Surface Passivation of Si Homojunction Solar Cells

In this section, the status quo and opportunities for ALD passivation layers for Si homojunction solar cells are discussed. First, the physical mechanisms of surface recombination and the basics of surface passivation are treated. Next, an overview of ALD Al_2O_3, which is widely used for the passivation of p- and p^+-type surfaces, is given. Moreover, also the requirements for deposition techniques in the HVM of silicon solar cells in industry are discussed together with novel high-throughput ALD reactors. In the final section, further opportunities and recent developments in the field of ALD for the passivation of Si are outlined. Examples include ALD layers for the passivation of n^+ Si, the passivation of surfaces with demanding topologies, and recent developments of novel, alternative passivation layers prepared by ALD.

2.2.1 Basics of Surface Passivation

2.2.1.1 The Physics of Surface Recombination

Upon absorption of light in Si, excess electrons and holes are created (Figure 2.2), with densities Δn and Δp, respectively, which increases their respective densities n and p from their equilibrium values n_0 and p_0 to $n = n_0 + \Delta n$ and $p = p_0 + \Delta p$. After generation, the excess charge carriers thermalize very rapidly (i.e., in $\sim 10^{-12}$ s) to the temperature of the Si lattice, T. Rather than by a single Fermi energy level, E_F, which is used to describe semiconductors in the dark, under illumination, the distributions of electrons and holes can be described by their own quasi-Fermi energies, E_{Fn} and E_{Fp}. Note that the carrier densities are still given by the energetic distance of the quasi Fermi level to the conduction and valence band, respectively. For Boltzmann statistics, the pn product is given by

$$p \cdot n = (p_0 + \Delta p)(n_0 + \Delta n) = n_i^2 \cdot \exp\left(\frac{E_{Fn} - E_{Fp}}{kT}\right) \quad (2.1)$$

with n_i the intrinsic carrier density and k the Boltzmann constant. The free energy per electron–hole pair is $E_{Fn} - E_{Fp}$ and corresponds to the implied voltage iV_{oc} in the cell, that is, $E_{Fn} - E_{Fp} = q \cdot iV_{oc}$, with q the elementary charge.

The quasi-Fermi level splitting induced by the generation of charge carriers is also the driver of processes that tend to restore thermodynamic equilibrium. In such processes, excess charge carriers recombine and their associated free energy is lost. To a certain extent, this recombination is unavoidable, because of intrinsic channels, which are direct (radiative) and Auger recombination. In these processes, the energy is transferred to a photon and/or phonons or a third carrier. In addition, recombination can take place via states in the band gap of Si, formed by extrinsic lattice defects or impurities. This pathway is named after Shockley, Read, and Hall (SRH), who formulated the theory describing this kind of recombination. In particular at the surface, as the silicon lattice terminates, a high density of Si dangling bonds is natively present, which form defect levels (e.g., P_{b0} defects) and therefore induce strong SRH recombination.

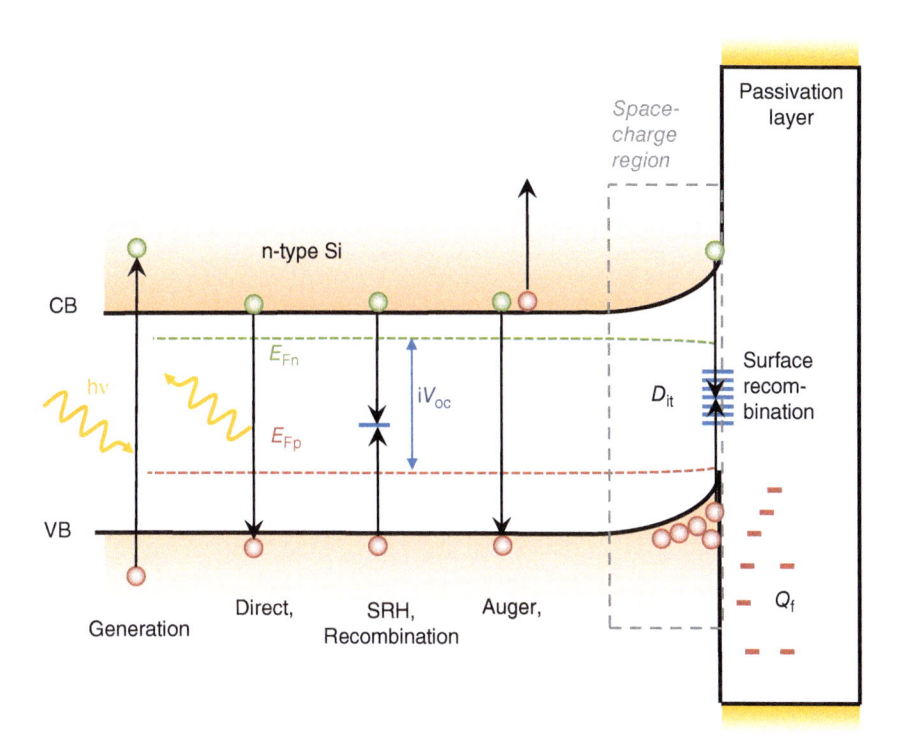

Figure 2.2 Schematic band diagram of the passivated silicon surface, indicating the generation of excess carriers by the absorption of light and the recombination thereof, which can take place directly, via a third carrier (in the Auger process), or via defect states (SRH recombination). At the surface, a high density of interface (defect) states D_{it} exists. A passivation layer reduces the D_{it}, and (in this example) reduces the electron concentration at the surface via a negative fixed charge density Q_f, which induces a space-charge region and upward band bending.

The recombination rate of charge carriers at the surface U_s (with units cm^{-2} s^{-1}) can be described via the extended SRH formalism, which considers a continuum of defects throughout the band gap of Si, with energy-dependent density, $D_{it}(E)$,

$$U_s = (p_s n_s - n_{i,\text{eff},s}^2) \cdot \int_{E_g} \frac{dE}{\frac{n_s + n_1(E)}{S_{p0}(E)} + \frac{p_s + p_1(E)}{S_{n0}(E)}} \tag{2.2}$$

with

$$S_{n0}(E) = v_t D_{it}(E) \sigma_n(E) \quad \text{and} \quad S_{p0}(E) = v_t D_{it}(E) \sigma_p(E)$$

In these expressions, v_t is the thermal velocity, $n_{i,\text{eff},s}$ is the *effective* intrinsic carrier concentration at the surface (taking into account band-gap narrowing and Fermi–Dirac statistics), n_s and p_s are the electron and hole concentrations at the surface, and $\sigma_n(E)$ and $\sigma_p(E)$ are the (energy-dependent) hole or electron capture cross sections, respectively, which are directly related to the physical processes of carrier capture by the defect states. The SRH densities n_ℓ and p_ℓ given by

$$n_1 = N_C \exp\left(-\frac{E_C - E_t}{kT}\right) \quad \text{and} \quad p_1 = N_V \exp\left(\frac{E_V - E_t}{kT}\right)$$

determine the effectiveness of the defects as recombination sites, where E_t is the energy level of the defect and N_C and N_V are the effective density of states of the conduction and valence band, respectively. In particular, defects near the center of the band gap of Si are most effective as recombination sites. In addition, when the capture rate of electrons by a defect equals the capture rate of holes, that is, when the following condition is satisfied,

$$\sigma_p \cdot p_s \approx \sigma_n \cdot n_s \tag{2.3}$$

the defect is most effective as a recombination site.

Even though U_s is the main physical parameter, which should be minimized, it is experimentally not directly accessible. For this reason, other figures of merit are usually used to assess the surface recombination (see Table 2.1), such as the effective minority carrier lifetime τ_{eff} and the implied open-circuit voltage iV_{oc}, both of which are ideally as high as possible. However, both of these parameters also include the influence of other recombination processes, such as those in the bulk of Si. To quantify only the surface recombination, the effective surface recombination velocity S_{eff} or surface saturation current density J_{0s} can be evaluated [23, 24]. The latter parameter J_{0s} (with units A cm^{-2}) offers the advantage that it (for flat quasi-Fermi levels) directly translates to the open-circuit voltage V_{oc} of the solar cell via

$$V_{oc} = \frac{nkT}{q} \ln\left(\frac{J_{sc}}{J_{0s} + J_{0,\text{others}}} + 1\right) \tag{2.4}$$

with n being the ideality factor, J_{sc} the short-circuit current density, and $J_{0,\text{others}}$ a parameter accounting for other recombination pathways, such as in the Si bulk. Due to the large surface-to-volume ratio in Si solar cells, the surface can be the dominant recombination source. With current trends toward thinner or higher quality Si wafers, bulk recombination is further reduced, making surface recombination even more important.

2.2.1.2 Surface Passivation

To minimize surface recombination, several strategies can be used. First of all, D_{it}, which is natively present, can be reduced by orders of magnitude by several approaches, which is referred to as *chemical passivation*. Chemical passivation can be obtained by depositing a thin film on the silicon surface, which binds to the Si dangling bonds. Moreover, (atomic) hydrogen released from the passivation scheme or the forming gas during a postdeposition annealing (PDA) can subsequently passivate the remaining defects. The most prominent examples of materials used for surface passivation of solar cells include thermally grown SiO_2 [25], plasma-enhanced chemical vapor deposited (PECVD) SiN_x [26], PECVD a-Si:H, and ALD Al_2O_3 [27, 28]. Naturally, to achieve chemical passivation, the passivation layers should also exhibit few states at energy levels that are relevant for charge-carrier recombination in Si. For this reason, materials with a wide band

Table 2.1 A selection of commonly used parameters to assess surface passivation quality.

Parameter	Symbol	Encompasses recombination in or at the Si	Assessed by	Definition
Effective minority carrier lifetime	τ_{eff}	Bulk, HDR, SCR, surface	PC, PL	$\tau_{\text{eff}} \equiv \dfrac{\Delta n}{U_{(\text{total})}}$
Surface recombination velocity (SRV)	S	Surface	Inaccessible (in most cases)	$S \equiv \dfrac{U_s}{\Delta n_s}$
Effective surface recombination velocity	S_{eff}	Surface, SCR	PC, PL	$S_{\text{eff}} \equiv \dfrac{U_s}{\Delta n_d}$
Implied open-circuit voltage	iV_{oc}	Bulk, HDR, SCR, surface	PC, PL	$iV_{\text{oc}} \equiv \dfrac{kT}{q} \ln\left(\dfrac{np}{n_i^2}\right)$
Surface saturation current density	J_{0s}	Surface	PC, PL	$J_{0s} \equiv qU_s\left(\dfrac{p_s n_s}{n_{i,\text{eff},s}^2} - 1\right)^{-1}$
("Emitter") Saturation current density	J_{0e}, J_0	HDR, SCR, surface	PC, PL, J–V, Suns-V_{oc}	$J_{0e} \equiv qU_{\text{HDR-surface}}\left(\dfrac{p_w n_w}{n_{i,\text{eff},w}^2} - 1\right)^{-1}$

Some parameters not only account for recombination at the surface but also encompass recombination in the highly doped region (HDR), in the Si bulk, or in the space-charge region (SCR) induced in the Si by the passivation scheme. The parameters are evaluated with respect to their average carrier densities in the Si bulk (n, p), the densities at the surface (n_s, p_s), the edge of the SCR (n_d, p_d), or the base side of the HDR (n_w, p_w). They can be assessed by photoconductance (PC), photoluminescence (PL), current–voltage (J–V), or Suns-V_{oc} measurements.

gap and low impurity content are most successful. Metals, on the contrary, have a large and continuous distribution of states and therefore act as catalyst for carrier recombination when brought in contact with Si.

Another strategy to reduce surface recombination is the suppression of either the electron or hole concentration at the surface (see Equation 2.2). Such suppression can be achieved via band bending in the Si, for instance, by a fixed charge density Q_f in the passivation layer and is commonly known as *field-effect passivation*. Thermally grown SiO_2 natively possesses a slight positive $Q_f \sim 10^{11}$ cm^{-2}, with SiN_x being a strong positive Q_f (typically $\sim 3 \times 10^{12}$ cm^{-2}), whereas ALD Al_2O_3 films typically exhibit a strong negative Q_f ($\sim 10^{12}$–10^{13} cm^{-2}) [29]. In Figure 2.3, a schematic overview of interface properties of various passivation schemes is given.

Naturally, the carrier densities at the Si surface are, besides band bending, largely dictated by the (local) *doping level* of the silicon N_s. The polarity of the fixed charge density in the dielectric is of importance when passivating doped surfaces, as Q_f can either further reduce or increase the minority carrier

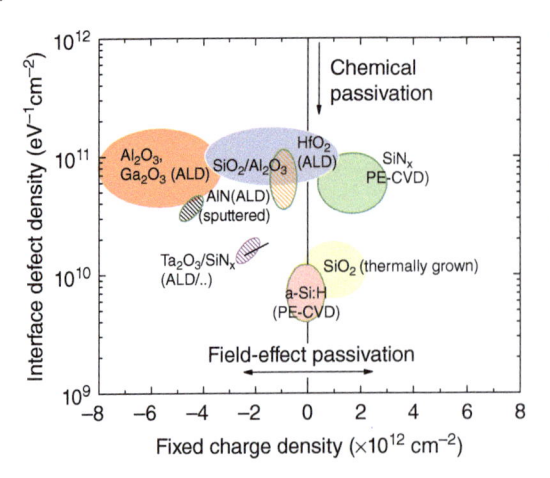

Figure 2.3 Schematic overview of typical interface defect densities and fixed charge densities of passivation schemes, adapted from Ref. [30]. Note that the actual interface properties might strongly depend on the processing conditions of the passivation layer.

density and, in this way, has a strong influence on the surface recombination. For instance, for heavily p-type doped surfaces (p^+ Si), electrons are the minority carriers. The passivation scheme for such a surface therefore ideally exhibits a negative Q_f to reduce the electron concentration even further. For n^+ Si surfaces, on the other hand, electrons are the *majority* carriers and their density is not reduced significantly by a negative Q_f. Even worse, the negative Q_f increases the minority carrier density, which increases surface recombination.

To further illustrate the effect of the doping level and fixed charge density on the surface recombination, as shown in Figure 2.4, the SRH equation is evaluated for a fixed level of chemical passivation. Interestingly, for very high doping levels (i.e., $N_s > 10^{20}$ cm^{-3}), the density and polarity of fixed charges in the range evaluated here have virtually no influence on the surface recombination, which is low in all cases. For lower N_s (i.e., in the range of ~10^{20} cm^{-3} or lower), the condition where the SRH recombination is the strongest (Equation 2.3) is met, but only when a passivation layer with the "wrong" charge polarity is used. Explicitly put, a high J_{0s} arises for a negative Q_f on n^+-type Si or for a positive Q_f on p^+ Si surfaces. Finally, for lower N_s (i.e., $N_s < 10^{18}$ cm^{-3}), the passivation quality is excellent, even in the case of a "wrong" charge polarity. In that case, the band bending induced by the fixed charge density brings the Si surface into inversion. Even though in that case the surface can be very well passivated, inversion layers are still undesirable in most solar cells. It is shown that they can form parasitic shunt pathways and possibly induce SRH recombination in the depletion region, which is formed near the surface, having an adverse impact on the efficiency of solar cells [35, 36]. In summary, it can be stated that for passivation schemes, a low D_{it} is always preferred, whereas, in general, the ideal Q_f depends on the surface doping concentration of the Si. Methods to control the Q_f of ALD passivation schemes are discussed in Section 2.2.4.

Finally, it should be noted that in case of very strong surface recombination, such as for poorly passivated surfaces or when metal contacts Si, the surface recombination rate is limited by the transport of excess charge carriers toward the surface. In this case, the presence of a highly doped (p^+ or n^+ Si) region can significantly reduce the transport of excess carriers toward the surface and, in

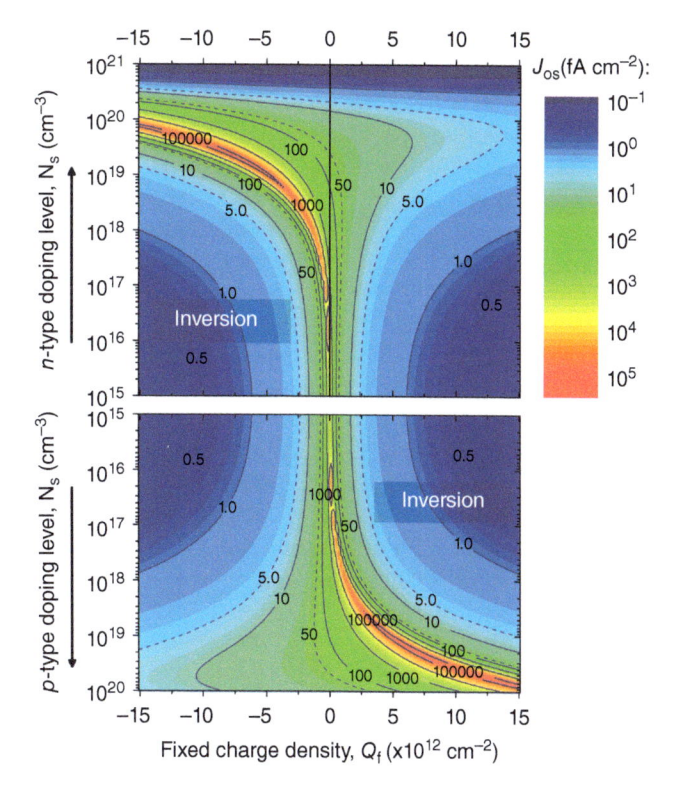

Figure 2.4 The effects of the fixed charge density and the surface doping concentration on the surface saturation current density for a fixed level of chemical passivation. The results are evaluated using Equation 2.2 for a single defect, with $S_{n0} = S_{p0} = 5000$ cm s^{-1}. The carrier densities are derived from the Girisch algorithm [31] using Fermi–Dirac statistics, $n_i = 9.65 \times 10^9$ cm^{-3} [32], a base injection level of $\Delta n = 1 \times 10^{15}$ cm^{-3}, and base doping level of $N_{base} = 1 \times 10^{15}$ cm^{-3}. For the band-gap narrowing of n- and p-type Si, the empirical models of Yan and Cuevas are used [33, 34].

this way, reduce recombination. Well-known examples are the p^+ Si and n^+ Si regions of Si homojunction solar cells (Figure 2.1), which "shield" the metal contacts and make them selective for the extraction of only one charge carrier type. For well-passivated surfaces, on the other hand, the transport of minority carriers through the highly doped region generally does not limit or affect the surface recombination rate. Then, the highly doped region is called *transparent*.

2.2.1.3 Compatibility with Si Homojunction Solar Cells

For implementation in Si solar cell manufacturing, passivation schemes should meet many requirements, from both device and processing perspectives. From an optical point of view, a high transparency to the solar spectrum is necessary when passivating the front or (for bifacial cells, which capture light from both sides) the rear surface. Moreover, a suitable refractive index n of ~2 at 2 eV is preferred for antireflection purposes, although, sometimes, the passivation scheme can be combined with a separate ARC. In contrast, a low refractive index is preferred when acting as dielectric mirror on the back side of the solar cell.

As solar cells are designed to last for decades, a long-term stability of the passivation scheme is also necessary. Passivation schemes can in particular suffer from light- or potential-induced degradation (LID or PID, respectively). An ultrahigh-temperature stability (i.e., up to 800 °C for some seconds) is required when screen- or stencil-printed metal fingers are "fired" through the passivation layer and ARC to contact the Si. In contrast, when the contacts are made at low temperatures by electroplating, this requirement is redundant. However, in the latter process, a pinhole-free thin film with a low leakage current is necessary to prevent undesired "ghost" plating of metal.

Finally, the implementation in solar cells can also impose (technological) requirements on the processing of passivation schemes. For instance, a large-area uniformity is desired, with solar cell dimensions of $156 \times 156 \, \text{mm}^2$ being the industrial standard. Moreover, a low-temperature process for the preparation of the passivation layer is preferred, not only from a cost perspective but also because high temperatures (for instance, used for thermal oxidation) can cause severe degradation of the Si bulk quality [37], in particular for multicrystalline Si. In addition to passivation of its surface, for multicrystalline Si, the passivation of defects located at grain boundaries by hydrogen can be required.

2.2.2 Surface Passivation by ALD Al_2O_3

The passivation of p-type Si and heavily doped p^+-type Si surfaces is very relevant for many high-efficiency solar cell concepts. Nonetheless, historically, before the introduction of ALD Al_2O_3, the passivation of such p-type Si surfaces was challenging. Thermal oxidation of Si yields excellent levels of chemical passivation with a very low positive Q_f, but its passivation of boron-doped surfaces is not stable over time [38]. Besides, thermal oxidation depletes the boron doping near the surface, and its high-temperature processing can have an adverse impact on the bulk lifetime [37, 39]. SiN_x, on the other hand, although very well suited as a passivation layer for n^+ Si and as ARC, is undesirable for the passivation of p-type Si surfaces due to its positive Q_f.

The first results of Si surface passivation for solar cells by Al_2O_3 were reported in the 1980s by Hezel and Jaeger [40, 41]. In this case, Al_2O_3 was prepared by CVD (pyrolysis). Nonetheless, the broad recognition of Al_2O_3 as an outstanding surface passivation layer was yet to come. It was not until 2006, before this was realized. This time, it was reported by Agostinelli *et al.* [27] and Hoex *et al.* [42] that Al_2O_3 provides excellent passivation on p-type Si surfaces. In these cases, the Al_2O_3 was prepared by ALD using trimethylaluminum (TMA) and H_2O or O_2 plasma as precursor. The excellent levels of surface passivation offered by the Al_2O_3 films were attributed to a combination of excellent chemical and field-effect passivation. In particular, ALD Al_2O_3 provides a very low D_{it} of $<10^{11} \, \text{cm}^{-2}$, while it distinctively exhibits a large *negative* Q_f of $10^{12}-10^{13} \, \text{cm}^{-2}$ [29]. Such interface properties were already reported in the first ALD Al_2O_3 experiments using TMA and H_2O in 1989 [43]. Due to its negative Q_f, Al_2O_3 is the ideal match for the passivation of p- and p^+ Si surfaces [44]. Shortly after its (re)discovery, the first solar cells with ALD Al_2O_3 demonstrated high efficiencies, for instance, PERC and PERL cells of 20.6% [45] and 23.2% [10] efficiency, respectively. Due to these advantages, ALD of Al_2O_3 has now found its way into solar cell manufacturing, as extensively discussed in Section 2.2.3. Moreover, the full potential of Al_2O_3 has likely not yet been realized, as is evident from, for example, recent results where Si solar cells with Al_2O_3 passivation layers demonstrated conversion efficiencies over 25% [3]. For an extensive overview on the topic of surface passivation by Al_2O_3, the reader is referred to [29]. Here, some key aspects related to surface passivation by ALD Al_2O_3 are outlined.

2.2.2.1 ALD of Al_2O_3 for Passivation

The most commonly applied and widely studied ALD process for Al_2O_3 is based on TMA as metal–organic precursor with either H_2O, O_3, or O_2 plasma as coreactant. The first half cycle can be described by the following reaction at the surface (*) for $n = 1$ or 2 [46]:

$$TMA : n^*OH + Al(CH_3)_3 \rightarrow {}^*O_n Al(CH_3)_{3-n} + nCH_4$$

After a purge step, the second ALD half cycle takes place, which in the case of thermal ALD can be described by the following ligand exchange reaction:

$$H_2O : {}^*Al(CH_3) + H_2O \rightarrow {}^*AlOH + CH_4$$

As a sufficient thermal budget is required for the latter to take place, alternatively highly reactive species are often used in the so-called energy-enhanced ALD processes, such as O_3 or O_2 plasma, when working at lower temperatures. Possible combustion-like reaction pathways are as follows:

$$O_2 \text{ plasma} : \quad {}^*Al(CH_3) + 4O \rightarrow {}^*AlOH + CO_2 + H_2O$$
$$O_3 : \quad {}^*Al(CH_3) + 4/3 O_3 \rightarrow {}^*AlOH + CO_2 + H_2O$$

The resulting Al_2O_3 films are amorphous, with a high optical band gap of ~6.4 eV. The latter makes Al_2O_3 films very suitable for application at the front surface of solar cells. Depending on the synthesis method, Al_2O_3 has a refractive index of ~1.55 to 1.65 at 2 eV, which is too low for application as an ARC, for which reason it is therefore often capped by SiN_x.

In the as-deposited state, the passivation quality provided by Al_2O_3 strongly depends on the oxidant chosen during the ALD process (Figure 2.5). When using water or ozone as oxidant, the passivation can be reasonable in the as-deposited state. However, for plasma-assisted ALD, the passivation is initially very poor. In fact, the surface is depassivated as a result of plasma damage, in particular due to vacuum ultraviolet radiation that is emitted in the plasma step [47, 48]. The D_{it} of as-deposited ALD Al_2O_3 films can be strongly reduced by a short PDA at temperatures between 400 and 450 °C, with comparable results for an anneal in forming gas (N_2/H_2) or N_2 ambient [49]. In actual Si solar cell processing, a dedicated PDA is often redundant because of the high temperatures used in the subsequent processing steps, such as the deposition of SiN_x or contact firing steps.

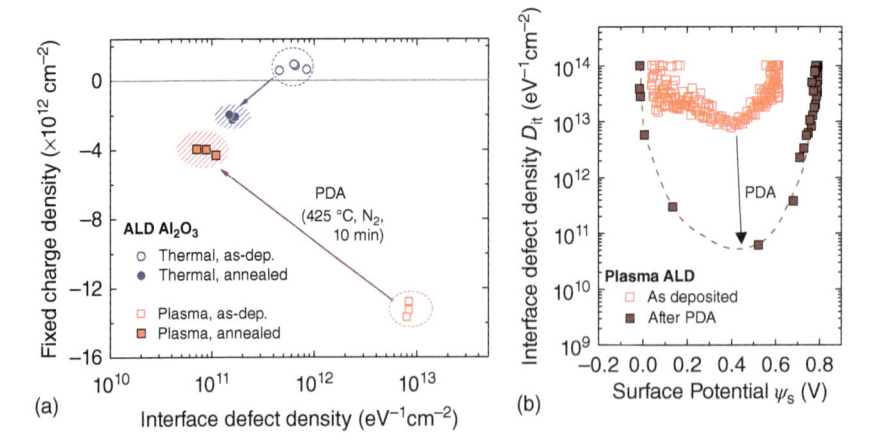

Figure 2.5 Interface properties as determined by corona oxide characterization of semiconductors (COCOS) of n-type Si (100) passivated by Al_2O_3 films prepared by plasma (i.e., using O_2 plasma) or thermal ALD (i.e., using H_2O) in as-deposited state and after postdeposition annealing at 425 °C in N_2 ambient. (a) After annealing, the D_{it} is significantly reduced for both deposition methods. (b) The D_{it} of the Si/Al_2O_3 interface is particularly reduced near mid-gap. Nandakumar of National University of Singapore is acknowledged for the COCOS measurements.

2.2.2.2 Hydrogenation of Interface Defects

Before the deposition of Al_2O_3, commonly, the SiO_2, which is natively present at the Si surface, is removed by dipping the Si in diluted hydrofluoric acid (HF). Nevertheless, after ALD, the Si/Al_2O_3 interface is in fact an Si/SiO_2/Al_2O_3 interface, as is commonly observed by, for example, transmission electron microscopy (TEM) (Figure 2.6) [42, 50].

Figure 2.6 High-resolution TEM image of the annealed Al_2O_3 layer prepared by plasma ALD on Si, showing the amorphous character of the Al_2O_3 layer. The SiO_2 interface layer has a thickness of 1.5 nm. (Hoex et al. 2006 [42]. Reproduced with permission of American Institute of Physics.)

Figure 2.7 Optical microscopy image of blisters formed in an Al_2O_3/SiN_x film after firing at 850 °C. The film stack passivates a Si surface with a random-pyramid texture. (Bordihn et al. 2014 [57]. Reproduced with permission of American Institute of Physics.)

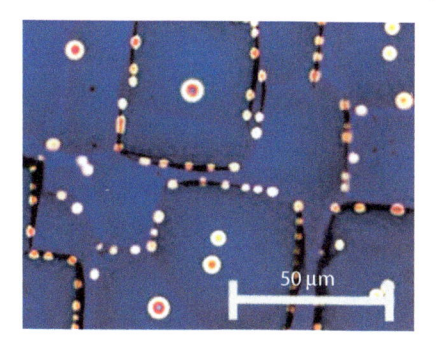

The remaining defects at the Si/SiO_2 interface are effectively passivated by hydrogen that is released from the Al_2O_3 during subsequent PDA. Evidence for this hydrogenation was found experimentally by the deposition of deuterated Al_2O_3, where deuterium present in the Al_2O_3 film diffused during PDA to the Si/SiO_2 interface [51]. The activation energy for the interface hydrogenation does not depend on the amount of hydrogen that is incorporated in the Al_2O_3 films, but the rate of hydrogenation does [52]. Therefore, the hydrogenation of interface defects is, to some extent, dependent on the amount of hydrogen present in the Al_2O_3 film and on its microstructure [53]. The best passivation properties are found for dense and hydrogen-containing Al_2O_3 films, such as, for ALD, Al_2O_3 films deposited at ~200 °C.

Interestingly, Al_2O_3 can also be used to passivate defects when it is not deposited directly onto the Si surface. For instance, the passivation of Si by materials such as HfO_2, TiO_2, or deposited SiO_2 layers could be strongly improved after capping by Al_2O_3 with a subsequent PDA [25, 51, 54, 55]. In fact, for SiO_2, it has even been shown that the Al_2O_3 capping layer can even be removed after PDA, without compromising the improved passivation properties [29, 51]. Nonetheless, capping by Al_2O_3 not only ensures the hydrogenation of interface defects during a PDA but also can improve the long-term and high-temperature stability of the surface passivation stack [51, 56].

Finally, it has to be noted that although the hydrogenation provided by the Al_2O_3 is preferred for passivation, the hydrogen can also lead to some adverse effects in solar cell manufacturing. For instance, a rapid increase in temperature (such as when contacts are "fired" through the dielectric) can induce blistering of the Al_2O_3 film (Figure 2.7) [58]. Most likely, hydrogen released from the Al_2O_3 at higher temperatures builds up under the Al_2O_3 film, which acts as a diffusion barrier, eventually causing a local delamination [58, 59]. In actual solar cell processing, the formation of blisters can be avoided by slowly releasing hydrogen by a more gradual increase in temperature during subsequent process steps. Furthermore, the formation of blisters can be avoided by using thin Al_2O_3 films (i.e., <10 nm in thickness) [59]. Interestingly, solar cell concepts exist where the blisters in the Al_2O_3 film are used to advantage, to form local point contacts to the Si [60].

2.2.2.3 Interface Engineering by Al_2O_3

From a scientific as well as from a processing perspective, it is interesting to see the effect of Al_2O_3 thickness on the surface passivation quality. The best passivation results are found for Al_2O_3 thicknesses >5 nm for plasma ALD [61] and >10 nm for thermal ALD [29]. For very thin Al_2O_3 layers, the level of chemical passivation achieved after PDA starts to deteriorate, which likely can be attributed due to a lack of interface hydrogenation during PDA [61]. Interestingly, it has been demonstrated that the fixed charge density of Al_2O_3, however, is located in very close vicinity of the Si–Al_2O_3 interface, that is, within the first 2 nm of the Al_2O_3 [61].

The latter is very relevant for Si solar cell processing. SiN_x is conventionally used as an ARC due to its suitable refractive index (tunable between 1.9 and 2.7), but when deposited directly on Si, it is not an ideal match for p-type or p^+-type Si due to its positive Q_f. By using Al_2O_3 as thin interlayer in Al_2O_3/SiN_x stacks, the total Q_f of the stacks becomes negative [62]. Therefore, Al_2O_3/SiN_x stacks are an ideal match for the passivation of the front p^+-type surfaces. The fact that very thin Al_2O_3 layers can be used makes ALD a viable technique for HVM, as further discussed in Section 2.2.3. It has to be stressed, however, that for ultra-thin Al_2O_3 layers (i.e., <10 nm) in Al_2O_3/SiN_x stacks, the interface properties such as Q_f and D_{it} are strongly dependent on the thickness of the Al_2O_3 film, the process conditions during film growth, and subsequent firing or annealing conditions [62].

2.2.2.4 Influence of the Surface Conditions on the Passivation Properties

It is common that the interface properties of passivation schemes are significantly dependent on the Si surface conditions, such as the doping level, crystallographic termination, and the surface topology [63, 64]. Such aspects have also been investigated for ALD Al_2O_3.

For highly doped p^+ Si surfaces with a wide range of surface doping concentrations, very low J_0 values are reported regardless of the oxidant used during ALD [44, 65–67]. In fact, Al_2O_3 passivates the p^+ Si surface so effectively that Auger recombination within the highly doped region is often the dominant recombination mechanism and the principal contributor to the measured J_0 [65, 68, 69]. Recently, it has even been shown by Black et al. that the interface parameters Q_f and S_{n0} for boron-doped Si surfaces passivated by Al_2O_3 prepared by thermal ALD are in fact *independent* of the surface doping concentration, at least up to $N_s \leq 3 \times 10^{19}$ cm^{-3} [67]. Nonetheless, care must still be taken when passivating p^+ Si, as, in some cases, a boron-rich layer (which is a boron–Si compound) can be formed on the Si surface during boron diffusion, which can inhibit the surface passivation by ALD Al_2O_3 and other passivation layers [65, 70]. Overall, as stated before, ALD Al_2O_3 is very effective for the passivation of p- and p^+-Si surfaces, virtually independent of the surface doping concentration.

For highly doped n^+ Si surfaces, the level of passivation provided by ALD Al_2O_3 is in many cases found to be compromised due to its negative Q_f [71–73], as could be expected on the basis of Figure 2.4. This holds in particular for surface doping levels in the range of 10^{18}–10^{20} cm^{-3}. Although on lowly doped n-type Si surfaces, excellent passivation results by Al_2O_3 are achieved, here the negative Q_f leads to inversion, which is undesirable in solar cells. Therefore, (novel) ALD-based schemes are also successfully being explored for the passivation of n- and n^+-type Si surfaces, as discussed further in Section 2.2.4.

Besides the doping level, the passivation of Si surfaces can strongly depend on other surface conditions, such as the crystallographic orientation. For ALD Al_2O_3, however, dedicated studies so far revealed marginal differences in the passivation of Si (100) and (111) surfaces [69]. On the other hand, the surface cleaning influences the final passivation properties of Al_2O_3 and should therefore be considered [74].

2.2.3 ALD in Solar Cell Manufacturing

2.2.3.1 Requirements for Manufacturing in the PV Industry

The success of ALD Al_2O_3 as a passivation layer on a lab scale was the incentive for the development of high-throughput ALD reactors, specifically designed for the PV industry. Even though high-throughput ALD reactors were readily used for manufacturing in the integrated-circuit (IC) industry, the demands set by the IC and PV industry are markedly different (Table 2.2). Whereas the requirements in terms of, for example, particle generation and film uniformity are relaxed for the PV industry compared to the IC industry, the former is rigorously more demanding in terms of throughput and costs. The production lines for crystalline Si solar cells are typically designed for a throughput of up to 1 wafer s^{-1}, whereas the costs for, for example, the deposition of Al_2O_3 are (depending on cell design) only allowed to be 0.03–0.05 $ per wafer in order to remain cost-effective.

For the deposition of thin films, ALD competes with physical vapor deposition (PVD), plasma-enhanced (PE) and atmospheric pressure (AP) chemical vapor deposition (CVD), and spray pyrolysis. In general, to compete with the aforementioned deposition techniques, ALD processes should offer additional advantages. For instance, they should be of low cost and preferably provide benefits in terms of spatial uniformity and solar cell efficiency. Furthermore, a high yield of the ALD process, efficient precursor usage, and low-temperature processing would aid in attaining a low cost of ownership. Apart from an optimized reactor design, another way to ensure an efficient precursor usage and optimized throughput is to minimize precursor dosing until the ALD half-reactions are at the onset of

Table 2.2 Comparison of typical film and equipment requirements by the integrated-circuit (IC) and photovoltaic (PV) industries.

	IC industry	PV industry
Number of process steps per device	200–400	15–20
Film uniformity (%)	>99	>96
Particle generation	Important	Irrelevant
Metal (Fe) contamination (cm^{-2})	<10^{10}	<10^{12}
Cost of ownership ($ per wafer)	3–10	0.03–0.05
Equipment cost (M$)	2–5	0.5–2.5
Equipment throughput (wafers h^{-1})	10–50	1000–3000
Equipment uptime	>95%	>95%
Wafer breakage	1 : 50 000	1 : 1000

Source: Adapted from Granneman et al. [75].

saturation. As long as the quality and homogeneity of the deposited films are still acceptable, this might be a valid approach. In addition, "solar grade" precursors can be used. These precursors are less pure and hence are of lower cost [76, 77]. When ALD films can be used on both sides of the solar cells (see e.g., Section 2.2.3), or when stringent requirements are set on the composition, uniformity, and thickness of the films or film stacks, such as for passivating contacts, ALD in particular might become the deposition method of choice.

2.2.3.2 High-Throughput ALD Reactors

Two distinct ALD methods are being used for HVM in the PV industry, that is, *temporal* and *spatial* ALD, which differ in the way the precursors are delivered to the solar cells (Figure 2.8).

With *temporal* ALD, the precursor and coreactant dosing steps are separated *in time* by purge or pump steps. Using temporal ALD, which is the most common form of ALD in research and development, a sufficiently high throughput for the PV industry can be achieved by placing many wafers (typically 500–1000 wafers) simultaneously in a batch reactor. As the wafers are placed back-to-back, parasitic deposition on the other side of the wafers (*wraparound*) is ideally avoided. As ALD uniquely is a surface-limited deposition method, homogeneous films can in principle be obtained on all wafers. Generally, in batch reactors, the purge, precursor, and reactant dosing times must be longer compared to single-wafer ALD tools, due to the very large surface areas that are exposed. Therefore, in batch ALD, the use of O_3 rather than H_2O as coreactant is often preferred due to its higher reactivity and because it allows for shorter purge times.

In *spatial* ALD (S-ALD), the precursor and coreactant dosing steps are separated not in *time*, as is the case for temporal ALD, but in *space*. The wafers can float on N_2 bearings through S-ALD reactors and move through separated precursor and coreactant zones (Figure 2.8), which are separated from each other by nitrogen gas bearings at (sub)atmospheric pressures. In this way, the lengthy purge times, which are otherwise required to separate the precursor and reactant gases, are avoided. As only the wafers encounter the spatially separated precursors, the ALD half-reactions take place only at the surface of the wafer and (ideally) not on the reactor walls. The spatial ALD concept is particularly interesting for *in-line* processing and can, because of the absence of purge steps, operate at atmospheric pressure. Moreover, in principle, no pumps or fast-switching valves are required. In Table 2.3, a selection of industrial aspects of temporal and spatial ALD are compared.

2.2.3.3 ALD Al_2O_3 in the PV Industry

Besides ALD, other synthesis methods of Al_2O_3 are suitable for the passivation of Si surfaces, such as PECVD or AP-CVD and sputtering [41, 67, 79–81]. In the PV industry, PECVD in particular competes with ALD as a deposition method for Al_2O_3, as PECVD reactors are already installed in most production lines for the deposition of SiN_x, whereas ALD is a relatively new technique in the field of PV manufacturing. Nonetheless, as discussed in Section 2.2.3, even very thin, homogeneous, films of Al_2O_3 of less than 2 nm in thickness could be sufficient for solar cells to benefit from its negative Q_f when combined with a SiN_x ARC or

2.2 Nanolayers for Surface Passivation of Si Homojunction Solar Cells | 61

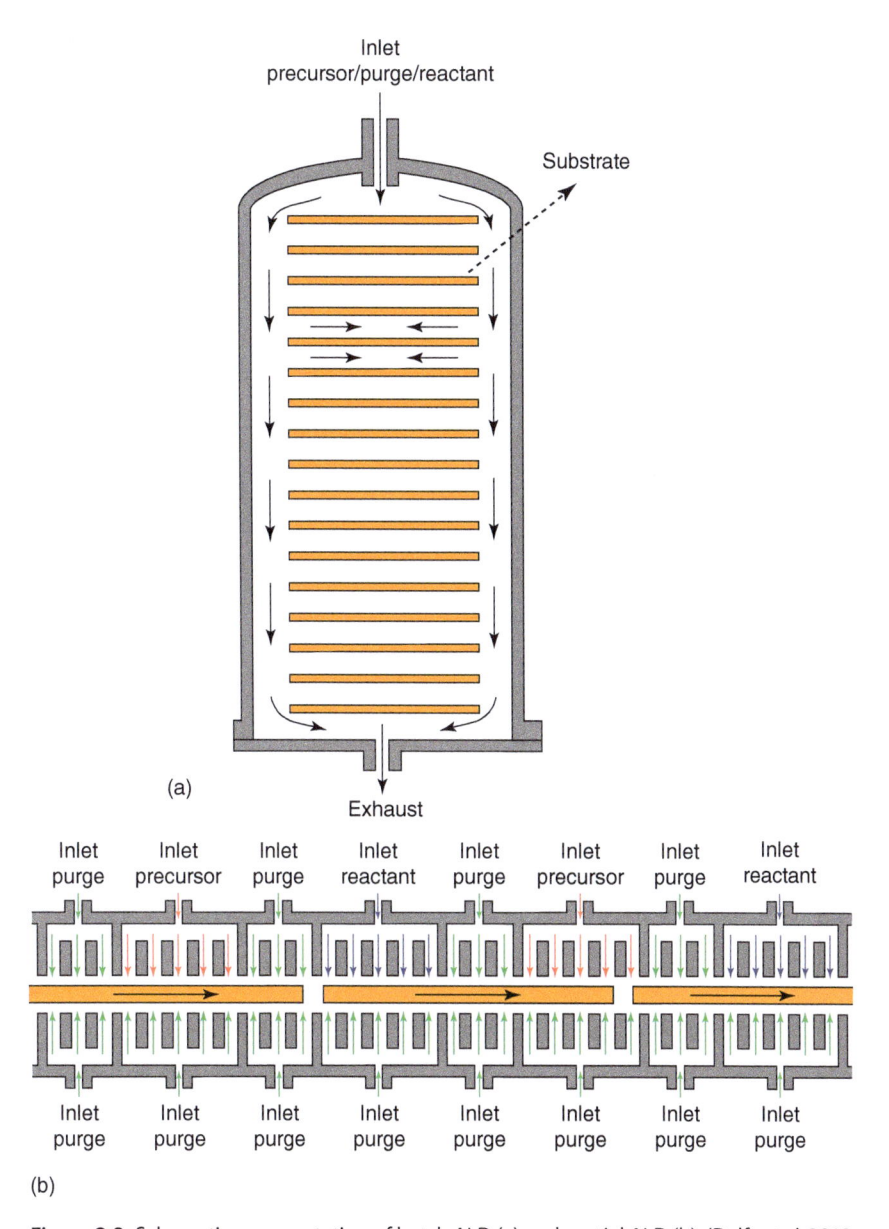

Figure 2.8 Schematic representation of batch ALD (a) and spatial ALD (b). (Delft *et al.* 2012 [78]. Reproduced with permission of Institute of Physics.)

a SiO_2 dielectric mirror. The fact that very thin Al_2O_3 layers can be used makes ALD in particular interesting for HVM. Moreover, in direct comparisons, the best passivation properties of Al_2O_3 are so far often achieved when it is prepared by ALD [67, 82].

For instance, Schmidt *et al.* [82] compared, in 2010, the Si surface passivation provided by single layers of Al_2O_3 prepared by sputtering, PECVD, spatial ALD,

Table 2.3 Comparison of industrial aspects for temporal and spatial ALD.

	Temporal ALD	Spatial ALD
Processing	Batch	In-line
Pressure	~1 Torr	Atmospheric or lower
Single side deposition by	Back-to-back wafer stacking	Precursor injection from one side of the wafer
Double-side deposition	Viable (no back-to-back stacking, half the throughput)	Viable (when precursor injection from both sides of the wafer)
Companies for Si PV	ASM, Beneq	Levitech, SoLayTec
Deposition on walls	Yes	No
Limit on throughput	Purge and dose times	Surface reaction kinetics
Deposition of stacks	Flexible	Possible
Turnaround time for approximately nanometer films	Long (~0.5–1 h)	Short (<1 min)

and temporal ALD. In Figure 2.9, a selection of results is given. ALD, with plasma ALD in particular, yielded the highest passivation performance before as well as after a high-temperature firing step. Importantly, the differences found in surface passivation quality translated well to differences in conversion efficiency of PERC solar cells, ranging from 20.1% for solar cells, where the Al_2O_3 was prepared by sputtering, to 21.4% for solar cells, where it was deposited by ALD and subsequently was capped by PECVD SiO_2 [82]. Note that research activities on the passivation by Al_2O_3 by various passivation methods are still ongoing. For Al_2O_3 films prepared by sputtering, for instance, the absence of hydrogen or the presence of sputter damage affected the surface passivation [29]. Indeed, the preparation of Al_2O_3 films by sputtering has since 2010 been significantly improved when using a hydrogen-containing ambient [83].

Currently, both spatial and temporal ALD systems are being piloted in the PV industry for the HVM of thin Al_2O_3 layers. Due to the excellent passivation of thermal ALD Al_2O_3 films, to date, no high-throughput reactors capable of plasma ALD have been developed for the Si PV industry yet. Very similar results in passivation are reported on lowly doped Si surfaces for spatial as well as temporal ALD of Al_2O_3 when combined with SiN_x capping layers [29, 75], even in direct comparison [82]. In addition, no significant differences in passivation quality by both deposition methods are found for either n^+ or p^+ Si surfaces, which is indicative of the robustness of ALD as a deposition method [84]. However, as it will become clear in the following sections, the different reactor designs have their own distinct advantages in terms of the preparation of doped films and stacks and in terms of processing complexity. In the end, cost-of-ownership and yield considerations are expected to play a decisive role in the final choice of reactor type or deposition method for the HVM in the PV industry.

Figure 2.9 Comparison between the passivation qualities in terms of upper limit of S_{eff} or SRV (where a low S_{eff} is preferred) of Al_2O_3 passivation layers prepared by spatial ALD, PECVD, and sputtering, before and after firing at 800 °C. (Schmidt et al. 2010 [82]. Reproduced with permission of EU PVSEC. Acknowledged by Katrin Aust.)

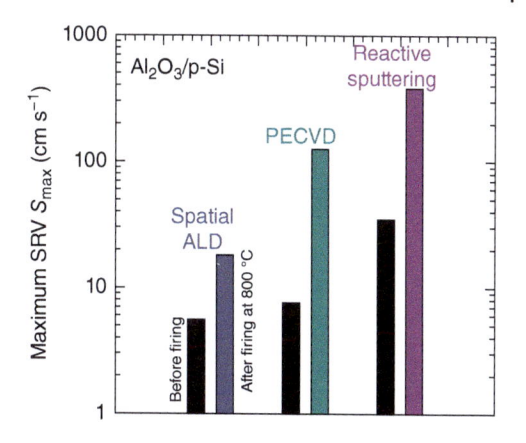

2.2.4 New Developments for ALD Passivation Schemes

With the passivation of p-type and p^+ Si surfaces by ALD being fully established, research incentives are currently also targeting the passivation of n^+ Si surfaces by ALD. Moreover, ALD passivation layers are being explored to enable novel light trapping schemes that otherwise suffer from a high surface recombination, such as "black Si" texturing. Finally, other new passivation materials prepared by ALD are also being explored, and ALD-based passivation schemes are being further tuned using doping or the fabrication of stacks. In this section, these recent developments in the field of surface passivation by ALD are outlined.

2.2.4.1 ALD Stacks for the Passivation of n^+ Si or n^+ and p^+ Si surfaces

The passivation of n^+ Si surfaces is very relevant for many high-efficiency solar cell designs (Figure 2.10). Unfortunately, as was mentioned in the previous sections, the negative Q_f makes Al_2O_3 less than ideal for this application. Additionally, , for example, for IBC solar cells, n^+ and p^+ Si regions are adjacent and preferably passivated simultaneously. By using passivation schemes without a significant Q_f, but with a high level of chemical passivation (i.e., a low D_{it}), effects associated with a "wrong" charge polarity, such as depletion region recombination or parasitic shunting, can ideally be avoided on both surface types. In such a

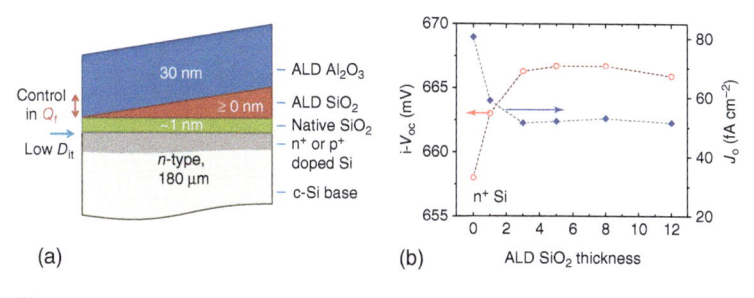

Figure 2.10 Schematic display of the SiO_2/Al_2O_3 passivation scheme (a). Passivation results obtained using SiO_2/Al_2O_3 stacks on an n^+ Si surface (having $N_s = 2 \times 10^{20}$ cm^{-3}) on a symmetrical lifetime sample (b). (Van De Loo et al. 2015 [11]. Reproduced with permission of Elsevier.)

"zero-charge" approach, rather than relying on field-effect passivation by a fixed charge, the high doping levels at the n^+ and p^+ Si surfaces could ensure the local reduction of the minority charge carrier density [11]. Finally, the simultaneous passivation of n^+ and p^+ Si surfaces by ALD could also be of interest from an (industrial) perspective, as it could allow for the simultaneous passivation of both sides of solar cells in a single ALD run. Fortunately, the negative Q_f of Al_2O_3-based layers can be tuned in several ways.

First, the Q_f of Al_2O_3 single layers is reduced using higher deposition temperatures (i.e., ~300–500 °C), although this approach comes at the expense of reduced chemical passivation [85, 86]. Furthermore, the fixed charge density of Al_2O_3 was found to be reduced by capping it with (PECVD) SiN_x in combination with a subsequent high-temperature (>800 °C) firing step [72]. Richter et al. demonstrated that such Al_2O_3/SiN_x stack is well capable of passivating n^+ surfaces with a wide range of surface doping concentrations (~10^{18}–2×10^{20} cm^{-3}) [72]. Nonetheless, in this stack, a negative Q_f of ~1.9×10^{12} cm^{-2} is still present [62], which is not ideal for implementation in solar cells due to inversion or depletion region effects.

Alternatively, ALD SiO_2/Al_2O_3 or HfO_2/Al_2O_3 stacks prepared at 200 °C (e.g., using Hf(NMeEt)$_4$ and H_2O for HfO_2, and $SiH_2(NEt_2)_2$ and O_2 plasma for SiO_2) can be used as passivation schemes with tunable Q_f [54, 56, 87]. The Al_2O_3 capping layer in these stacks ensures the hydrogenation of the Si–SiO_2 or Si–HfO_2 interface during annealing, resulting in excellent levels of chemical passivation, with D_{it} values <10^{11} cm^{-2} [54, 56]. The very thin (i.e., 2–4 nm) SiO_2 or HfO_2 interlayer in turn prevents the formation of a negative Q_f in the Al_2O_3 [54, 56, 87]. For thicker SiO_2 interlayers (>4 nm), the overall charge polarity of SiO_2/Al_2O_3 stacks even becomes positive (depending on the preparation method of SiO_2), as the SiO_2 layer contains a very low positive Q_f in the bulk [54, 56, 87]. Overall, the "digital" thickness control and the excellent homogeneity of the interlayers provided by ALD enable a precise control of Q_f in SiO_2/Al_2O_3 and HfO_2/Al_2O_3 stacks [11, 54, 56]. Recently, it was demonstrated that ALD SiO_2/Al_2O_3 stacks outperform the passivation of n^+ Si surfaces provided by single layers of Al_2O_3 or by fired Al_2O_3/SiN_x stacks [11].

As already mentioned, the possibility of passivating n^+ Si with an ALD-based passivation scheme opens up new opportunities for ALD reactors in HVM. For S-ALD, Al_2O_3 can be deposited on p^+ Si, whereas simultaneously, SiO_2/Al_2O_3 stacks could be deposited on n^+ Si side of the solar cell using the bottom side of the spatial ALD reactor. For batch ALD, both sides of solar cells could be passivated in a single deposition run. In fact, SiO_2/Al_2O_3 stacks have already successfully been scaled up in industrial batch ALD reactors [11].

2.2.4.2 ALD for the Passivation of Surfaces with Demanding Topologies

Most high-efficiency solar cells have, at least at the front surface, a random-pyramid (RP) texture (created by wet chemical etching) in combination with an ARC, to ensure good light trapping and correspondingly a high short-circuit current density (Figure 2.11a). The RP texture typically results in an increase of ~1.7 in surface recombination for Al_2O_3-based passivation schemes compared to planar surfaces [89], which can be attributed to the increase in surface area. Alternatively, the front surface can be etched by reactive ions, which creates

2.2 Nanolayers for Surface Passivation of Si Homojunction Solar Cells | 65

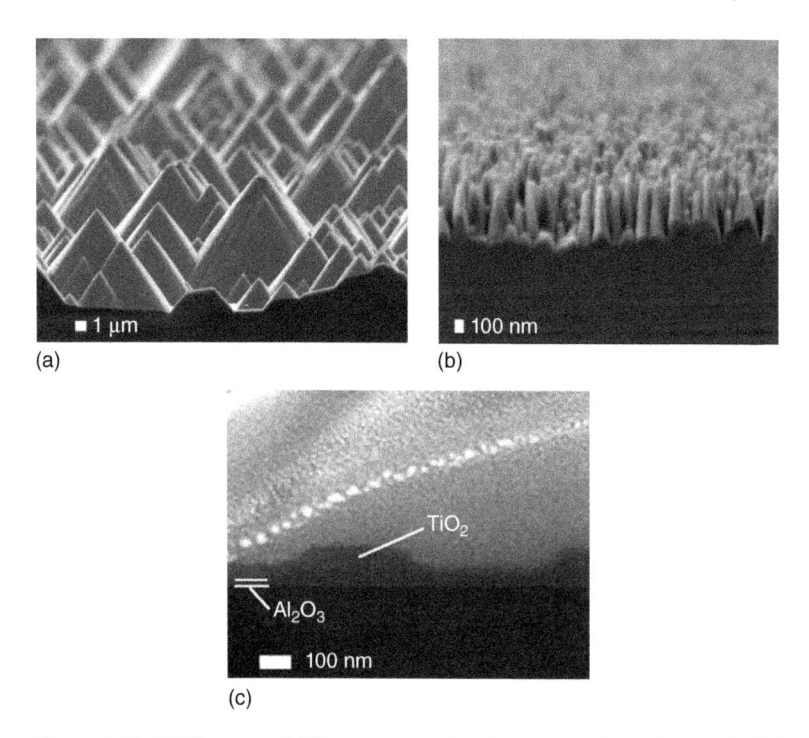

(a)

(b)

(c)

Figure 2.11 SEM images of different approaches that are used to enhance the light absorption in Si solar cells. In (a), the current industrial standard method for monocrystalline Si is used, where random pyramids are created by wet chemical etching of the Si. On the RP-textured surface, subsequently, the passivation layer and ARC are deposited. In (b), a "black Si" texture is shown, which is formed by reactive ion etching and subsequently is passivated by a 30-nm Al_2O_3 film prepared by plasma ALD (the authors would like to acknowledge Ingenito from Delft University of Technology for the black Si synthesis). (c) A cross section of a TiO_2 resonator is shown. The TiO_2 is deposited by electron beam evaporation on a 30 nm Al_2O_3 film prepared by plasma ALD, which enables the surface passivation. (Panel (c): Spinelli *et al.* 2013 [88]. Reproduced with permission of American Institute of Physics.)

a very rough surface topology, referred to as "black silicon" (Figure 2.11b). Black Si exhibits excellent light trapping properties, even without ARC [90, 91]. Compared to RP-textured Si, black Si shows particularly a high absorption in the short-wavelength range. Moreover, its absorption is less dependent on the angle of incidence compared to RP-textured Si [91].

Despite these benefits, the surface area of the black Si surface is significantly increased, compared to planar substrates, typically by a factor in the range 7–14 [92, 93]. This strong increase in surface area imposes stringent requirements on the passivation quality of the black Si surface. For a long time, solar cells with black Si texture only achieved efficiencies <18.2%. Only recently, it has been reported in IBC cells comprising black Si with a conversion efficiency of 22.1% [91]. In this case, the black Si surface was passivated by Al_2O_3 deposited by thermal ALD [91]. Conceivably, a full coverage of Al_2O_3 films over the pillars of the black Si (see, e.g., the TEM results shown in Figure 2.12) is crucial for the surface passivation, making ALD an ideal candidate. Moreover, it is commonly

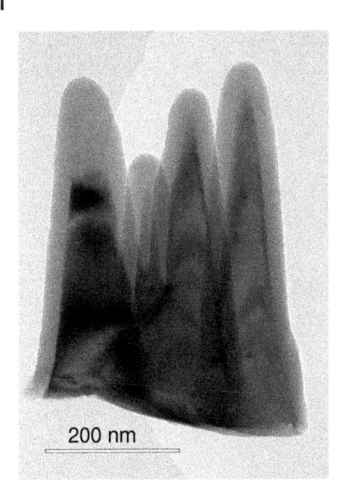

Figure 2.12 Bright-field TEM image of black Si pillars shown in Figure 2.11b. A conformal Al_2O_3 layer of 30 nm thickness is deposited by plasma ALD.

observed that the recombination rate of charge carriers at the black Si surfaces is much lower than what would be expected on the basis of the large surface area [92–95]. An important reason for this observation is that the field-effect passivation is markedly more effective for black Si textures than for planar or RP-textured surfaces [92, 95]. More specifically, the fixed charge density in the passivation layer can bring the needles of the black Si texture almost completely into inversion or accumulation, and in this way, it effectively suppresses the surface recombination. Finally, for application in solar cells, more heavily doped black Si surface textures are also relevant. Whereas first results for Al_2O_3 on p^+ doped black Si are promising [96], for n^+-doped black Si surfaces, the SiO_2/Al_2O_3 or HfO_2/Al_2O_3 ALD stacks, discussed in the previous section, could be interesting candidates.

Besides black Si, ALD Al_2O_3 films have been used for the surface passivation of other textures with an even more demanding surface topology, such as Si nanowires or hierarchical textures [90, 97]. For such topologies, multiple precursor dosing steps are even used during ALD to ensure a good conformality of the film over the nanostructure [90]. Interestingly, light-trapping approach has also been developed, in which *flat* Si surfaces could be used [88]. In this approach, the flat surface was passivated by ALD Al_2O_3. On top of the passivation layer, nano-sized TiO_2 resonators are deposited, which enhance the light trapping in the Si (Figure 2.11c) [88]. This approach not only makes the texturing of the surface and the use of an ARC redundant but also does not adversely affect the surface passivation provided by the ALD Al_2O_3 films and is therefore very promising for application in solar cells.

2.2.4.3 Novel ALD-Based Passivation Schemes

After a tremendous success of ALD Al_2O_3, the surface passivation properties of various other materials prepared by ALD have also been explored, including TiO_2, Ta_2O_5, and Ga_2O_3. A selection of passivation results and corresponding interface properties for these materials is given in Table 2.4. Recently, Cuevas and coworkers identified Ga_2O_3 [100] and Ta_2O_5 [98] prepared by ALD (the

Table 2.4 Selection of optimal surface passivation conditions provided by several materials prepared by ALD.

Material or stack	Metal precursor	Co-reactant	T_{dep} (°C)	GPC (Å)	PDA (°C, ambient)	D_{it} (×10^{11} eV^{-1} cm^{-2})	Q_f (×10^{12} cm^{-2})	S_{eff} (cm s^{-1})	References
Al_2O_3	$AlMe_3$	O_2 plasma	200	1.1	450, N_2	0.8	−5.6	2.8	[47]
	$AlMe_3$	O_3	200	0.9	400, N_2	1.0	−3.4	6.0	[47]
	$AlMe_3$	H_2O	200	1.1	350, N_2	0.4	−1.3	4.0	[47]
SiO_2	$SiH_2(NEt_2)_2$	O_2 plasma	200	1.2	400, N_2	10	0.6–0.8	25^a	[55]
SiO_2/Al_2O_3	$SiH_2(NEt_2)_2$	O_2 plasma	200	1.2	400, N_2	1	−5.8 to 0.6	3	[55]
HfO_2/Al_2O_3	$Hf(NMeEt)_4$	H_2O	150	1.1	350, N_2/H_2	<1	−(4–1)	<1	[54]
Ta_2O_5	$Ta_2(OEt)_{10}$	H_2O	250	0.3	No	n.a.	−1.8	~467	[98]
Ta_2O_5/SiN_x	$Ta_2(OEt)_{10}$	H_2O	250	0.3	No	n.a.	−1.0	3.2	[98]
TiO_2	$TiCl_4$	H_2O	100	0.6	200–250, N_2, light soaking	n.a.	n.a.	2.8	[99]
Ga_2O_3	$GaMe_3$	O_3	250	0.2	350, H_2/Ar	n.a.	n.a.	6.5	[100]

The SiN_x layers, which are sometimes used as capping layers, are prepared by PECVD. Note that S_{eff} depends on the doping level of the bulk [24]. Therefore, only S_{eff} data based on polished, float-zone, n-type Si wafers with a base resistivity of 1–5 Ω cm are included here. For each process, the substrate temperature T_{dep} and growth per cycle (GPC) are indicated.
n.a., Not available. ^{a}Unstable in time.

latter in combination with PECVD SiN$_x$ capping layer) as materials that achieve excellent surface passivation. Although outstanding surface passivation results have been demonstrated, these materials do not yet offer apparent benefits in terms of processing complexity, cost, or passivation quality compared to, for example, Al$_2$O$_3$ films or SiO$_2$/Al$_2$O$_3$ stacks. Nonetheless, these novel materials might become very useful in the emerging field of passivating contacts, as discussed in Section 2.4.

Other passivation materials can be interesting because of their refractive index of ~2 at 2 eV, which makes them also suitable as ARCs. TiO$_2$ was, for instance, traditionally used before SiN$_x$ as ARC in Si solar cells, although it was gradually replaced by SiN$_x$ due to the poor passivation quality of the former. However, Liao et al. [99] recently demonstrated excellent surface passivation by TiO$_2$ using thermal ALD using TiCl$_4$ and H$_2$O as precursors. The passivation of TiO$_2$ could be achieved after a PDA in combination with a subsequent light-soaking step [99]. Interestingly, TiO$_2$ is also potentially suitable as a part of a passivating contact.

Besides the deposition of single layers, ALD is well-suited to prepare stacks or doped films in a well-controlled way. This ability can potentially be used to carefully tune the electrical or optical properties of the passivation layer or scheme. For instance, using materials with different refractive indices, it is possible to make double-layered or graded ARCs or Bragg mirrors to enhance light trapping in solar cells. A passivation scheme prepared by ALD, which is also suitable as ARC, could be a stack of Al$_2$O$_3$/ZnO, where even improvements in passivation properties were found after capping the Al$_2$O$_3$ by ZnO [22]. The interface properties such as D_{it} and Q_f of ALD Al$_2$O$_3$ or SiO$_2$ layers could also be improved by using doping [101, 102]. For instance, it was found that Al$_2$O$_3$ doped with TiO$_2$ has a slightly more negative Q_f compared to single layers of Al$_2$O$_3$ [101].

Finally, ALD also offers the possibility of obtaining nanolaminates or alloys, which potentially also could be used to improve the passivation properties and can even add new functionalities to it. For example, it has been reported that ALD TiO$_2$–Al$_2$O$_3$ nanolaminates have an increased conductivity compared to a single layer of Al$_2$O$_3$, albeit at the expense of a reduced passivation quality [102]. Moreover, Al$_2$O$_3$–TiO$_2$ nanolaminates are successfully used to enhance the damp-heat stability of the surface passivation scheme, as they act as a moisture barrier layer [103]. Al$_2$O$_3$–TiO$_2$ "alloys" have also been explored (deposited using an Al$_2$O$_3$:TiO$_2$ ALD cycle ratio of 1 : 1), resulting in a slight improvement in surface passivation [104]. Overall, it has been demonstrated that the precise control of film growth and film composition offered by ALD opens up numerous possibilities to further develop ALD-based passivation schemes.

2.3 Transparent Conductive Oxides for Si Heterojunction Solar Cells

In this section, the role of ALD in preparing both ZnO- and In$_2$O$_3$-based TCOs for SHJ solar cells is reviewed. First, the basics of TCOs and important aspects with respect to implementation in SHJ solar cells are reviewed. From this, it follows that the upcoming high-mobility In_2O_3-based TCOs are especially promising for use on the front side of the solar cell due to their excellent conductivity and transparency. Consequently, these TCOs have played a key role in achieving the recent record efficiencies for SHJ cells. ZnO-based TCOs are mostly promising as a low-cost alternative at the rear side of the solar cell, since the optoelectronic requirements at the rear side are less stringent, whereas the work function of ZnO is also more suited for the electron-collecting side.

ALD processes of both ZnO- and In_2O_3-based TCOs are discussed. In the subsection on ALD of doped ZnO-based films (ZnO:X, X = Al, B, Ga, etc.), the high control over the doping level offered by the use of dopant supercycles is discussed [105, 106]. The challenge of dopant clustering, which is inherent to the supercycle approach, is addressed, together with several strategies to mitigate this clustering. For (doped) In_2O_3, an overview of existing ALD processes is given, with special attention to the ALD process of high-mobility H-doped In_2O_3, given the very promising properties it can provide. Finally, recent developments in the HVM of ALD TCOs, with a focus on (doped) ZnO, are briefly touched upon.

2.3.1 Basics of TCOs in SHJ Solar Cells

2.3.1.1 Lateral Conductivity

The first requirement of a TCO is that it should exhibit a low resistivity in order to prevent excessive Ohmic losses during lateral charge transport to the metal grid. The resistivity ρ is determined by the carrier density N_e and the mobility μ_e of the charge carriers by $(N_e \mu_e e)^{-1}$, where e is the elementary charge. A typical sheet resistance for the front TCO with a thickness of 75 nm is $\sim 40\,\Omega\,\gamma^{-1}$, which translates to a resistivity of 0.3 mΩ cm [107]. Note that the TCO should preferably also serve as an antireflection coating to maximize the incoupling of light, which more or less fixes the thickness to 75 nm, given that TCOs typically exhibit a refractive index of around 2. In Figure 2.13, the dashed isolines of constant resistivity show the typical N_e ($>1 \times 10^{20}$ cm^{-3}) and μ_e values that are needed to achieve such resistivity values.

TCOs typically have a high band gap of $\sim$3 eV and therefore a very low intrinsic carrier density N_e. Nonetheless, thin films of In_2O_3 and ZnO are often unintentionally n-type doped by the presence of doubly charged oxygen vacancies (V_O^{2+}) and singly charged H$^+$, which leads to a typical N_e up to a level of $\sim 10^{19}$ cm^{-3} or higher. In order to achieve a sufficiently low resistivity, TCOs are often intentionally further n-type doped by other elements such as Sn for In_2O_3 and Al, Ga, and B for ZnO to increase N_e to the order of 10^{20}–10^{21} cm^{-3}.

The carrier mobility μ_e is limited by the scattering of charge carriers. An intrinsic, unavoidable scattering process is phonon scattering. Depending on the quality of the (typically) polycrystalline TCO, extrinsic scattering processes can also be significant, for example, in terms of crystallographic defects such as grain boundaries and other impurities. Nonetheless, typically, the dominant scattering process for carrier density values in the range of interest ($>1 \times 10^{20}$ cm^{-3}) is the Coulombic scattering that arises from the introduction of ionized dopants,

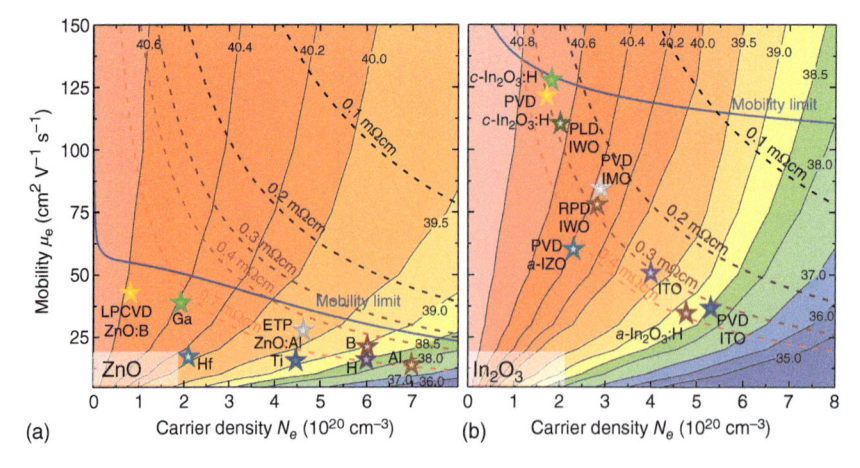

Figure 2.13 Contour plots of the simulated J_{sc} (in mA cm^{-2}) of SHJ solar cells featuring a ZnO-based (a) or In$_2$O$_3$-based (b) TCO as a function of carrier density and mobility. Dashed lines are isolines of constant resistivity. The simulated cell structure consists of a textured wafer with 5 nm of intrinsic a-Si:H, 10 nm of p-type a-Si:H and 75 nm of TCO on the front side of the solar cell. Photon currents were simulated using OPAL2 [108], and the optical constants of In$_2$O$_3$ and ZnO were taken from ellipsometry measurements [106, 109]. In order to calculate the effect of free-carrier absorption for varying carrier density and mobility, the contribution of the Drude oscillator ε_{Drude} to the modeled dielectric function was varied accordingly. The interband absorption was assumed to remain constant, together with a constant m^* of $0.23m_e$ for In$_2$O$_3$ and $0.4m_e$ for ZnO, respectively. The thick solid line is the mobility limit as a function of the carrier density. For ZnO, this was calculated using the Masetti et al. model [110] with the most recent parameters of [111]. The mobility limit for In$_2$O$_3$ was calculated using the mobility limits due to both phonon and ionized impurity scattering [109, 112]. For ZnO, literature values (found in Table 2.5) of ALD films with various dopant atoms are displayed, as well as ZnO:B obtained by low-pressure CVD and ZnO:Al by expanding thermal plasma CVD [113]. For In$_2$O$_3$, values for amorphous (a-In$_2$O$_3$:H) and crystallized indium oxide(c-In$_2$O$_3$:H) (prepared by ALD [109]), ITO [106], IMO [114], and amorphous IZO [115] (prepared by sputtering) and IWO (prepared by reactive plasma deposition (RPD) [116] and pulsed laser deposition (PLD) [117]) are shown. ALD processes are denoted in bold [118].

known as ionized impurity scattering (IIS). For homogeneously dispersed ionized dopants, the mobility limit due to IIS, μ_{ii}, can be, as calculated by Pisarkiewicz et al. [112],

$$\mu_{ii} = \frac{3(\varepsilon_r \varepsilon_0)^2 h^3}{Z^2 m^{*2} e^3} \cdot \frac{N_e}{N_i} \cdot \frac{1}{F_{ii}^{np}(\xi_0)} \tag{2.5}$$

In this equation, h is Planck's constant, ε_0 and ε_r are the vacuum and relative permittivity, respectively, and m^* is the effective electron mass. Z is the charge state of the ionized impurity, and N_i is the concentration of ionized impurities. $F_{ii}^{np}(\xi_0)$ is the screening function for IIS in a degenerate semiconductor, which depends on the carrier density through the factor ξ_0 [112],

$$\xi_0 = \sqrt[3]{3\pi^2} \cdot \frac{\varepsilon_r \varepsilon_0 h^2 N_e^{1/3}}{m^* e^2}$$

For N_e values $>1 \times 10^{20}$ cm^{-3}, IIS limits the mobility of ZnO to approximately <50 cm^2 V^{-1} s^{-1} and that of In$_2$O$_3$ to <150 cm^2 V^{-1} s^{-1} [119, 120]. In this respect,

Table 2.5 Selection of reported low-temperature ALD processes of doped ZnO using DEZ as the Zn source.

Dopant	Doping precursor	T_{dep} (°C)	Doping level (at.%)	N_e (10^{20} cm^{-3})	μ_e (cm^2 V^{-1} s^{-1})	ρ (mΩ cm)	References
Al	AlMe$_3$	200	1.9	1.4	14.3	3.1	[134]
	AlMe$_3$ [a]	170	—	4.3	7	2.1	[135]
	AlMe$_3$ [b]	200	7	8	—	—	[136]
	AlMe$_2$(O^{i}Pr)	250	4.6	10	6	1.1	[137]
	AlMe$_2$(O^{i}Pr)	200	—	0.7–7	13.4–15.6	0.7–6.7	[106]
B	B(O^{i}Pr)$_3$	150–240	1.6	<3	<12	2.2–3.5	[138]
	B$_2$H$_6$	150	—	~6	~20	0.64	[139]
Ti	Ti(OCHMe$_2$)$_4$	200	1.6	2.9	20.4	1.05	[140]
		—	4.5	15	0.9	[141]
Ga	GaMe$_3$	210	—	~2	25–40	0.8	[142]
Hf	Hf(NMeEt)$_4$	220	1.7	2.1	17	1.6	[143]
H	H$_2$ plasma	200	—	6	15	0.7	[144]

a) This process uses dehydroxylation to decrease the amount of TMA deposited.
b) This process uses functionalization by alkyl alcohols to decrease the amount of TMA deposited.

In$_2$O$_3$ is at a clear advantage, which is mostly caused by a lower effective mass of the electrons of ~0.2–0.3m_e as compared to ~0.4–0.5m_e for ZnO. The limits of the carrier mobility as a function of carrier density in ZnO- and In$_2$O$_3$-based TCOs, which for high carrier densities is mainly caused by IIS, are shown by the blue solid lines in Figure 2.13.

2.3.1.2 Transparency

Besides having a low resistivity, TCOs should provide excellent transparency for the photon energy range of interest for solar cells (~1.12–3.5 eV). Due to their high band gap ($E_g > 3$ eV), ZnO and In$_2$O$_3$ are in principle very transparent up to that photon energy, as can be seen by the spectral absorption coefficient of nominally undoped ZnO in Figure 2.14a. Above the band gap of ~3 eV, a strong increase in optical absorption in the ZnO is observed as expected. For higher doping, the onset of absorption shifts to higher photon energies, which corresponds to an increase in optical band gap. This effect is known as the Burstein–Moss (BM) shift: as can be seen in Figure 2.14b, the Fermi level E_F of ZnO is close to the conduction band since it is (almost) degenerate by the unintentional doping. By increased doping of the TCO, the Fermi level is raised further into the conduction band. This leads to an increase in the optical band gap, since the occupied states in the bottom of the conduction band are unavailable for optical transitions from the top of the valence band. In this respect, high doping of the TCO is beneficial since the transparency is extended to higher photon energies. This advantage is, however, relatively small, since very few photons are present in the solar spectrum at these high photon energies, as can be seen in Figure 2.2a.

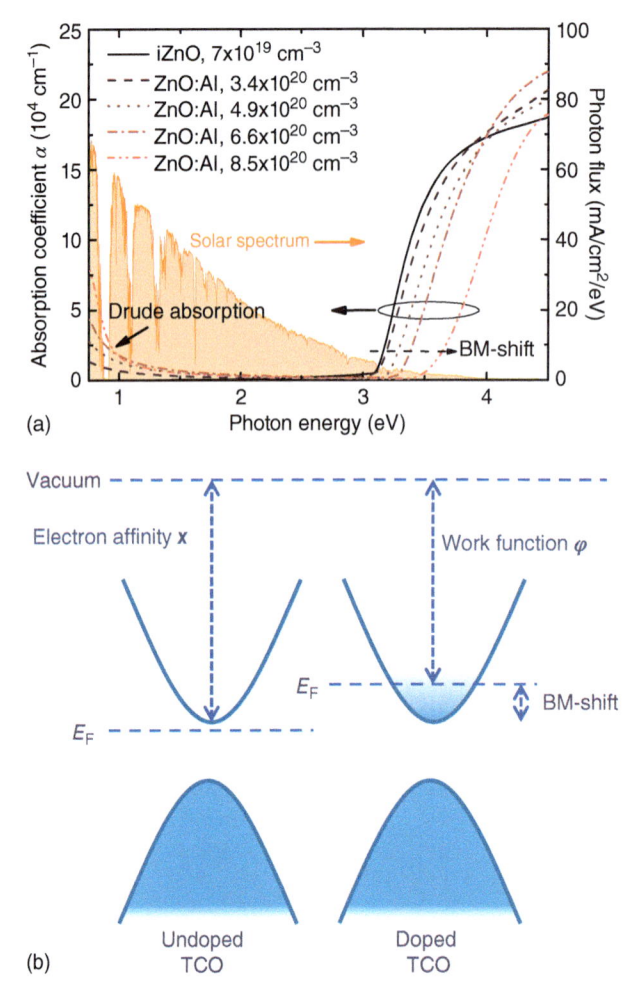

Figure 2.14 (a) Absorption coefficient of Al-doped ZnO layers of varying carrier density prepared by thermal ALD at 200 °C using DEZ and DMAI as Zn and Al source, respectively. The increasing Drude contribution and Burstein–Moss (BM) shift with doping level are indicated. (b) Schematic band diagram of an unintentionally doped TCO (a) and an intentionally doped TCO (b). The electron affinity χ denotes the energetic distance from the conduction band edge to the vacuum level, whereas the work function φ is the distance from the Fermi level E_F to the vacuum.

The free carriers generated by doping lead to increased free carrier absorption (FCA) at low photon energies due to an increase in Drude absorption, as can be seen in Figure 2.14a. This is a very detrimental effect, since the solar spectrum contains the most photons at low photon energy. Additionally, this increased Drude contribution reduces the refractive index n at low photon energies, leading to enhanced free-carrier reflection (FCR) due to a refractive index mismatch.

2.3 Transparent Conductive Oxides for Si Heterojunction Solar Cells

The Drude contribution to the complex dielectric function $\varepsilon_{\text{Drude}}$ is determined by the plasma frequency ω_p and the scatter frequency ω_τ [121]:

$$\varepsilon_{\text{Drude}}(\omega) = -\frac{\omega_p^2}{\omega^2 + i\omega\omega_\tau}, \quad \omega_p = \sqrt{\frac{e^2 N_e}{\varepsilon_0 m^*}}, \quad \omega_\tau = \frac{e}{m^* \mu_e} \quad (2.6)$$

where ω_p denotes the onset frequency of the Drude contribution, whereas the damping term ω_τ determines the broadening around the onset frequency. From these equations, it can be seen that a high mobility is advantageous in two ways. First, it reduces the required N_e for a low resistivity and thereby it reduces ω_p. At the same time, a high mobility (or low ω_τ) reduces the Drude contribution above the plasma frequency because of a reduced broadening.

The effect of the Drude contribution on the J_{sc} of an SHJ solar cell due to FCA and FCR is shown for both ZnO- and In$_2$O$_3$-based TCOs in the simulated contour plots of Figure 2.13. The figure shows that a high μ_e and a low N_e are key to achieving a high J_{sc}. On the other hand, both high μ_e and N_e are desirable for a low resistivity. In this respect, the higher mobility limit of In$_2$O$_3$ compared to ZnO allows for higher J_{sc} as well as lower resistivity values. Nonetheless, the contour plot shows a lower sensitivity of the J_{sc} for N_e and μ_e in the case of ZnO in comparison to In$_2$O$_3$. This is mostly due to the higher effective electron mass in ZnO, which leads to lower ω_p and ω_τ for given N_e and μ_e, respectively.

For comparison, a selection of carrier density and mobility values taken from literature reports (which is discussed in the next section) is shown in Figure 2.13, where the boldfaced labels denote ALD processes. The ZnO-based processes show mobility values well below the mobility limit, which indicates that scattering processes related to material quality (e.g., grain boundary and neutral impurity scattering) play an important role. Nonetheless, ALD is at least on par with other deposition methods, especially since the reported films deposited by expanding thermal plasma (ETP) and low-pressure CVD (LPCVD) were much thicker (>500 nm).

For the In$_2$O$_3$-based processes, various high-mobility TCOs have appeared that greatly outperform sputtered ITO. In particular, crystallized In$_2$O$_3$:H (c-In$_2$O$_3$:H) is very promising because of the low resistivity (<0.3 mΩ cm) and very high J_{sc} due to μ_e, which is very close to the mobility limit. This material was initially developed by sputtering [122], but recently, an ALD process has also been developed as will be discussed later [4, 109]. Nonetheless, it has proven to be difficult to make a good metal–TCO contact for this material, and the H-doped material is less stable under operating conditions, which, however, can be remedied by the use of bilayers [123, 124]. Recently, sputtered amorphous Zn-doped indium oxide (IZO) has also gained interest, as it combines a relatively high μ_e (~60 cm^2 V^{-1} s^{-1}) with good metal–TCO contact properties and stability [115, 124]. In addition, Mo-doped In$_2$O$_3$ (IMO) and W-doped In$_2$O$_3$ (IWO) are high-mobility TCOs, and SHJ solar cells based on the latter TCO in combination with copper metallization have reached efficiencies exceeding 22% [116].

2.3.1.3 Compatibility with SHJ Solar Cells

It is required that the produced TCO and the processing thereof are compatible with the SHJ solar cell design and processing. The restrictions in terms of TCO processing are mostly dictated by the ultrathin intrinsic and doped a-Si:H layers. Especially, the *p*-type doped a-Si:H layer is very temperature sensitive, and this puts an upper limit on the processing temperature of 200 °C [107]. Additionally, these ultrathin layers are prone to plasma-induced damage. For example, sputter deposition of TCOs leads to plasma-induced damage of the underlying a-Si:H films, thereby reducing the level of surface passivation [106, 125]. Although postdeposition annealing can (partially) recover this damage, the microstructure of the a-Si:H layers is irreversibly altered [125].

In addition to the aforementioned considerations, the work function φ of the TCO is of importance. The front TCO contacting the a-Si:H(p) layer preferably has a φ equal to or exceeding the φ of the a-Si:H(p) layer (~5.3 eV), whereas the rear TCO contacting the a-Si:H(n) layer preferably has a φ equal to or below the φ of the a-Si:H(n) layer (~4.2 eV) [126]. A mismatch in φ leads to a Schottky contact between the doped a-Si:H/TCO layer, and the associated depletion region can extend into the Si wafer and reduce the band bending, thereby seriously affecting the fill factor and even the open-circuit voltage [127, 128]. Although this effect can be mitigated by increasing the thickness and doping level of a-Si:H layers, this comes at the expense of enhanced parasitic light absorption and defect density in especially the a-Si:H(p) layer [126, 129]. As can be seen in the schematic of Figure 2.14b, the φ of a degenerate TCO is determined by the electron affinity χ and the doping level through the Burstein–Moss shift, $\chi - \Delta E_{BM} = \phi$ (ignoring band-gap narrowing) [130]. Therefore, control over the work function of the TCO through its doping level is of importance for the optimization of the doped a-Si:H/TCO contact [106, 126, 129]. Since the electron affinity χ of ZnO (~4.4 eV) is lower than that of In_2O_3 (~5.0 eV), ZnO seems more suited to contact the a-Si:H(n) layer, whereas lowly doped In_2O_3 is more suited for the front a-Si:H(p) layer [130]. Together with the less stringent requirements on transparency at the rear side of the cell, ZnO-based TCOs are therefore more likely to be an effective low-cost alternative for indium-based TCOs at this side of the cell [131]. In fact, it has been recently shown that doped ZnO can replace the rear ITO without conversion efficiency loss [131].

2.3.2 ALD of Transparent Conductive Oxides

2.3.2.1 ALD of Doped ZnO

The by far most studied ALD process for ZnO is the process based on diethyl zinc (DEZ, $ZnEt_2$) and water, which yields high growth rates (typically >1.5 Å per cycle) at temperatures <200 °C [132]. The process can be described by the following reaction [133]:

$$DEZ: \quad {}^*ZnOH + Zn(C_2H_5)_2 \rightarrow {}^*ZnOZn(C_2H_5) + C_2H_6$$
$$H_2O: \quad {}^*ZnOZn(C_2H_5) + H_2O \rightarrow {}^*ZnOH + C_2H_6$$

Unintentionally doped ZnO made by ALD can have an electron density up to $\sim 10^{19}$ cm^{-3} due to the presence of oxygen vacancies and/or H dopants.

2.3 Transparent Conductive Oxides for Si Heterojunction Solar Cells | 75

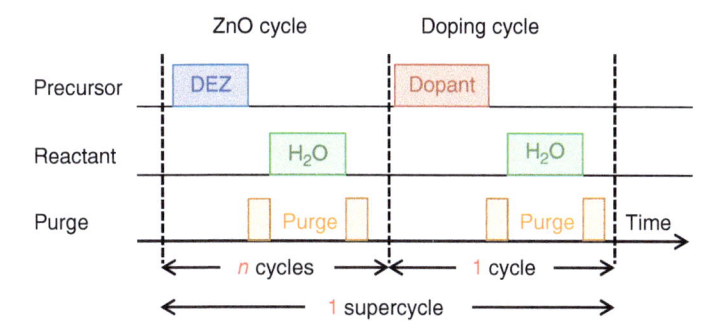

Figure 2.15 Schematic representation of the ALD supercycle principle. In ALD supercycles used for doping of ZnO, n cycles of the ZnO ALD process are followed by 1 cycle of a process containing a dopant element (Al, B, Ga, etc.). This supercycle is repeated until the desired film thickness is reached. (Adapted from Ref. [46].)

Nonetheless, the resistivity of ZnO is typically in the high $10^{-3}\,\Omega\,\text{cm}$ regime, which necessitates cationic doping in order to reach the (low) $10^{-4}\,\Omega\,\text{cm}$ regime required for SHJ solar cell applications.

The most prevalent approach for doping of ZnO is by Al using TMA, although other dopant precursors and atoms such as B, Ga, Ti, Hf, and even H have gained significant interest over the past years. An overview of low-temperature ALD processes of doped ZnO can be found in Table 2.5.

The introduction of dopants in the ZnO matrix is most commonly achieved using the so-called ALD supercycles. The principle of an ALD supercycle is shown in Figure 2.15. In such an ALD supercycle, an integer number n of ZnO cycles is followed by one dopant cycle. By repeating such supercycles, a structure as shown in Figure 2.16a is obtained, in which the dopants lie in distinct planes. This is distinctly different from CVD and PVD methods, in which the dopants

Figure 2.16 (a) Schematic representation of the superstructure obtained when using ALD supercycles. The cycle ratio n controls the *vertical* spacing of the dopants. (b) Demonstration of the accurate control over the carrier density in ZnO that is possible using the supercycle approach. The process used employs DMAI as dopant precursor and DEZ as Zn precursor, at a deposition temperature of 200 °C.

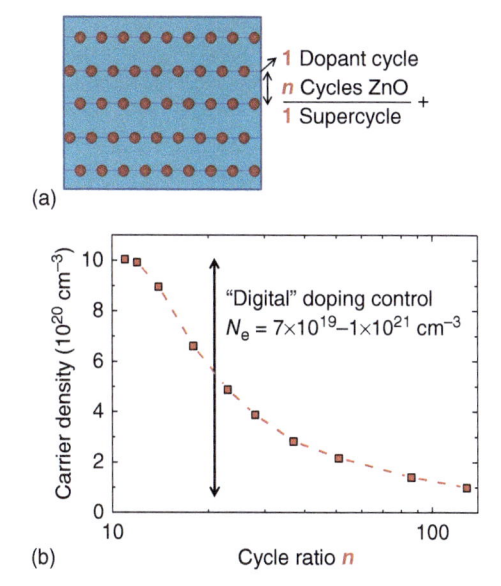

are randomly incorporated into the lattice. The vertical spacing of dopants can be accurately controlled by the number of ZnO cycles between the dopant (e.g., Al_2O_3) steps, that is, the cycle ratio n. Therefore, using such a supercycle approach, the carrier density can be controlled with great accuracy by the cycle ratio n, as shown in Figure 2.4b.

For a proper control of the *vertical* spacing of the dopants, it is important to be aware of deviations from steady-state ALD behavior that can occur when switching ALD processes in the supercycle approach. For example, a reduction in growth rate has been observed for ZnO after a doping step using either TMA or DMAI, which only recovers after ~4 ZnO cycles [137, 145]. This reduction has been attributed to the coexistence of AlOH* and ZnOH* surface species after the Al_2O_3 step, which leads to proton transfer from the AlOH* group to the relatively basic ZnOH* group, resulting in a reduced density of surface OH groups [137]. Additionally, for the case of TMA, it has been observed that ZnO can be etched during the TMA exposure step by the following reaction:

$$^*ZnOH + Al(CH_3)_3 \rightarrow {}^*Al(CH_3)OH + Zn(CH_3)_2$$

Such effects show that the growth per supercycle (GPSC) can vary from what would be expected from linear addition of the growth rates of the comprising ALD cycles.

The supercycle approach presented inherently results in planes of high dopant density in the superstructure, which potentially leads to clustering of dopants. Such clustering can have several detrimental effects. First, clustering leads to a reduced doping efficiency, which enhances neutral impurity scattering by inactive dopants and limits the maximum achievable doping level. Secondly, the dopant cycle can interrupt the grain growth of the ZnO, thereby enhancing grain boundary scattering [105]. Finally, for a given dopant density, IIS is minimized when the dopants are isotropically distributed. Therefore, besides control over the *vertical* spacing, control over the *lateral* spacing of dopants is highly desired for optimization of doped ZnO layers.

Several approaches have been proposed in the literature to reduce this dopant clustering by reducing the number of dopant atoms deposited per cycle. Wu et al. have shown that by replacing TMA as dopant precursor by a bulkier precursor, such as DMAI, the lateral distance between Al atoms can be increased due to enhanced steric hindrance, as is schematically shown in Figure 2.17 [137]. Because of the decrease in Al atoms deposited per dopant cycle from ~1.1 at. nm^{-2} to ~0.3 at. nm^{-2}, the percentage of Al atoms in the film that are active as dopants, that is, the doping efficiency, increases from ~10% to almost 60% [137]. Due to this, a maximum carrier density level up to 10^{21} cm^{-3} could be achieved, as opposed to a maximum of ~4×10^{20} cm^{-3} for TMA.

Besides the use of steric hindrance, the amount of Al deposited in the dopant step can be reduced by reducing the amount of hydroxyl sites available for TMA chemisorption. A rather simple approach was used by Park and Heo, who prolonged the purge time after the water dose in order to reduce the amount of hydroxyl groups via dehydroxylation reactions [135]. Another approach is to dose the TMA immediately after the last DEZ dose, that is, to omit the H_2O dose. Since less number of hydroxyl sites are available for TMA chemisorption, the

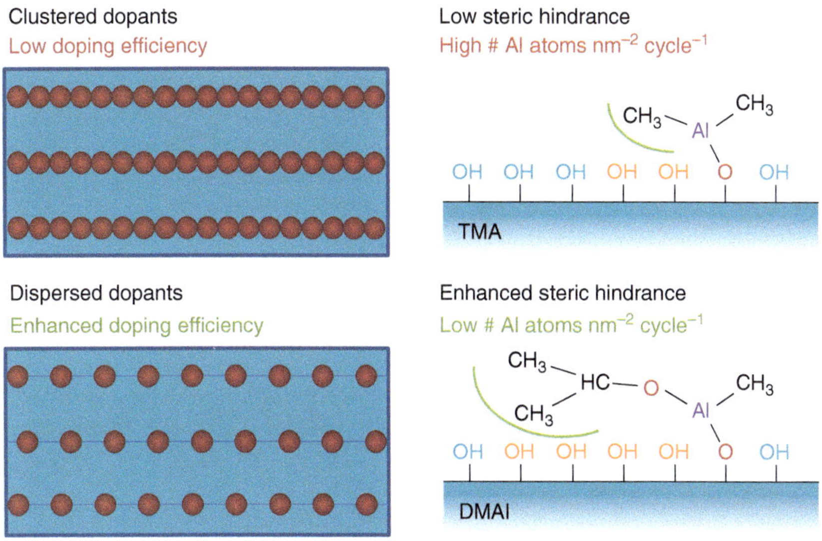

Figure 2.17 Schematic of the correlation between dopant clustering and doping efficiency. One approach for reduced dopant clustering is the use of dopant precursors with bulky ligands that lead to steric hindrance, as shown for TMA and DMAI in the figure.

Al incorporation is reduced and the doping efficiency is enhanced [145]. Finally, functionalization of the surface by exposure to alkyl alcohols prior to TMA exposure can be used to reduce the number of sites available for chemisorption of TMA. The alkoxide or alkoxylate surface groups that are formed are subsequently removed during the oxidant step [136].

2.3.2.2 Beyond Al Doping: Doping by B, Ti, Ga, Hf, and H

Although Al doping has been the by far most studied dopant for ZnO, other dopant atoms such as B, Ti, Ga, Hf, and even H have gained significant interest as well. Boron-doping of ZnO, well known in the field of low-pressure CVD, has been demonstrated using both B_2H_6 and triisopropyl borate ($B(O^iPr)_3$ or TIB) as B-precursors. Using B_2H_6, Sang et al. reported on ZnO:B with a promisingly low resistivity of 0.64 mΩ cm obtained at a low deposition temperature of 150 °C (Table 2.5) [139]. In addition, the carrier mobility, 20 cm^2 V^{-1} s^{-1}, was quite respectable. Despite these promising results, few further reports on the use of B_2H_6 as ALD dopant can be found in the literature. This could very well be due to the high toxicity of B_2H_6 in conjunction with its extremely high vapor pressure, which renders controlled dosing difficult [138]. Recently, $B(O^iPr)_3$ has emerged as a promising alternative B-precursor. At a deposition temperature of 200 °C, a low resistivity of 0.9 mΩ cm could be achieved. Similar to the case of DMAI, the $B(O^iPr)_3$ precursor benefits from its bulky ligands in reducing the amount of B dopants deposited per cycle [138].

As can be seen from Table 2.5, also the use of Ti and Ga as dopant has been shown to enable resistivity values <1 mΩ cm. Interestingly, Thomas et al. have demonstrated that it is also possible to dope ZnO with H by interleaving ZnO cycles by H_2 plasma treatments [144]. In this way, a resistivity of 0.7 mΩ cm was achieved at a reasonably high mobility of 15 cm^2 V^{-1} s^{-1}.

To conclude this section on doped ZnO, several groups have shown that resistivity values well below 1 mΩ cm can be achieved at low temperatures ($\leq$200 °C). The key to achieving efficient doping is reduction of dopant clustering through the use of clever approaches to reduce the number of dopant atoms deposited per cycle. Even though values typical for conventional sputtered ITO (~0.4 mΩ cm) cannot be reached, considerations regarding cost and material availability could make doped ZnO an effective alternative to ITO, especially at the rear side of an SHJ solar cell.

2.3.2.3 ALD of In_2O_3

Although historically, Sn-doped In_2O_3 (ITO) has been the most widely used TCO for SHJ solar cells, relatively less literature on ALD of In_2O_3-based TCOs seems to be available compared to ZnO. In Table 2.6, a brief overview of ALD processes of (doped) In_2O_3 is given.

In 1995, Asikainen demonstrated ALD of both In_2O_3 and ITO using a halide $InCl_3$ precursor and achieved a very low resistivity of 0.25 mΩ cm by Sn doping [146]. However, the process featured a low GPC of 0.2 Å and required a high

Table 2.6 Selection of reported ALD processes for (doped) In_2O_3.

Dopant	Precursors	T_{dep} (°C)	N_e (×10^{20} cm^{-3})	μ_e (cm^2 V^{-1} s^{-1})	ρ (mΩ cm)	GPC (Å)	References
—	$InCl_3$, H_2O	500	0.25	72	3	0.2	[146]
Sn	$InCl_3$, $SnCl_2$, H_2O	500	5.2	47	0.25	0.2	[146]
—	$In(acac)_3$, H_2O	160–255	—	—	30–6×10^4	0.15–0.25	[147]
—	$In[(^iPrN)_2CN(CH_2Me)]_3$, H_2O	230–300	—	—	—	0.45	[148]
—	$InMe_3$, H_2O	200–250	0.27	84	2.8	~0.39	[149]
—	DMLDMIn, H_2O	300	0.75	28.7	1.6	0.6	[150]
—	$In(TMHD)_3$, O_2 plasma	100–400	—	—	2.5–18	0.14	[151]
—	InCp, O_3	200–450	—	—	16	1.3–2.0	[152]
Sn	InCp, O_3, TDMASn, H_2O_2	275	4	50	0.3	1.1–1.7	[153]
H?	InCp, H_2O+O_2	100–250	0.8–4.5	38–111	0.34–2.5	1.0–1.6	[154]
H[a]	InCp, H_2O+O_2	100	1.8	138	0.27	1.2	[4, 109]

a) The films were postcrystallized at 150–200 °C.

deposition temperature of 500 °C. At lower temperatures, the use of In(acac)$_3$, In[(^{i}PrN)$_2$CN(CH$_2$Me)]$_3$, InMe$_3$, DMLDMIn, In(TMHD)$_3$, and InCp has been reported. In(acac)$_3$ and In[(^{i}PrN)$_2$CN(CH$_2$Me)]$_3$ have a somewhat higher GPC, but the optoelectronic properties have so far not been reported. InMe$_3$ and H$_2$O can yield a reasonable resistivity (~3 mΩ cm) at an intermediate temperature (200–250 °C) and a GPC of ~0.39 Å.

In 2006, Elam et al. demonstrated an ALD process for In$_2$O$_3$ based on indium cyclopentadienyl (InCp) and ozone at temperatures 200–450 °C [152]. Promisingly, a high GPC of 1.3–2.0 Å was achieved, and a very low resistivity of 0.3 mΩ cm was reached at 275 °C by Sn doping using TDMASn and H$_2$O$_2$. A few years later, Libera et al. showed that although H$_2$O and O$_2$ by themselves do not yield growth in combination with InCp, a combination of H$_2$O and O$_2$ as oxidants yields a high GPC (1.0–1.6 Å) at very low deposition temperatures (100–250 °C) [154]. They proposed that both oxidants are needed for growth since they serve different roles, that is, H$_2$O eliminates the Cp ligand and O$_2$ oxidizes the surface In from oxidation state +1 to +3. Moreover, a low resistivity of 0.34 mΩ cm was achieved for an amorphous film at 100 °C. The highest mobility value of 111 cm^2 V^{-1} s^{-1} was achieved at 140 °C, which is around the amorphous–polycrystalline growth transition temperature.

Macco et al. have demonstrated that the ALD process by Libera et al. using InCp and a combination of O$_2$ and H$_2$O actually unintentionally yields H-doped In$_2$O$_3$ (In$_2$O$_3$:H), where an amorphous film deposited at 100 °C has a H content of 4.2 at.% [4]. By low-temperature postdeposition annealing at 150–200 °C, solid-phase crystallization of the film occurs, which yields a low resistivity of 0.27 mΩ cm at a record-high electron mobility of ~138 cm^2 V^{-1} s^{-1} and a relatively low carrier density of 1.8×10^{20} cm^{-3}. This combination leads to negligible FCA in the photon energy range relevant for SHJ solar cells [4]. In fact, the quality of crystallized layers is such that only phonon and IIS processes play a role, meaning that the mobility is at its fundamental limit, as can also be seen in Figure 2.13b [109]. The excellent optoelectronic properties in combination with the low temperature processing and high growth rate make this process very interesting for SHJ solar cell applications. To the authors' knowledge, there are currently no ALD processes reported for IZO, IMO, and IWO, which is a clear opportunity for further development.

2.3.3 High-Volume Manufacturing of ALD TCOs

Although ALD is not yet used in the industry for the preparation of TCOs for SHJ solar cells, the ALD approach can potentially offer some key benefits over the most commonly used sputtering method. First, due to the absence of harsh plasma conditions during ALD, plasma damage (e.g., during sputtering) to the substrate is avoided. Therefore, recent studies have focused on the use of ALD as a "soft deposition" method to deposit TCOs in SHJ solar cells and have demonstrated improved passivation [106, 155]. Additionally, as discussed in the section on ALD of doped ZnO, ALD allows for a high level of control over the doping level of the TCO and thereby its work function. This greatly facilitates the optimization of the doped a-Si:H/TCO contact by varying the doping level and potentially even the doping profile [106, 126, 129].

Table 2.7 Selection of reported S-ALD results of both intrinsic and doped ZnO using DEZ as the Zn source.

Dopant	Doping precursor	Growth rate (nm^{-1} s^{-1})	T_{dep} (°C)	N_e (10^{20} cm^{-3})	μ_e (cm^2 V^{-1} s^{-1})	ρ (mΩ cm)	References
—	—	0.6	200			1–2 × 10^5	[160, 161]
—	—	~1	75–250	0.2–0.7	14–30	4–150	[159]
Al	TMA	0.2	200	5	6	2	[162]
Al	DMAI	~1.5	250	—	—	0.46	[163]
In	InMe$_3$	0.1	200	6	3	3	[164]
Ga	GaMe$_3$	0.4	250	—	—	2	[165]

Moreover, after the recent introduction of high-throughput ALD reactors for HVM of Al$_2$O$_3$ in the PV industry, the deposition of (doped) ZnO by such reactors, most notably spatial ALD (S-ALD), is also being explored [78, 156–159]. As can be seen in Table 2.7, S-ALD processes have been reported that combine high deposition rates exceeding 1 nm s^{-1} with rather good material properties. In particular, Ellinger et al. showed that a very low resistivity (<0.5 mΩ cm) can be obtained at high growth rates (~1.5 nm s^{-1}), on par with typical growth rates obtained with sputtering, at an intermediate deposition temperature (250 °C).

A key difference between temporal and spatial ALD of doped TCOs is that in spatial ALD, the dopants can, apart from using the supercycle approach, also be introduced by *premixing* or by *coinjection* with the other precursor, due to a homogeneous delivery of the precursor to the substrate. In such approaches, both precursors compete for reactive surface sites. As a result, the amount of dopant incorporation depends, for example, on the partial pressures of both precursors and can even depend on exposure times [162]. One might thus say that some level of control, which is typical for the supercycle approach, is lost when switching to premixed or coinjected precursors. Nonetheless, the good material properties that have been reported and the ability to even successfully deposit other multicomponent oxides such as InGaZnO demonstrate that this is not necessarily a drawback for the industrial application of precursor mixing or coinjection in S-ALD [166].

2.4 Prospects for ALD in Passivating Contacts

In this section, the upcoming field of passivating contacts and the possible role of ALD therein are discussed. First, the basic principles and requirements of passivating contacts and some of its concepts are outlined. Subsequently, examples of passivating tunnel and carrier-selective oxides and the use of ALD for preparing such oxides are reviewed.

2.4.1 Basics of Passivating Contacts

A passivating contact is typically a stack of thin films on the Si absorber, which passivates the Si surface and simultaneously acts as a selective membrane for either holes or electrons. Examples include the traditional SHJ cell and the TOPCon concept, as discussed in the introduction.

2.4.1.1 How to Make a Passivating Contact?

To briefly illustrate the working principle and the merits of passivating contacts, schematic band diagrams of example strategies to produce carrier-selective contacts are shown in Figure 2.18. All diagrams consider the Si under illumination, which leads to excess charge carriers. Since in a metal (or TCO), there can be no quasi-Fermi level splitting, the two quasi-Fermi levels must converge at the contact. Since a gradient in a Fermi level represents a force, this leads to a current of both electrons (J_n) and holes (J_p) toward the metal:

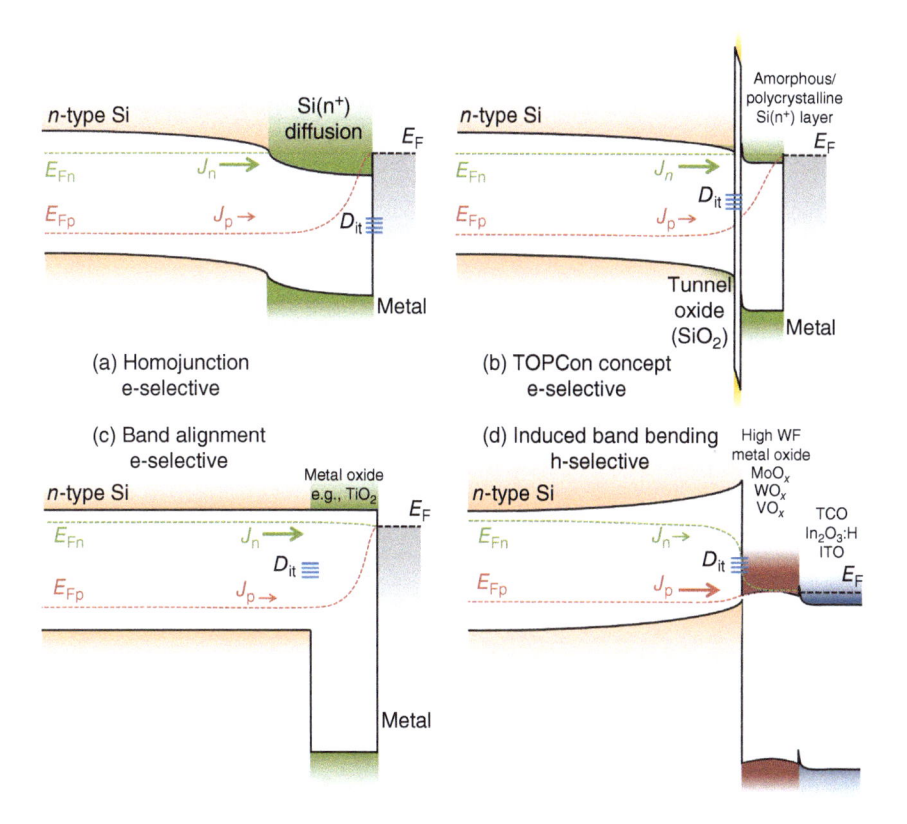

$$J_n = en\mu_n \nabla E_{Fn}, \quad J_p = ep\mu_p \nabla E_{Fp} \tag{2.7}$$

Figure 2.18 Schematic band diagrams of various approaches for obtaining carrier selective contacts. All band diagrams refer to the situation under illumination and are not drawn to scale. (a) A conventional electron-selective contact made by n^+-type doping. (b) An electron-selective contact obtained by the TOPCon concept, consisting of a tunnel oxide and a thin (partially) crystalline n^+-Si film. (c) An electron-selective contact that is realized through *band alignment* of the Si with a metal oxide film. (d) A hole-selective contact that is realized through *induced band bending* by a high-work-function metal oxide film. Often concepts (c) and (d) also employ separate ultrathin passivation layers, but these were not drawn for the sake of simplicity.

In these equations, μ_n and μ_p are the electron and hole mobilities, respectively. In addition, note that all these quantities in principle depend on the spatial coordinate x.

In order for a contact to be *selective*, the region or film(s) in between the Si and metal contact must, besides providing passivation, induce a strong *asymmetry* in the electron and hole currents to the metal contact. In order to understand how this asymmetry can be achieved, one should realize that Equation 2.7 is basically Ohm's law for electrons and holes [167]. If the conductivity for a charge carrier is high (i.e., a high product of mobility and carrier concentration), there will be a low gradient in the quasi-Fermi level toward the contact for a given current (i.e., little voltage drop). Therefore, in the case of a passivating contact, the quasi-Fermi level of the carrier that is to be extracted should be as flat as possible (i.e., low resistance), whereas the other quasi-Fermi level should show a high degree of bending (i.e., high resistance). This is markedly different from an ideal passivation layer on Si where both quasi-Fermi levels are flat (high resistance to both carriers), and in case of a Si/metal contact, in principle, both quasi-Fermi levels will bend (low resistance to both carriers).

In Figure 2.18a, the conventional method for making an electron-selective contact ($J_n \gg J_p$) is shown, that is, by heavily doping the Si. The selectivity for electrons arises from the high resistance for holes in the heavily doped n^+ Si region. Note that the holes actually experience a strong force toward the metal, as seen by the strong gradient of the Fermi level, E_{Fp}. Nevertheless, the hole current J_p in this region is very low since the large energetic distance between the valence band and E_{Fp} ensures a low density of holes p and hence a high resistivity for holes. Although contacts based on such homojunctions can be very selective, the V_{oc} of such devices is typically limited by Auger recombination occurring in the highly doped region. This drawback is avoided by using passivating contacts, examples of which are depicted in Figure 2.18b–d.

In Figure 2.18b, the TOPCon concept is shown. Here, the n^+ doping of the (partially) crystalline Si layer provides selectivity to extract electrons in a manner quite similar to the doped region of (a). A very thin SiO_2 tunnel oxide of ~1.4 Å, typically prepared by a nitric acid oxidation step (NAOS), enables chemical passivation and acts as a diffusion barrier for dopants. Note that the typical thickness of the doped Si layer (few tens of nanometers) is much less than a typical doped region (~0.5 µm) and thus leads to much less Auger recombination.

In Figure 2.18c, an electron-selective contact is formed by *band alignment*: a wide-band-gap material is deposited on Si, with little (or ideally no) conduction band offset. In this way, the hole current J_p is greatly reduced by the large valence band offset. Again, the strong gradient in E_{Fp} shows that the holes experience a strong force toward the metal, but the large energetic distance between the valence band and E_{Fp} ensures a low density of holes p in the metal oxide film and thereby a low hole current J_p. Note that, for simplicity, it is assumed here that no band bending occurs (i.e., no fixed charge and equal work functions of the n-type Si and the metal oxide film are assumed).

In Figure 2.18d, a hole-selective contact is formed by *induced band bending*: in this example, the high work function ($\gtrsim 5.5$ eV) of a metal oxide (such as MoO_x, WO_x, or VO_x) induces a strong upward bending at the n-type Si surface, leading

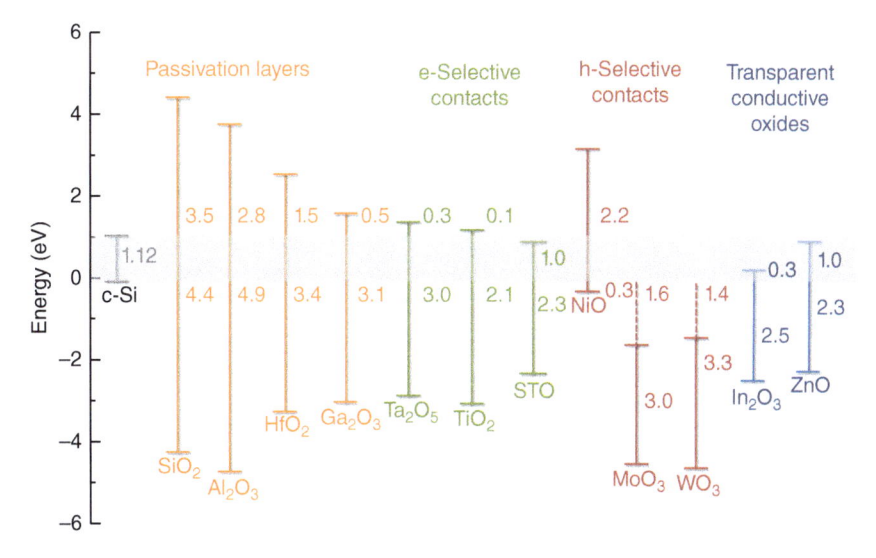

Figure 2.19 Schematic of the band offsets of a selection of oxides with Si. Offsets are denoted in electronvolt. (Adapted from Refs [168] and [169].)

to inversion. Whereas the band bending reduces the electron concentration at the surface, it facilitates a high hole current J_p to the metal.

In Figure 2.19, an overview of the band offsets with Si of (a selection of) oxides that are of interest for the formation of a passivating contact is given. It should be noted that these values can vary considerably depending on the exact processing conditions, and the values are therefore indicative. Ta_2O_5, TiO_2, and strontium titanate (STO) are of interest as electron-selective contacts due to their small conduction band offset. Similarly, NiO is of interest as a hole-selective contact due to its small conduction band offset [18, 170], whereas MoO_3 and WO_3 form hole-selective contacts by induced band bending [17, 171].

For the sake of simplicity, separate passivation layers were not shown in the band diagrams in Figure 2.18. Nonetheless, since the oxides used for selectivity generally do not offer (excellent) passivation, many passivating contact schemes employ a-Si:H or ultrathin (<2 nm) tunnel oxide layers such as Al_2O_3 and SiO_2 in between the Si and the carrier-selective layer for interface passivation.[1]

In Figure 2.19, typically used oxide passivation layers are also shown. As discussed in Section 2.2, such oxides should have a low interface defect density D_{it}, whereas the presence of a fixed charge density Q_f can be either beneficial or detrimental to the passivation quality, depending on the doping of the wafer. When applied to passivating contacts, additional requirements for such layers come into play. Since the oxides should allow for tunneling, the D_{it} should be low even for ultrathin (<2 nm) layers. The presence of fixed charge is also of

---

1 Besides providing interface passivation, these layers are thought to aid in selectivity: if, for example, a-Si:H is added in between the Si and high WF metal oxide of Figure 2.18d, most of the drop in E_{Fn} will occur in the a-Si:H instead of in the Si. Since the mobility in a-Si:H is orders of magnitudes lower than in Si, this will lead to a reduced J_n toward the metal contact, according to Equation 2.7.

importance, as the band bending induced by fixed charge can affect the selectivity, analogous to the case shown in Figure 2.18d. For example, the high negative fixed charge of Al_2O_3 makes Al_2O_3 more suited for hole-selective contacts rather than for electron-selective contacts. Finally, the band offsets with Si play a role as well since tunnel probabilities are inversely and exponentially proportional to the band offsets. For example, for SiO_2, the asymmetry in the valence band (4.4 eV) and conduction band (3.5 eV) offset makes it such that electrons tunnel much easier compared to holes [172]. Therefore, asymmetry in the band offsets can also aid in selectivity.

2.4.1.2 Requirements of a Passivating Contact

To assess the potential of various passivating contact schemes, it is instructive to discuss the two main figures of merit of a passivating contact:

- The contact resistance ρ_c for the charge carrier type the contact should be selective to.
- The recombination current of the other charge carrier type to the metal contact, which can be characterized by J_0.

To a first order approximation, the contact resistance influences the FF, whereas the recombination current limits the V_{oc}. In Figure 2.20, a contour plot of the maximum Si solar cell efficiency is displayed as a function of the ρ_c and J_0 values of the rear contact. This calculation assumes a full-area rear contact with no other loss mechanisms (no optical losses, no other recombination anywhere in the cell, and no other resistive losses) and thus represents the upper bound of the solar cell efficiency set by the rear contact.

As can be seen from Figure 2.20, high-efficiency (>25%) devices require both low ρ_c (<1 $\Omega\,cm^{-2}$) and J_0 (<100 fA cm^{-2}) values, a region that can be defined as a criterion for being a passivating contact. For comparison, a typical Si/metal contact in a p-type Al-BSF concept has a very low contact resistivity (~5 $m\Omega\,cm^2$), but the high J_0 (>500 fA cm^{-2}) severely limits the cell efficiency. On the other hand, an Al_2O_3 passivation layer can yield a very low J_0 (<10 fA cm^{-2}) but is insulating. For this reason, many solar cell designs (e.g., PERC and PERL) employ a local metal contacting scheme: by making local metal contacts to the silicon, a trade-off is made between passivated regions of low J_0 ($J_{0,pass}$) and contacting regions that have a high J_0 ($J_{0,cont}$) but a low ρ_c. The effective J_0 and ρ_c of a locally contacted rear are determined by the contact area fraction f by the relations

$$J_{0,eff} = fJ_{0,cont} + (1-f)J_{0,pass} \quad \text{and} \quad \rho_{c,eff} = \rho_c/f$$

However, local contacting might add to processing complexity and additionally induces resistive FF losses in the bulk of Si due to the required lateral transport of carriers therein, as shown in Figure 2.21 [3].

Figure 2.20 also shows the literature values for various passivating contact concepts, as well as for various rear sides of cells using partial metallization. The classical SHJ concept based on (doped) a-Si:H layers is probably the best-known example of a passivating contact. The rear contact of the current-record solar cell of Kaneka (25.1% efficiency) combines a very low J_0 value of 12 fA cm^{-2} with a low contact resistivity of 30 $m\Omega\,cm^2$ [14]. The TOPCon concept and

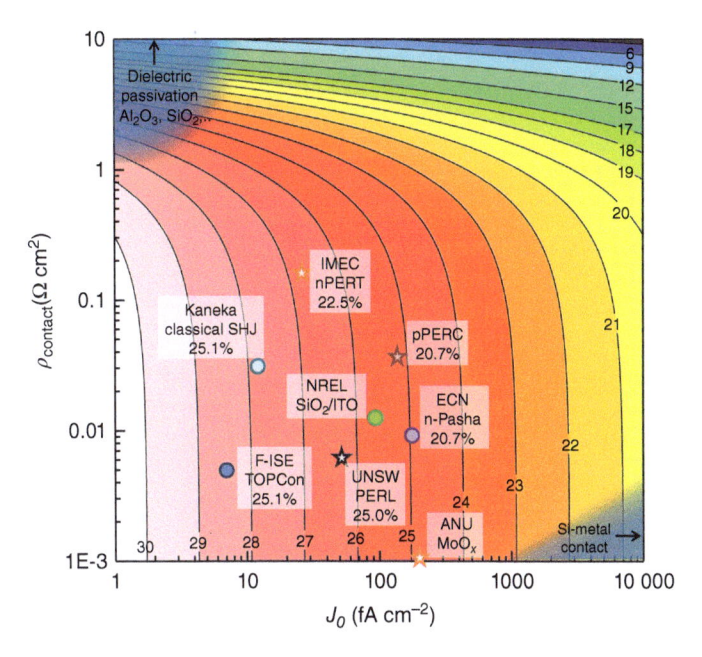

Figure 2.20 Contour plot of the calculated upper limit of efficiency of a solar cell featuring a full-area passivating contact, as a function of J_0 and ρ_c. The calculation is done similarly as in Refs [173] and [174] and assumes no other recombination channels (surface nor bulk), no shunting, and no optical losses (i.e., a J_{sc} of 44 mA cm^{-2}). For comparison, data points for various reported structures/cells are shown, along with the efficiencies of the full devices. These include an SHJ cell of Kaneka (2015, private communication) [14], the TOPCon concept of F-ISE [3, 175], the SiO$_2$/ITO stack of NREL [173], the TiO$_2$ cell of ANU [176], the UNSW PERL [177], a p-type PERC cell [177], the IMEC nPERT [178, 179], the ECN nPasha (2015, private communication) [178], and a p-type Si/MoO$_x$ contact [180]. Hole-selective contacts are denoted by star-shaped symbols, whereas electron-selective contacts are denoted by circular symbols. Concepts employing a full-area rear contact are noted in bold. For the PERL cell, the J_0 was estimated using the reported surface recombination velocity in Ref. [177] and case 3 in Ref. [24]. For solar cell concepts that use a partial rear contact, the J_0 and ρ_c values have been corrected for the contact area fraction f.

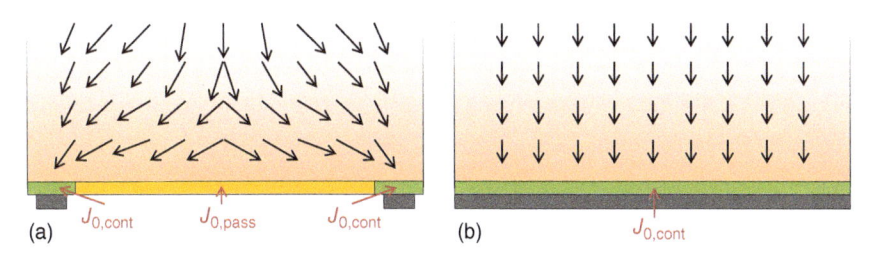

Figure 2.21 Schematic showing the current flow pattern in (a) a locally contacted cell and (b) a solar cell with a full-area rear contact. (Adapted from Ref. [3].)

the SiO$_2$/ITO stack of [173] (discussed as follows) are both electron-selective contacts that use a tunnel oxide. As can be seen, both concepts have a very low ρ_c, demonstrating that efficient transport can occur through such oxides. Moreover, the TOPCon concept shows that very low J_0 values can be achieved with tunnel oxides. Due to the excellent properties of such passivating contacts, full-area contacts can be employed while still having a low J_0. This has the benefit of having one-dimensional charge transport in the solar cell, thereby reducing resistive losses due to lateral transport, as well as being very straightforward from a processing point of view. As can be seen, the solar cell concepts employing partial rear metallization are also capable of reaching low ρ_c values, but the local metal contacts inevitably lead to higher J_0 values. As discussed further, other metal oxide films are promising for passivating contact formation as well, such as TiO$_2$ for electron-selective contacts and MoO$_x$, WO$_x$, and VO$_x$ for hole-selective contacts. Evaporated MoO$_x$ directly on p-type Si has been reported to have a very low contact resistivity (1 mΩ cm^2) and an intermediate J_0 of ~200 fA cm^{-2}, as can be seen in Figure 2.20 [180]. Because of the low ρ_c and intermediate J_0, such contacts are best used in a partial rear contacting scheme, as shown by a reported 20.4% efficient solar cell using a 5% rear contact area fraction [181]. Moreover, MoO$_x$ has been used as a replacement of the a-Si:H(p) layer in a conventional SHJ solar cell. Due to the passivating properties of the a-Si:H layer, an impressive 22.5% efficiency has been reported for a full-area front contact [182]. The J_0 and ρ_c values of the a-Si:H/MoO$_x$/TCO contact were not reported for this solar cell.

2.4.2 ALD for Passivating Contacts

As was discussed in the previous section, (stacks of) various thin films are of interest to serve as passivating contact, which in principle can be prepared by ALD. Moreover, the use of ALD may offer distinct advantages, for example, in terms of processing/doping control and the easy manufacturing of stacks. Nonetheless, this new field has yet to be fully explored. Therefore, in this section, the possible role of ALD in preparing tunneling oxides and carrier-selective oxides is discussed, addressing both the few examples of ALD already shown in this field and the future prospects.

2.4.2.1 ALD for Tunneling Oxides

For tunnel oxides, the contact resistivity (i.e., ρ_c) and level of surface passivation (i.e., J_0) are extremely dependent on the thickness of the oxide. Therefore, it is expected that the submonolayer thickness control over a large surface area offered by ALD can be a key enabler in this respect. It is worth mentioning that ultrathin ALD Al$_2$O$_3$ has also been used in between the metal contacts and the highly doped region of the Si wafer in PERC cells. Although, strictly speaking, this is not a passivating contact, it has been shown that only two cycles of Al$_2$O$_3$ enhance the passivation (12 mV increase in V_{oc}) without a significant increase in contact resistivity [183].

A tunneling ALD Al$_2$O$_3$ film has been successfully used in an Al$_2$O$_3$/ZnO(:Al) stack to make a hole-selective contact, fully prepared by ALD [21, 22]. Such a stack achieves selectivity toward holes by the negative fixed charge in the Al_2O_3, which leads to accumulation of holes at the Si surface, which is analogous to the use of a high-work-function metal oxide as shown in Figure 2.18d. The holes subsequently recombine with the electrons in the conduction band of the ZnO(:Al). Interestingly, such a stack takes advantage of the fact that the fixed charge in Al_2O_3 has been observed to be *interfacial*, that is, it resides at the Si–Al_2O_3 interface, and is thus persistent even for ultrathin films [184]. Additionally, the position of the Fermi level in the TCO, which is readily tuned in ALD by control of the doping level, was found to be crucial for the working of such a contact. Nonetheless, although this work nicely demonstrates a proof of concept, the reported high J_0 ($>10^4$ fA cm^{-2}) and intermediate ρ_c ($>1\,\Omega$ cm^2) values hinder a high efficiency.

Besides Al_2O_3, many passivating contacts employ SiO_2 as tunnel oxide. Young et al. have shown that a stack of thin SiO_2 and sputtered ITO can also make an electron-selective contact through energetic lineup of the Fermi levels of the Si and the heavily degenerate ITO [173]. Promisingly, a low J_0 of 92.5 fA cm^{-2} and a contact resistivity ρ_c of only 11.5 mΩ cm^2 were achieved, as indicated in Figure 2.20. Remarkably, the optimal SiO_2 thickness prior to ITO sputtering was found to be 4.5 nm, much more than would be expected on the basis of a tunneling process. This has been attributed to intermixing of the SiO_2/ITO layer by the energetic ions coming from the plasma [173]. In this respect, ALD of TCOs (In_2O_3:H, ZnO) [4] could be a much better controlled process due to the absence of plasma-related damage [106].

Finally, it is worth noting that the use of ultrathin ALD metal oxides as tunneling layers has already been explored in the field of organic PV [185, 186]. Specifically, ALD layers of Ga_2O_3 and Ta_2O_5 have been used successfully in such cells. These two materials have recently been shown to provide excellent Si surface passivation, also making these materials highly interesting for passivating contact formation for Si solar cells [98, 100].

2.4.2.2 ALD for Electron-Selective Contacts

An electron-selective contact has been prepared by an ultrathin (1–4 nm) layer of TiO_2 by Avasthi et al. [19]. The selectivity of this film is achieved through band alignment as shown in Figure 2.18c. Although the work of Avasthi et al. demonstrates the electron selectivity of TiO_2, the rather simple device structure and lack of passivation severely limited the performance of the device. Promisingly, as discussed earlier, Liao et al. have demonstrated that it is possible to achieve excellent surface passivation using a thermal ALD process using $TiCl_4$ and H_2O at 100 °C to deposit TiO_2, which suggests that this concept could be optimized further [99]. In a separate piece of work, another very low surface recombination velocity of 16 cm s^{-1} was observed for a carefully prepared TiO_2/Si heterojunction, showing that this interface can be highly passivating [187]. Yang et al. have subsequently demonstrated that 4.5 nm of TiO_2 prepared by ALD (using Ti(O^{i}Pr)$_4$ and H_2O at 230 °C) can also yield a relatively low contact resistivity of ~0.25 Ω cm^2 and a J_0 of 25 fA cm^{-2} [176]. When combining this ALD TiO_2 with a 1.5-nm SiO_2 interlayer, they demonstrated an impressive 20.5% efficiency for their champion cell.

As can be seen from Figure 2.19, Ta_2O_5 and STO also have the proper band alignment to serve as an electron-selective contact. As already pointed out in

Section 2.2.4, ALD Ta_2O_5 can provide excellent surface passivation when capped with SiN_x [98]. Therefore, the use of a dedicated passivation layer in between the Si and Ta_2O_5 can potentially be avoided. Nonetheless, the observed negative fixed charge of $\sim 10^{12}$ cm^{-3} observed might hinder the working as a selective electron contact.

STO is mostly known in the semiconductor industry for its very high dielectric constant. It has been shown experimentally that the conduction band offset of STO on both *n*- and *p*-type Si is negligible (~ 0.1 eV) and does not change significantly if a very thin SiO_2 interlayer is applied (<1.2 nm) [188]. More importantly, DFT calculations predict that the conduction band offset is highly dependent on the initial layer of the STO thin film: the desired negligible conduction band offset

Table 2.8 Selection of ALD processes reported in the literature of potential carrier-selective oxides.

Metal oxide	Metal precursor	Reactant	T_{dep} (°C)	GPC (Å)	References
Electron-selective oxides					
TiO_2	$Ti(O^iPr)_4$	H_2O	150–300	0.2–0.3	[191]
	$TiCl_4$	H_2O	100	0.6	[99]
	$Ti(Cp^*)(OMe)_3$	O_2 plasma	50–300	0.5	[192]
	$Ti(NMe_2)_4$	H_2O	25–325	0.5–1.4	[193]
Ta_2O_5	$Ta_2(OEt)_{10}$	H_2O	250	0.3	[98]
	$Ta(NMe_2)_5$	O_2 plasma	100–250	0.8–0.9	[192]
	$Ta(NMe_2)_5$	H_2O/O_3	200–300	0.9/1.1	[194]
STO	$C_p(Me)_5Ti(OMe)_3/$ $Sr(^iPr_3Cp)_2DME$	O_2 plasma	150–350	2.3–2.6$^{a)}$	[190]
	$Ti(O^iPr)_4/Sr(thd)_2$	H_2O plasma	250	0.6$^{b)}$	[195]
Hole-selective oxides					
MoO_x	$(N^tBu)_2(NMe_2)_2Mo$	O_3	100–300	0.3–2.4	[196]
	$(N^tBu)_2(NMe_2)_2Mo$	O_2 plasma	50–350	0.8–1.9	[20, 197]
	$Mo(CO)_6$	O_3	152–172	0.8	[198]
WO_x	$(N^tBu)_2(NMe_2)_2W$	H_2O	300–350	0.4–1.0	[199]
	$W(CO)_6$	O_3	195–205	0.2	[200]
	$WH_2(^iPrCp)_2$	O_2 plasma	300	0.9	[201]
VO_x	$V(NEtMe)_4$	H_2O	125–200	0.8	[202]
	$VO(OPr)_3$	H_2O	170–190	1.0	[203]
	$VO(OPr)_3$	O_2/H_2O plasma	50–200	0.7	[204]
NiO_x	$Ni(Et_2Cp)_2$	O_3	150–300	0.4–0.9	[205]
	$Ni(thd)_2$	H_2O	260	0.4	[206]
	$Ni(Cp)_2$	H_2O	165	—	[207]
	$Ni(dmamp)_2$	H_2O	120	0.8	[208]

a) The reported GPC is the growth per supercycle for a [SrO]/[TiO_2] cycle ratio of 1:3.
b) The reported GPC is the growth per supercycle for a [SrO]/[TiO_2] cycle ratio of 1:1.

Figure 2.22 False-colored cross-sectional TEM image of a stack of amorphous silicon, ALD MoO$_x$, and crystallized ALD hydrogen-doped indium oxide (c-In$_2$O$_3$:H). (Adapted from Ref. [20].)

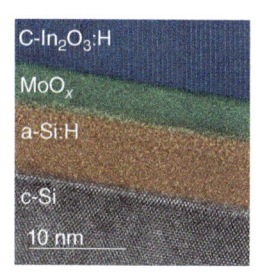

(0.1–0.2 eV) occurs when the initial layer of the STO film consists of SrO, whereas a higher offset of 1.2–1.3 eV has been predicted when the initial layer consists of TiO$_2$ [189]. Since both the interfacial and bulk compositions can be controlled accurately by ALD by choosing the appropriate initial cycle (TiO$_2$ or SrO) and cycle ratio [190], respectively, it can be expected that ALD is very well suited for the preparation of such oxides (Table 2.8).

2.4.2.3 ALD for Hole-Selective Contacts

Molybdenum oxide (MoO$_x$) is well known in the organic PV literature as a hole transport material. However, it has only very recently been demonstrated that evaporated MoO$_x$ can replace the hole-selective a-Si:H(p) layer at the front of a standard SHJ solar cell [16, 180, 209]. The working principle of such a selective hole contact is thought to be based on the high work function of MoO$_x$ (~6.6 eV), as shown in Figure 2.18d. Promisingly, the reduced optical losses enable a substantial enhancement in photocurrent of 1.9 mA cm^{-2}. Currently, the highest reported efficiency for a MoO$_x$-based SHJ is already 22.5%, which is very promising, given the novelty of this approach [182].

Recently, it has also been shown that MoO$_x$ can be deposited by plasma-enhanced ALD using (N^{t}Bu)$_2$(NMe$_2$)$_2$Mo and O$_2$ plasma at temperatures down to 50 °C [197]. Additionally, this ALD MoO$_x$ layer was implemented in an a-Si:H/MoO$_x$/ALD In$_2$O$_3$:H stack (Figure 2.22), and a high level of passivation in combination with a high optical transparency was demonstrated [20]. In other work, initial solar cell results based on ALD MoO$_x$ have been reported, although the efficiency (~11%) is not yet on par with its evaporated counterpart [210].

For the other aforementioned hole-selective materials of WO$_x$, VO$_x$, and NiO$_x$, there have been no reports yet on the use of ALD to make carrier-selective contacts with these materials. Since all these materials can be obtained by ALD, it is likely that this will be explored in the near future. The assessment of the passivation quality, possibly in combination with an a-Si:H layer or a tunnel oxide, and the carrier-selectivity should be the main focus when screening the ALD approaches for the fabrication of passivating contacts.

2.5 Conclusions and Outlook

In the field of Si PV, ALD of Al$_2$O$_3$ resulted in a breakthrough in the passivation of p-type Si surfaces. As a result, it is currently incorporated in solar cells with high efficiencies over 25%, and it enables challenging concepts, such as solar cells with black Si surface textures. Due to these successes, in the last few years, high-throughput reactors based on temporal and spatial ALD have successfully been developed. These reactors can meet the stringent demands of HVM in PV industries in terms of throughput and cost, and ALD is competitive with other techniques for the deposition of Al_2O_3. In addition, other materials prepared by ALD have successfully been explored for the passivation of Si. Examples include HfO_2, SiO_2, Ga_2O_3, Ta_2O_5, and TiO_2, which altogether can passivate a variety of doped surfaces. The passivation by SiO_2/Al_2O_3 stacks has recently even been successfully scaled up using batch ALD.

Apart from surface passivation, the potential of ALD to prepare TCOs, such as doped ZnO and In_2O_3 films, has been recognized. Key advantages of ALD as deposition method of TCOs include a very precise control of film properties, in particular when using extrinsic dopants, such as for doped ZnO. Moreover, its soft nature does not induce damage on sensitive a-Si:H passivation layers. For In_2O_3 TCOs prepared by ALD, the electron mobility is record-high and reaches even the fundamental limit. The latter allows for the best possible trade-off between conductivity and transparency. Altogether, these merits make ALD very promising as deposition method for TCOs in Si solar cell manufacturing, although its potential is yet to be demonstrated on a solar cell level. Moreover, the industrial viability of ALD to prepare TCOs in solar cell manufacturing remains to be determined, although the deposition of doped ZnO films by high-throughput spatial ALD reactors has recently been achieved.

Finally, an interesting, emerging field of research governs passivating contacts. In this field, very thin films or stacks of metal oxides (i.e., a thickness of 1–80 nm) should meet many requirements in terms of surface passivation, carrier selectivity to Si, a low contact resistance, and so on. Fortunately, ALD is ideally suited to deposit such stacks in a precisely controlled way. Moreover, the knowledge gained from ALD of TCOs and passivating films can be combined in this field. For instance, stacks of Al_2O_3/ZnO are pioneered as a hole-selective contact to Si, whereas SiO_2/In_2O_3 stacks are promising as an electron-selective passivating contact. With high-throughput ALD reactors available, it is likely that if such passivating contact schemes have come to full development, they can be prepared in a single deposition run, even on both sides of the solar cells at once. This could yield a significant process simplification in Si solar cell manufacturing and would underline the potential of ALD in the field of PVs.

References

1 Fraunhofer ISE (2015) Photovoltaics Report, *Freiburg 17 November* 2015. https://www.ise.fraunhofer.de/de/downloads/pdf-files/aktuelles/photovoltaics-report-in-englischer-sprache.pdf (accessed 18 November 2016)

2 Agostinelli, G.G., Vitanov, P., Alexieva, Z., Harizanova, A., Dekkers, H.F.W., De Wolf, S., and Beaucarne, G. (2004) Proceedings of the 19th European Photovoltaic Solar Energy Conference and Exhibition, Paris, France, June 7–11, 2004, pp. 2529–2532.

3 Glunz, S.W., Feldmann, F., Richter, A., Bivour, M., Reichel, C., Steinkemper, H., Benick, J., and Hermle, M. (2015) Proceedings of the 31st European Photovoltaic Solar Energy Conference and Exhibition, Hamburg, p. 259.
4 Macco, B., Wu, Y., Vanhemel, D., and Kessels, W.M.M. (2014) *Phys. Status Solidi RRL*, **8**, 987.
5 Green, M.A., Emery, K., Hishikawa, Y., Warta, W., and Dunlop, E.D. (2014) *Prog. Photovoltaics Res. Appl.*, **22**, 701.
6 Cousins, P.J., Smith, D.D., Luan, H.C., Manning, J., Dennis, T.D., Waldhauer, A., Wilson, K.E., Harley, G., and Mulligan, W.P. (2010) Conference Record IEEE Photovoltaic Specialists Conference p. 275.
7 Yamaguchi, T., Ichihashi, Y., Mishima, T., Matsubara, N., and Yamanishi, T. (2014) *IEEE J. Photovoltaics*, **4**, 1433.
8 Smith, D.D., Cousins, P., Westerberg, S., De Jesus-Tabajonda, R., Aniero, G., and Shen, Y.-C. (2014) *IEEE J. Photovoltaics*, **4**, 1465.
9 Richter, A., Hermle, M., and Glunz, S.W. (2013) *IEEE J. Photovoltaics*, **3**, 1184.
10 Benick, J., Hoex, B., van de Sanden, M.C.M., Kessels, W.M.M., Schultz, O., and Glunz, S.W. (2008) *Appl. Phys. Lett.*, **92**, 253504.
11 van de Loo, B.W.H., Knoops, H.C.M., Dingemans, G., Janssen, G.J.M., Lamers, M.W.P.E., Romijn, I.G., Weeber, A.W., and Kessels, W.M.M. (2015) *Sol. Energy Mater. Sol. Cells*, **143**, 450.
12 Taguchi, M., Yano, A., Tohoda, S., Matsuyama, K., Nakamura, Y., Nishiwaki, T., Fujita, K., and Maruyama, E. (2014) *IEEE J. Photovoltaics*, **4**, 96.
13 Nandakumar, N., Dielissen, B., Garcia-alonso, D., Liu, Z., and Kessels, W.M.M. (2014) Technical Proceedings of the 6th World Conference on Photovoltaic Energy Conversion, pp. 1–2.
14 Adachi, D., Hernández, J.L., and Yamamoto, K. (2015) *Appl. Phys. Lett.*, **107**, 233506.
15 Moldovan, A., Feldmann, F., Zimmer, M., Rentsch, J., Benick, J., and Hermle, M. (2015) *Sol. Energy Mater. Sol. Cells*, **142**, 123.
16 Battaglia, C., de Nicolás, S.M., De Wolf, S., Yin, X., Zheng, M., Ballif, C., and Javey, A. (2014) *Appl. Phys. Lett.*, **104**, 113902.
17 Gerling, L.G., Mahato, S., Morales-vilches, A., Masmitja, G., Ortega, P., Voz, C., Alcubilla, R., and Puigdollers, J. (2015) *Sol. Energy Mater. Sol. Cells*, **3**, 1.
18 Islam, R. and Saraswat, K.C. (2014) 40th IEEE Photovoltaic Specialists Conference, PVSC 2014, p. 285.
19 Avasthi, S., McClain, W.E., Man, G., Kahn, A., Schwartz, J., and Sturm, J.C. (2013) *Appl. Phys. Lett.*, **102**, 203901.
20 Macco, B., Vos, M.F.J., Thissen, N.F.W., Bol, A.A., and Kessels, W.M.M. (2015) *Phys. Status Solidi RRL*, **9**, 393.
21 Smit, S., Garcia-Alonso, D., Bordihn, S., Hanssen, M.S., and Kessels, W.M.M. (2014) *Sol. Energy Mater. Sol. Cells*, **120**, 376.
22 Garcia-Alonso, D., Smit, S., Bordihn, S., and Kessels, W.M.M. (2013) *Semicond. Sci. Technol.*, **28**, 082002.
23 del Alamo, J.A. and Swanson, R.M. (1984) *IEEE Trans. Electron Devices*, **31**, 1878.
24 McIntosh, K.R. and Black, L.E. (2014) *J. Appl. Phys.*, **116**, 014503.

25 Dingemans, G., van de Sanden, M.C.M., and Kessels, W.M.M. (2011) *Phys. Status Solidi RRL*, **5**, 22.
26 Hoex, B., Peeters, F.J.J., Erven, A.J., Bijker, M.D., Kessels, W.M.M., and De Sanden, M.C.M. (2006) 4th World Conference on Photovoltaic Energy Conversion, pp. 1036–1039.
27 Agostinelli, G., Delabie, A., Vitanov, P., Alexieva, Z., Dekkers, H.F.W., De Wolf, S., and Beaucarne, G. (2006) *Sol. Energy Mater. Sol. Cells*, **90**, 3438.
28 Hoex, B., Schmidt, J., Pohl, P., van de Sanden, M.C.M., and Kessels, W.M.M. (2008) *J. Appl. Phys.*, **104**, 044903.
29 Dingemans, G. and Kessels, W.M.M. (2012) *J. Vac. Sci. Technol., A*, **30**, 040802.
30 Cuevas, A., Allen, T., Bullock, J., Wan, Y., Yan, D., and Zhang, X. (2015) Proceedings of 42nd IEEE Photovoltaic Specialists Conference, New Orleans.
31 Girisch, R.B.M., Mertens, R., and Keersmaecker, R.F. (1988) *IEEE Trans. Electron Devices*, **35**, 203.
32 Altermatt, P.P., Schenk, A., Geelhaar, F., and Heiser, G. (2003) *J. Appl. Phys.*, **93**, 1598.
33 Yan, D. and Cuevas, A. (2013) *J. Appl. Phys.*, **114**, 044508.
34 Yan, D. and Cuevas, A. (2014) *J. Appl. Phys.*, **116**, 194505.
35 Dauwe, S., Mittelstadt, L., Metz, A., and Hezel, R. (2002) *Prog. Photovoltaics Res. Appl.*, **10**, 271.
36 Dirnstorfer, I., Simon, D.K., Jordan, P.M., and Mikolajick, T. (2014) *J. Appl. Phys.*, **116**, 044112.
37 Cousins, P.J. and Cotter, J.E. (2006) *Sol. Energy Mater. Sol. Cells*, **90**, 228.
38 Thomson, A.F. and McIntosh, K.R. (2009) *Appl. Phys. Lett.*, **95**, 052101.
39 Grove, A.S., Leistiko, O., and Sah, C.T. (1964) *J. Appl. Phys.*, **35**, 2695.
40 Jaeger, K. and Hezel, R. (1985) 18th IEEE Photovoltaic Specialists Conference, p. 1752.
41 Hezel, R. and Jaeger, K. (1989) *J. Electrochem. Soc.*, **136**, 518.
42 Hoex, B., Heil, S.B.S., Langereis, E., Van De Sanden, M.C.M., and Kessels, W.M.M. (2006) *Appl. Phys. Lett.*, **89**, 042112.
43 Higashi, G.S. and Fleming, C.G. (1989) *Appl. Phys. Lett.*, **55**, 1963.
44 Hoex, B., Schmidt, J., Bock, R., Altermatt, P.P., Van De Sanden, M.C.M., and Kessels, W.M.M. (2007) *Appl. Phys. Lett.*, **91**, 112107.
45 Schmidt, J., Merkle, A., Brendel, R., Hoex, B., van de Sanden, M.C.M., and Kessels, W.M.M. (2008) *Prog. Photovoltaics Res. Appl.*, **16**, 461.
46 Knoops, H.C.M., Potts, S.E., Bol, A.A., and Kessels, W.M.M. (2015) *Handbook of Crystal Growth*, 2nd edn, Elsevier, pp. 1101–1134.
47 Dingemans, G., Terlinden, N.M., Pierreux, D., Profijt, H.B., van de Sanden, M.C.M., and Kessels, W.M.M. (2011) *Electrochem. Solid-State Lett.*, **14**, H1.
48 Profijt, H.B., Kudlacek, P., van de Sanden, M.C.M., and Kessels, W.M.M. (2011) *J. Electrochem. Soc.*, **158**, G88.
49 Dingemans, G., Seguin, R., Engelhart, P., Van De Sanden, M.C.M., and Kessels, W.M.M. (2010) *Phys. Status Solidi RRL*, **4**, 10.
50 Hoex, B., Bosman, M., Nandakumar, N., and Kessels, W.M.M. (2013) *Phys. Status Solidi RRL*, **7**, 937.

51 Dingemans, G., Beyer, W., van de Sanden, M.C.M., and Kessels, W.M.M. (2010) *Appl. Phys. Lett.*, **97**, 152106.
52 Richter, A., Benick, J., Hermle, M., and Glunz, S.W. (2014) *Appl. Phys. Lett.*, **104**, 061606.
53 Dingemans, G., Einsele, F., Beyer, W., van de Sanden, M.C.M., and Kessels, W.M.M. (2012) *J. Appl. Phys.*, **111**, 093713.
54 Simon, D.K., Jordan, P.M., Dirnstorfer, I., Benner, F., Richter, C., and Mikolajick, T. (2014) *Sol. Energy Mater. Sol. Cells*, **131**, 72.
55 Dingemans, G., van Helvoirt, C.A.A., Pierreux, D., Keuning, W., and Kessels, W.M.M. (2012) *J. Electrochem. Soc.*, **159**, H277.
56 Dingemans, G., Terlinden, N.M., Verheijen, M.A., van de Sanden, M.C.M., and Kessels, W.M.M. (2011) *J. Appl. Phys.*, **110**, 093715.
57 Bordihn, S., Mertens, V., Müller, J.W., and (Erwin) Kessels, W.M.M. (2014) *J. Vac. Sci. Technol., A*, **32**, 01A128.
58 Richter, S.W.G.A., Henneck, S., Benick, J., Hörteis, M., and Hermle, M. (2010) 25th European Photovoltaic Solar Energy Conference and Exhibition, Valencia, Spain, pp. 1453–1459.
59 Hennen, L., Granneman, E.H.A., and Kessels, W.M.M. (2012) 38th IEEE Photovoltaic Specialists Conference (IEEE, 2012), pp. 001049–001054.
60 Vermang, B., Goverde, H., Uruena, A., Lorenz, A., Cornagliotti, E., Rothschild, A., John, J., Poortmans, J., and Mertens, R. (2012) *Sol. Energy Mater. Sol. Cells*, **101**, 204.
61 Terlinden, N.M., Dingemans, G., van de Sanden, M.C.M., and Kessels, W.M.M. (2010) *Appl. Phys. Lett.*, **96**, 112101.
62 Schuldis, D., Richter, A., Benick, J., Saint-Cast, P., Hermle, M., and Glunz, S.W. (2014) *Appl. Phys. Lett.*, **105**, 231601.
63 Altermatt, P.P., Schumacher, J.O., Cuevas, A., Kerr, M.J., Glunz, S.W., King, R.R., Heiser, G., and Schenk, A. (2002) *J. Appl. Phys.*, **92**, 3187.
64 Baker-Finch, S.C. and McIntosh, K.R. (2011) *IEEE J. Photovoltaics*, **1**, 59.
65 Mok, K.R.C., van de Loo, B.W.H., Vlooswijk, A.H.G., Kessels, W.M.M., and Nanver, L.K. (2015) *IEEE J. Photovoltaics*, **5**, 1310.
66 Liao, B., Stangl, R., Ma, F., Mueller, T., Lin, F., Aberle, A.G., Bhatia, C.S., and Hoex, B. (2013) *J. Phys. D: Appl. Phys.*, **46**, 385102.
67 Black, L.E., Allen, T., McIntosh, K.R., and Cuevas, A. (2014) *J. Appl. Phys.*, **115**, 093707.
68 Liao, B., Stangl, R., Ma, F., Hameiri, Z., Mueller, T., Chi, D., Aberle, A.G., Bhatia, C.S., and Hoex, B. (2013) *J. Appl. Phys.*, **114**, 094505.
69 Liang, W., Weber, K.J., Suh, D., Phang, S.P., Yu, J., McAuley, A.K., and Legg, B.R. (2013) *IEEE J. Photovoltaics*, **3**, 678.
70 Phang, S.P., Liang, W., Wolpensinger, B., Kessler, M.A., and MacDonald, D. (2013) *IEEE J. Photovoltaics*, **3**, 261.
71 Hoex, B., van de Sanden, M.C.M., Schmidt, J., Brendel, R., and Kessels, W.M.M. (2012) *Phys. Status Solidi RRL*, **6**, 4.
72 Richter, A., Benick, J., Kimmerle, A., Hermle, M., and Glunz, S.W. (2014) *J. Appl. Phys.*, **116**, 243501.
73 Bordihn, S., Dingemans, G., Mertens, V., and Kessels, W.M.M. (2013) *IEEE J. Photovoltaics*, **3**, 925.

74 Bordihn, S., Engelhart, P., Mertens, V., Kesser, G., Köhn, D., Dingemans, G., Mandoc, M.M., Müller, J.W., and Kessels, W.M.M. (2011) *Energy Procedia*, **8**, 654.
75 Granneman, E.H.A., Kuznetsov, V.I., and Vermont, P. (2014) *ECS Trans.*, **61**, 3.
76 Dingemans, G. and Kessels, W.M.M. (2010) 25th European Photovoltaic Solar Energy Conference and Exhibition, Valencia, Spain, pp. 1083–1090.
77 Nandakumar, N., Lin, F., Dielissen, B., Souren, F., Gay, X., Gortzen, R., Duttagupta, S., Aberle, A.G., and Hoex, B. (2013) 28th European Photovoltaic Solar Energy Conference and Exhibition, pp. 1105–1107.
78 van Delft, J.A., Garcia-Alonso, D., and Kessels, W.M.M. (2012) *Semicond. Sci. Technol.*, **27**, 074002.
79 Miyajima, M.K.S., Irikawa, J., and Yamada, A. (2008) 23rd European Photovoltaic Solar Energy Conference and Exhibition, Valencia, Spain, pp. 1029–1032.
80 Black, L.E., Allen, T., Cuevas, A., McIntosh, K.R., Veith, B., and Schmidt, J. (2014) *Sol. Energy Mater. Sol. Cells*, **120**, 339.
81 Li, T.-T. and Cuevas, A. (2009) *Phys. Status Solidi RRL*, **3**, 160.
82 Schmidt, J., Werner, F., Veith, B., Zielke, D., Bock, R., Tiba, V., Poodt, P., Roozeboom, F., Li, T.A., Cuevas, A., and Brendel, R. (2010) *Industrially Relevant Al2O3 Deposition Techniques for the Surface Passivation of Si Solar Cells*. 25th European Photovoltaic Solar Energy Conference and Exhibition, Valencia, Spain, pp. 1130–1133.
83 Zhang, X. and Cuevas, A. (2013) *Phys. Status Solidi RRL*, **7**, 619.
84 Saynova, D.S., Romijn, I.G., Cesar, I., Lamers, M.W.P.E., Gutjahr, A., Dingemans, G., Knoops, H.C.M., van de Loo, B.W.H., Kessels, W.M.M., Siarheyeva, O., Granneman, E.H.A., Gautero, L., Borsa, D.M., Venema, P.R., and Vlooswijk, A.H.G. (2013) 28th European Photovoltaic Solar Energy Conference and Exhibition, pp. 1188–1193.
85 Dingemans, G. and Kessels, W.M.M. (2011) *ECS Trans.*, **41**, 293.
86 Duttagupta, S., Ma, F., Lin, S., Mueller, T., Aberle, A.G., and Hoex, B. (2013) *IEEE J. Photovoltaics*, **3**, 1163.
87 Terlinden, N.M., Dingemans, G., Vandalon, V., Bosch, R.H.E.C., and Kessels, W.M.M. (2014) *J. Appl. Phys.*, **115**, 033708.
88 Spinelli, P., Macco, B., Verschuuren, M.A., Kessels, W.M.M., and Polman, A. (2013) *Appl. Phys. Lett.*, **102**, 233902.
89 Richter, A., Benick, J., and Hermle, M. (2013) *IEEE J. Photovoltaics*, **3**, 236.
90 Wang, W.-C., Lin, C.-W., Chen, H.-J., Chang, C.-W., Huang, J.-J., Yang, M.-J., Tjahjono, B., Huang, J.-J., Hsu, W.-C., and Chen, M.-J. (2013) *ACS Appl. Mater. Interfaces*, **5**, 9752.
91 Savin, H., Repo, P., von Gastrow, G., Ortega, P., Calle, E., Garín, M., and Alcubilla, R. (2015) *Nat. Nanotechnol.*, **10**, 624.
92 von Gastrow, G., Alcubilla, R., Ortega, P., Yli-Koski, M., Conesa-Boj, S., Fontcuberta i Morral, A., and Savin, H. (2015) *Sol. Energy Mater. Sol. Cells*, **142**, 29.
93 Allen, T., Bullock, J., Cuevas, A., Baker-Finch, S., and Karouta, F. (2014) IEEE 40th Photovoltaic Specialist Conference (PVSC) 562.

94 Otto, M., Kroll, M., Käsebier, T., Salzer, R., Tünnermann, A., and Wehrspohn, R.B. (2012) *Appl. Phys. Lett.*, **100**, 1.
95 Repo, P., Haarahiltunen, A., Sainiemi, L., Yli-Koski, M., Talvitie, H., Schubert, M.C., and Savin, H. (2013) *IEEE J. Photovoltaics*, **3**, 90.
96 Repo, P., Benick, J., Von Gastrow, G., Vähänissi, V., Heinz, F.D., Schön, J., Schubert, M.C., and Savin, H. (2013) *Phys. Status Solidi RRL*, **7**, 950.
97 Liu, Y., Das, A., Lin, Z., Cooper, I.B., Rohatgi, A., and Wong, C.P. (2014) *Nano Energy*, **3**, 127.
98 Wan, Y., Bullock, J., and Cuevas, A. (2015) *Appl. Phys. Lett.*, **106**, 201601.
99 Liao, B., Hoex, B., Aberle, A.G., Chi, D., and Bhatia, C.S. (2014) *Appl. Phys. Lett.*, **104**, 253903.
100 Allen, T.G. and Cuevas, A. (2014) *Appl. Phys. Lett.*, **105**, 031601.
101 Benner, F., Jordan, P.M., Richter, C., Simon, D.K., Dirnstorfer, I., Knaut, M., Bartha, J.W., and Mikolajick, T. (2014) *J. Vac. Sci. Technol., B*, **32**, 03D110.
102 Simon, D.K., Jordan, P.M., Knaut, M., Chohan, T., Mikolajick, T., and Dirnstorfer, I. (2015) IEEE 42nd Photovoltaic Specialists Conference (IEEE, 2015).
103 Suh, D. (2015) *Phys. Status Solidi RRL*, **9**, 344.
104 Repo, P., Talvitie, H., Li, S., Skarp, J., and Savin, H. (2011) *Energy Procedia*, **8**, 681.
105 Wu, Y., Hermkens, P.M., van de Loo, B.W.H., Knoops, H.C.M., Potts, S.E., Verheijen, M.A., Roozeboom, F., and Kessels, W.M.M. (2013) *J. Appl. Phys.*, **114**, 024308.
106 Macco, B., Deligiannis, D., Smit, S., van Swaaij, R.A.C.M.M., Zeman, M., and Kessels, W.M.M. (2014) *Semicond. Sci. Technol.*, **29**, 122001.
107 De Wolf, S., Descoeudres, A., Holman, Z.C., and Ballif, C. (2012) *Green*, **2**, 7.
108 McIntosh, K.R. and Baker-Finch, S.C. (2012) 38th IEEE Photovoltaic Specialists Conference (IEEE, 2012), pp. 000265–000271.
109 Macco, B., Knoops, H.C.M., and Kessels, W.M.M. (2015) *ACS Appl. Mater. Interfaces*, **7**, 16723.
110 Masetti, G., Severi, M., and Solmi, S. (1983) *IEEE Trans. Electron Devices*, **30**, 764.
111 Ellmer, K. and Mientus, R. (2008) *Thin Solid Films*, **516**, 4620.
112 Pisarkiewicz, T., Zakrzewska, K., and Leja, E. (1989) *Thin Solid Films*, **174**, 217.
113 Sharma, K., Williams, B.L., Mittal, A., Knoops, H.C.M., Kniknie, B.J., Bakker, N.J., Kessels, W.M.M., Schropp, R.E.I., and Creatore, M. (2014) *Int. J. Photoenergy*, **2014**, 1.
114 Yoshida, Y., Wood, D.M., Gessert, T.A., and Coutts, T.J. (2004) *Appl. Phys. Lett.*, **84**, 2097.
115 Morales-Masis, M., Martin De Nicolas, S., Holovsky, J., De Wolf, S., and Ballif, C. (2015) *IEEE J. Photovoltaics*, **5**, 1340.
116 Yu, J., Bian, J., Duan, W., Liu, Y., Shi, J., Meng, F., and Liu, Z. (2016) *Sol. Energy Mater. Sol. Cells*, **144**, 359.
117 Newhouse, P.F., Park, C.-H., Keszler, D.A., Tate, J., and Nyholm, P.S. (2005) *Appl. Phys. Lett.*, **87**, 112108.

118 Macco, B., Knoops, H.C.M., Vos, M.F.J., Kuang, Y., Verheijen, M.A., and Kessels, W.M.M. (2015) 30th European Photovoltaic Solar Energy Conference and Exhibition.
119 Preissler, N., Bierwagen, O., Ramu, A.T., and Speck, J.S. (2013) *Phys. Rev. B*, **88**, 085305.
120 Ellmer, K. (2012) *Nat. Photonics*, **6**, 809.
121 Knoops, H.C.M., van de Loo, B.W.H., Smit, S., Ponomarev, M.V., Weber, J.-W., Sharma, K., Kessels, W.M.M., and Creatore, M. (2015) *J. Vac. Sci. Technol., A*, **33**, 021509.
122 Koida, T., Fujiwara, H., and Kondo, M. (2007) *Jpn. J. Appl. Phys.*, **46**, L685.
123 Barraud, L., Holman, Z.C., Badel, N., Reiss, P., Descoeudres, A., Battaglia, C., De Wolf, S., and Ballif, C. (2013) *Sol. Energy Mater. Sol. Cells*, **115**, 151.
124 Tohsophon, T., Dabirian, A., De Wolf, S., Morales-Masis, M., and Ballif, C. (2015) *APL Mater.*, **3**, 116105.
125 Demaurex, B., De Wolf, S., Descoeudres, A., Holman, Z.C., and Ballif, C. (2012) *Appl. Phys. Lett.*, **101**, 171604.
126 Ritzau, K.-U., Bivour, M., Schröer, S., Steinkemper, H., Reinecke, P., Wagner, F., and Hermle, M. (2014) *Sol. Energy Mater. Sol. Cells*, **131**, 9.
127 Rößler, R., Leendertz, C., Korte, L., Mingirulli, N., and Rech, B. (2013) *J. Appl. Phys.*, **113**, 144513.
128 Bivour, M., Schröer, S., and Hermle, M. (2013) *Energy Procedia*, **38**, 658.
129 Tomasi, A., Sahli, F., Fanni, L., Seif, J.P., De Nicolas, S.M., Holm, N., Geissbühler, J., Paviet-Salomon, B., Löper, P., Nicolay, S., De Wolf, S., and Ballif, C. (2016), IEEE J. Photovoltaics **6**, 17.
130 Klein, A., Körber, C., Wachau, A., Säuberlich, F., Gassenbauer, Y., Harvey, S.P., Proffit, D.E., and Mason, T.O. (2010) *Materials (Basel)*, **3**, 4892.
131 Carroy, G.R.P., Muñoz, D., Ozanne, F., Valla, A., and Mur, P. (2015) 30th European Photovoltaic Solar Energy Conference and Exhibition, pp. 359–364.
132 Tynell, T. and Karppinen, M. (2014) *Semicond. Sci. Technol.*, **29**, 043001.
133 Elam, J.W. and George, S.M. (2003) *Chem. Mater.*, **15**, 1020.
134 Lee, D.-J., Kim, H.-M., Kwon, J.-Y., Choi, H., Kim, S.-H., and Kim, K.-B. (2011) *Adv. Funct. Mater.*, **21**, 448.
135 Park, H.K. and Heo, J. (2014) *Appl. Surf. Sci.*, **309**, 133.
136 Yanguas-Gil, A., Peterson, K.E., and Elam, J.W. (2011) *Chem. Mater.*, **23**, 4295.
137 Wu, Y., Potts, S.E., Hermkens, P.M., Knoops, H.C.M., Roozeboom, F., and Kessels, W.M.M. (2013) *Chem. Mater.*, **25**, 4619.
138 Garcia-Alonso, D., Potts, S.E., van Helvoirt, C.A.A., Verheijen, M.A., and Kessels, W.M.M. (2015) *J. Mater. Chem. C*, **3**, 3095.
139 Sang, B., Yamada, A., and Konagai, M. (1997) *Sol. Energy Mater. Sol. Cells*, **49**, 19.
140 Lee, D.-J., Kim, K.-J., Kim, S.-H., Kwon, J.-Y., Xu, J., and Kim, K.-B. (2013) *J. Mater. Chem. C*, **1**, 4761.
141 Ye, Z.-Y., Lu, H.-L., Geng, Y., Gu, Y.-Z., Xie, Z.-Y., Zhang, Y., Sun, Q.-Q., Ding, S.-J., and Zhang, D.W. (2013) *Nanoscale Res. Lett.*, **8**, 108.
142 Ott, A.W. and Chang, R.P.H. (1999) *Mater. Chem. Phys.*, **58**, 132.

143 Geng, Y., Xie, Z.-Y., Yang, W., Xu, S.-S., Sun, Q.-Q., Ding, S.-J., Lu, H.-L., and Zhang, D.W. (2013) *Surf. Coat. Technol.*, **232**, 41.
144 Thomas, M.A., Armstrong, J.C., and Cui, J. (2013) *J. Vac. Sci. Technol., A*, **31**, 01A130.
145 Na, J.-S., Peng, Q., Scarel, G., and Parsons, G.N. (2009) *Chem. Mater.*, **21**, 5585.
146 Asikainen, T. (1995) *J. Electrochem. Soc.*, **142**, 3538.
147 Nilsen, O., Balasundaraprabhu, R., Monakhov, E.V., Muthukumarasamy, N., Fjellvåg, H., and Svensson, B.G. (2009) *Thin Solid Films*, **517**, 6320.
148 Gebhard, M., Hellwig, M., Parala, H., Xu, K., Winter, M., and Devi, A. (2014) *Dalton Trans.*, **43**, 937.
149 Lee, D.-J., Kwon, J.-Y., Il Lee, J., and Kim, K.-B. (2011) *J. Phys. Chem. C*, **115**, 15384.
150 Kim, D., Nam, T., Park, J., Gatineau, J., and Kim, H. (2015) *Thin Solid Films*, **587**, 1.
151 Ramachandran, R.K., Dendooven, J., Poelman, H., and Detavernier, C. (2015) *J. Phys. Chem. C*, **119**, 11786.
152 Elam, J.W., Martinson, A.B.F., Pellin, M.J., and Hupp, J.T. (2006) *Chem. Mater.*, **18**, 3571.
153 Elam, J.W., Baker, D.A., Martinson, A.B.F., Pellin, M.J., and Hupp, J.T. (2008) *J. Phys. Chem. C*, **112**, 1938.
154 Libera, J.A., Hryn, J.N., and Elam, J.W. (2011) *Chem. Mater.*, **23**, 2150.
155 Demaurex, B., Seif, J.P., Smit, S., Macco, B., Kessels, W.M.M., Geissbuhler, J., De Wolf, S., and Ballif, C. (2014) *IEEE J. Photovoltaics*, **4**, 1387.
156 Hoye, R.L.Z., Muñoz-Rojas, D., Nelson, S.F., Illiberi, A., Poodt, P., Roozeboom, F., and MacManus-Driscoll, J.L. (2015) *APL Mater.*, **3**, 040701.
157 Munoz-Rojas, D. and MacManus-Driscoll, J. (2014) *Mater. Horiz.*, **1**, 314.
158 Illiberi, A., Poodt, P., Bolt, P.J., and Roozeboom, F. (2014) *Chem. Vap. Depos.*, **20**, 234.
159 Illiberi, A., Roozeboom, F., and Poodt, P. (2012) *ACS Appl. Mater. Interfaces*, **4**, 268.
160 Nandakumar, N., Dielissen, B., Garcia-Alonso, D., Liu, Z., and Kessels, W.M.M., (2014), in *Proc. 6th World Conf. Photovolt. Energy Convers.*, pp. 2–3.
161 Nandakumar, N., Dielissen, B., Garcia-Alonso, D., Liu, Z., Roger, G., Kessels, W.M.M.E., Aberle, A.G., and Hoex, B. (2015) *IEEE J. Photovoltaics*, **5**, 1462.
162 Illiberi, A., Scherpenborg, R., Wu, Y., Roozeboom, F., and Poodt, P. (2013) *ACS Appl. Mater. Interfaces*, **5**, 13124.
163 Ellinger, C.R. and Nelson, S.F. (2014) *Chem. Mater.*, **26**, 1514.
164 Illiberi, A., Scherpenborg, R., Roozeboom, F., and Poodt, P. (2014) *ECS J. Solid State Sci. Technol.*, **3**, P111.
165 Nandakumar, N., Hoex, B., Dielissen, B., Garcia-Alonso, D., Gortzen, R., Kessels, W.M.M., Fin, L., Aberle, A.G., and Mueller, T. (2015) Presented at the 25th Asia Photovoltaic Solar Energy Conference and Exhibition.
166 Illiberi, A., Cobb, B., Sharma, A., Grehl, T., Brongersma, H., Roozeboom, F., Gelinck, G., and Poodt, P. (2015) *ACS Appl. Mater. Interfaces*, **7**, 3671.
167 Wurfel, U., Cuevas, A., and Wurfel, P. (2015) *IEEE J. Photovoltaics*, **5**, 461.

168 Robertson, J. (2000) *J. Vac. Sci. Technol., B*, **18**, 1785.
169 Stradins, P., Essig, S., Nemeth, W., Lee, B.G., Young, D., Norman, A., Liu, Y., Luo, J., Warren, E., Dameron, A., Lasalvia, V., Page, M., and Ok, Y. (2014) 6th World Conference on Photovoltaic Energy Conversion.
170 Islam, R., Shine, G., and Saraswat, K.C. (2014) *Appl. Phys. Lett.*, **105**, 182103.
171 Bivour, M., Temmler, J., Steinkemper, H., and Hermle, M. (2015) *Sol. Energy Mater. Sol. Cells*, **142**, 34.
172 Ng, K.K. and Card, H.C. (1980) *J. Appl. Phys.*, **51**, 2153.
173 Young, D.L., Nemeth, W., Grover, S., Norman, A., Lee, B.G., and Stradins, P. (2014) IEEE 40th Photovoltaic Specialists Conference (IEEE, 2014), pp. 1–5.
174 Khanna, A., Mueller, T., Stangl, R.A., Hoex, B., Basu, P.K., and Aberle, A.G. (2013) *IEEE J. Photovoltaics*, **3**, 1170.
175 Feldmann, F., Bivour, M., Reichel, C., Hermle, M., and Glunz, S.W. (2014) *Sol. Energy Mater. Sol. Cells*, **120**, 270.
176 Yang, X., Zheng, P., Bi, Q., and Weber, K. (2016) *Sol. Energy Mater. Sol. Cells*, **150**, 32–38.
177 Fell, A., McIntosh, K.R., Altermatt, P.P., Janssen, G.J.M., Stangl, R., Ho-Baillie, A., Steinkemper, H., Greulich, J., Muller, M., Min, B., Fong, K.C., Hermle, M., Romijn, I.G., and Abbott, M.D. (2015) *IEEE J. Photovoltaics*, **5**, 1250.
178 Cornagliotti, E., Uruena, A., Aleman, M., Sharma, A., Tous, L., Russell, R., Choulat, P., Chen, J., John, J., Haslinger, M., Duerinckx, F., Dielissen, B., Gortzen, R., Black, L., and Szlufcik, J. (2015) *IEEE J. Photovoltaics*, **5**, 1366.
179 Urueña, M.H.A., Aleman, M., Cornagliotti, E., Sharma, A., Deckers, J., Tous, J.S.L., Russell, R., John, J., Yao, Y., Söderström, T., and Duerinckx, F. (2012) 31st European Photovoltaic Solar Energy Conference and Exhibition, vol. **33**, p. 81.
180 Bullock, J., Cuevas, A., Allen, T., and Battaglia, C. (2014) *Appl. Phys. Lett.*, **105**, 232109.
181 Bullock, J., Samundsett, C., Cuevas, A., Yan, D., Wan, Y., and Allen, T. (2015) 42nd IEEE Photovoltaic Specialists Conference, vol. **5**, p. 1591.
182 Geissbühler, J., Werner, J., Martin de Nicolas, S., Barraud, L., Hessler-Wyser, A., Despeisse, M., Nicolay, S., Tomasi, A., Niesen, B., De Wolf, S., and Ballif, C. (2015) *Appl. Phys. Lett.*, **107**, 081601.
183 Zielke, D., Petermann, J.H., Werner, F., Veith, B., Brendel, R., and Schmidt, J. (2011) *Phys. Status Solidi RRL*, **5**, 298.
184 Werner, F., Veith, B., Zielke, D., Kühnemund, L., Tegenkamp, C., Seibt, M., Brendel, R., and Schmidt, J. (2011) *J. Appl. Phys.*, **109**, 113701.
185 Chandiran, A.K., Nazeeruddin, M.K., and Graetzel, M. (2014) *Adv. Funct. Mater.*, **24**, 1615.
186 A.K. Chandiran, N. Tetreault, R. Humphry-baker, F. Kessler, C. Yi, K. Nazeeruddin, and M. Grätzel, **12**, 3941 (2012).
187 Sahasrabudhe, G., Rupich, S.M., Jhaveri, J., Berg, A.H., Nagamatsu, K., Man, G., Chabal, Y.J., Kahn, A., Wagner, S., Sturm, J.C., and Schwartz, J. (2015) *J. Am. Chem. Soc.*, **137**, 14842. doi: 10.1021/jacs.5b09750
188 Chambers, S.A., Liang, Y., Yu, Z., Droopad, R., and Ramdani, J. (2001) *J. Vac. Sci. Technol., A*, **19**, 934.

189 Först, C.J., Ashman, C.R., Schwarz, K., and Blöchl, P.E. (2004) *Nature*, **427**, 53.
190 Longo, V., Leick, N., Roozeboom, F., and Kessels, W.M.M. (2012) *ECS J. Solid State Sci. Technol.*, **2**, N15.
191 Ritala, M., Leskela, M., Niinisto, L., Haussalo, P., and Niinist, L. (1993) *Chem. Mater.*, **5**, 1174.
192 Potts, S.E., Keuning, W., Langereis, E., Dingemans, G., van de Sanden, M.C.M., and Kessels, W.M.M. (2010) *J. Electrochem. Soc.*, **157**, P66.
193 Xie, Q., Jiang, Y.-L., Detavernier, C., Deduytsche, D., Van Meirhaeghe, R.L., Ru, G.-P., Li, B.-Z., and Qu, X.-P. (2007) *J. Appl. Phys.*, **102**, 083521.
194 Kim, M.K., Kim, W.H., Lee, T., and Kim, H. (2013) *Thin Solid Films*, **542**, 71.
195 Kwon, O.S., Kim, S.K., Cho, M., Hwang, C.S., and Jeong, J. (2005) *J. Electrochem. Soc.*, **152**, C229.
196 Bertuch, A., Sundaram, G., Saly, M., Moser, D., and Kanjolia, R. (2014) *J. Vac. Sci. Technol., A*, **32**, 01A119.
197 Vos, M.F.J., Macco, B., Thissen, N.F.W., Bol, A.A., and (Erwin) Kessels, W.M.M. (2016) *J. Vac. Sci. Technol., A*, **34**, 01A103.
198 Diskus, M., Nilsen, O., and Fjellvåg, H. (2011) *J. Mater. Chem.*, **21**, 705.
199 Liu, R., Lin, Y., Chou, L.-Y., Sheehan, S.W., He, W., Zhang, F., Hou, H.J.M., and Wang, D. (2011) *Angew. Chem. Int. Ed.*, **50**, 499.
200 Malm, J., Sajavaara, T., and Karppinen, M. (2012) *Chem. Vap. Depos.*, **18**, 245.
201 Song, J., Park, J., Lee, W., Choi, T., Jung, H., Lee, C.W., Hwang, S., Myoung, J.M., Jung, J., Kim, S., Lansalot-matras, C., and Kim, H. (2013) *ACS Nano*, 7, 11333.
202 Blanquart, T., Niinistö, J., Gavagnin, M., Longo, V., Heikkilä, M., Puukilainen, E., Pallem, V.R., Dussarrat, C., Ritala, M., and Leskelä, M. (2013) *RSC Adv.*, **3**, 1179.
203 Boukhalfa, S., Evanoff, K., and Yushin, G. (2012) *Energy Environ. Sci.*, **5**, 6872.
204 Musschoot, J., Deduytsche, D., Van Meirhaeghe, R.L., and Detavernier, C. (2009) 216th Electrochemical Society Meeting, pp. 29–37.
205 Lu, H.L., Scarel, G., Li, X.L., and Fanciulli, M. (2008) *J. Cryst. Growth*, **310**, 5464.
206 Lindahl, E., Ottosson, M., and Carlsson, J.-O. (2009) *Chem. Vap. Depos.*, **15**, 186.
207 Chae, J., Park, H.-S., and Kang, S. (2002) *Electrochem. Solid-State Lett.*, **5**, C64.
208 Yang, T.S., Cho, W., Kim, M., An, K.-S., Chung, T.-M., Kim, C.G., and Kim, Y. (2005) *J. Vac. Sci. Technol., A*, **23**, 1238.
209 Battaglia, C., Yin, X., Zheng, M., Sharp, I.D., Chen, T., McDonnell, S., Azcatl, A., Carraro, C., Ma, B., Maboudian, R., Wallace, R.M., and Javey, A. (2014) *Nano Lett.*, **14**, 967.
210 Ziegler, J., Mews, M., Kaufmann, K., Schneider, T., Sprafke, A.N., Korte, L., and Wehrspohn, R.B. (2015) *Appl. Phys. A*, **120**, 811.

3

ALD for Light Absorption

Alex Martinson

Argonne National Laboratory, Materials Science Division, 9700 South Cass Avenue, Lemont, IL 60439, USA

3.1 Introduction to Solar Light Absorption

Independent of photovoltaic (PV) technology, the first step in the intricate process of converting solar energy to direct current electricity is the absorption of photons. Although not simple, solar light absorption is an important step toward the ultimate goal of high conversion efficiency. Overall solar power conversion efficiency (η_p) is directly proportional to the light absorption efficiency (η_a), charge separation efficiency (η_{cs}), and charge extraction efficiency (η_{ce}) according to

$$\eta_p = \eta_a \cdot \eta_{cs} \cdot \eta_{ce} \tag{3.1}$$

Therefore, any *inefficiency* in each of three primary steps is multiplicatively propagated through to the final PV performance.

The task of solar light absorption is made further challenging by the spectrum of the Sun's radiation, which is close to that of a blackbody with temperature ∼5800 K and, as such, emits radiation with significant power across the ultraviolet, visible, and infrared spectra, Figure 3.1. Given the wide spectrum of photon energies incident upon the Earth, their random polarization, and their direct and indirect (scattered) nature, the efficient collection of solar energy is not trivial.

The challenge of solar light absorption is perhaps most clearly illustrated by illustrating the fraction of the solar spectrum that is absorbed by the most common PV absorber – crystalline silicon, Figure 3.1. A significant fraction of the UV, visible, and NIR photons with wavelength as long as 1100 nm (1.1 eV) is absorbed by Si, but not all of them. Reflectivity, insufficient absorption near the band gap, and subband-gap transmission all factor into the effective η_a and will depend upon the material properties and morphological details as discussed as follows.

Even in the UV, where silicon's absorptivity is the strongest, the incoming light is rejected from the surface due to reflectance. Specular (mirror-like) reflections are described by Fresnel's reflection laws, which depend on the angle, wavelength,

Atomic Layer Deposition in Energy Conversion Applications, First Edition. Edited by Julien Bachmann.
© 2017 Wiley-VCH Verlag GmbH & Co. KGaA. Published 2017 by Wiley-VCH Verlag GmbH & Co. KGaA.

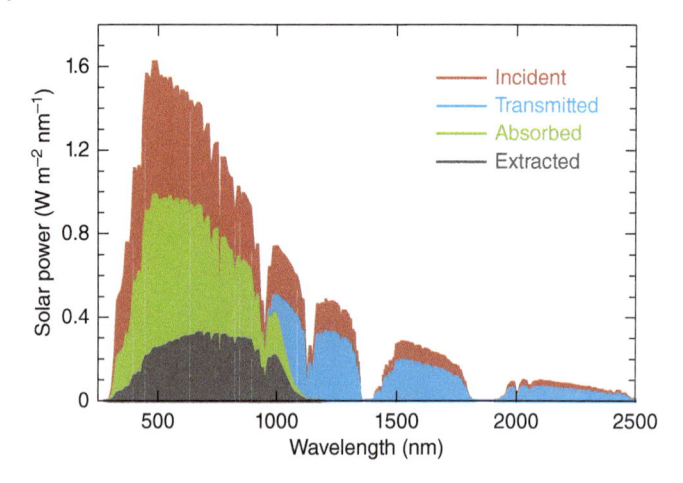

Figure 3.1 The spectral power of our Sun incident upon the Earth at 37° latitude (red). The approximate solar power absorbed by a 100-micron-thick silicon wafer after accounting for reflectivity and absorption coefficient is overlaid (green). The power extracted as electricity (gray) is estimated from perfect photon-to-electron efficiency but is reduced by the modest voltage of Si photovoltaics (0.64 V) relative to the absorbed photon energy (4.5–1.1 eV).

polarization, and the difference in refractive index at the interface.

$$R_s = \left| \frac{n_1 \cos \Theta_i - n_2 \cos \Theta_t}{n_1 \cos \Theta_i + n_2 \cos \Theta_t} \right|^2, \quad R_p = \left| \frac{n_1 \cos \Theta_t - n_2 \cos \Theta_i}{n_1 \cos \Theta_t + n_2 \cos \Theta_i} \right|^2$$

At normal incidence ($\Theta_i = \Theta_t = 0$), the cosine term tends to 1 such that both R_s and R_p can be simplified to $|(n_1 - n_2)/(n_1 + n_2)|^2$. Therefore, if a planar semiconductor such as Si ($n_2 = 4.5$) averaged over the visible light wavelengths) is simply interfaced with air ($n_1 = 1.0$), then the average reflectivity is ~40%. This can be visualized in Figure 3.1, where ~60% of the incident power is absorbed with the remaining 40% being lost almost solely to reflection. Clearly, this is a significant effect that must be considered in any serious attempt to maximize light absorption. As such, antireflection treatments in the form of texturing and antireflective coatings are commonly employed in order to minimize the effect of reflectivity. ALD is a notably expedient approach to deposit advanced antireflection coatings, the most advanced of which require precise multilayer thickness control and graded refractive index tuning [1, 2].

Solar light that is not rejected due to reflectivity must be absorbed by the desired material in order to contribute to PV performance. To first order, photons with energy greater than the semiconductor band gap are absorbed with intensity (absorptivity) proportional to the density of states separated by a given photon energy and further modified by the spectroscopic selection rules. If the minimal-energy state in the conduction band and maximal-energy state in the valance band occur at the same crystal momentum (k-vector) in the Brillouin zone, then the semiconductor is referred to as having a direct band gap. In these materials, photons can be directly absorbed without passing through an intermediate state and transferring momentum to the crystal lattice.

This intermediate step is, however, necessary for indirect semiconductors, which have minimal and maximal energy states offset in the Brillouin zone. This effect is important for PVs as indirect semiconductors (e.g., silicon) typically require hundreds of micrometers of thickness to absorb a significant fraction of incident light in a single pass, in contrast to direct semiconductors (e.g., CdTe or CIGS), which often require only one or two micrometers of thickness. A finite but insufficient absorption of photons is observed between 1000 and 1100 nm as shown in Figure 3.1.

The third major solar light absorption deficiency is transmission of photons with energy less than that of the band gap – as shown in Figure 3.1 – for wavelengths greater than 1100 nm. As no interband transitions are feasible at subgap energies, no useful absorption events are possible. The fact that semiconductors with smaller gap are capable of adsorbing a larger fraction of the solar spectrum seems to imply this as a useful approach to more efficient solar energy conversion. However, under normal PV operation, the useful device voltage – the energy at which photons are extracted – is proportional to the band gap. This trade-off between maximizing light absorption (and therefore current) and maximizing the potential of each photon (and therefore voltage) is famously captured in the detailed balance limit for PVs as first calculated by William Shockley and Hans Queisser. The result, Figure 3.2, reveals that a theoretical optimum for the semiconductor band gap exists in order to optimize the overall power efficiency (η_p) according to

$$\eta_p = J_{sc} \cdot V_{oc} \cdot FF \tag{3.2}$$

It is within this framework that we consider the ALD of semiconductor thin films with band gaps ranging from 2.5 to 0.9 eV.

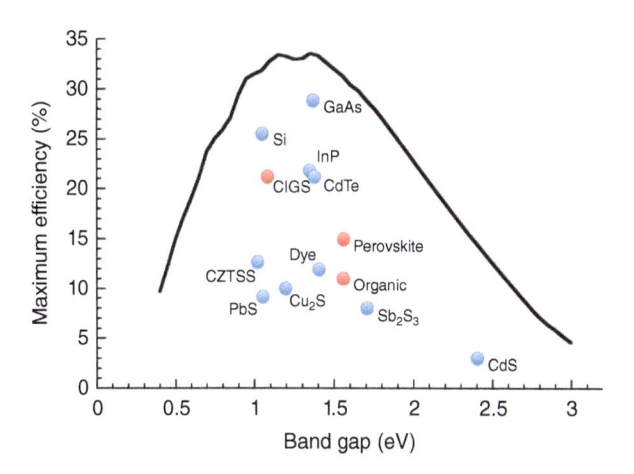

Figure 3.2 Shockley–Queisser single-junction efficiency limit, with record photovoltaic efficiency plotted against band gap. Materials for which ALD processes have been reported are highlighted in blue. (Record efficiencies compiled from Green, Dasgupta, and references therein.)

3.2 Why ALD for Solar Light Absorbers?

The fact that solar absorbing materials *can* be produced through ALD processes does not motivate *why* an ALD approach to PV might be pursued. ALD espouses many meritorious qualities including tight thickness control, relatively dense polycrystalline films, and a modest vacuum requirement. However, these advantages are not exclusive to ALD – consider chemical vapor deposition or closed-space evaporation. Indeed, in several aspects important to PV fabrication, ALD is lacking relative to other deposition options. Foremost, ALD is a thin-film deposition method, with a practical thickness ranging from 0.1 to 100 nm for most processes. There is no intrinsic limit on thickness, and reports of ALD-grown films with greater than 1 μm are no longer uncommon but are simply slow to grow. A conservative estimate of common dose (0.1 s) and purge times (8 s) during each cycle alongside typical growth rates (0.1 nm per cycle) reveals that >2 h depositions are required for even 100 nm thin films. Compare this to the 300–1000-nm-thick absorber thickness found in the most efficient "thin-film" PVs. However, overnight depositions are feasible in many nonproduction settings. ALD is also uniquely amenable to batch coating, as detailed as follows.

Despite its relatively slow growth rate in time, ALD possesses several unique advantages that make it an increasingly important method for scientific study and potentially for commercial production of solar absorbers. The most exceptional characteristics with relevance to light absorption are considered as follows.

3.2.1 Uniformity and Precision of Large-Area Coatings

The intrinsic uniformity of both thickness and composition renders ALD coatings exceptionally strong candidates for PV absorbers. Unlike many physical deposition methods, which require a source target size comparable to the substrate, ALD utilizes gas-phase precursors that may rapidly diffuse to all points on all substrates. While the diffusion of precursors may not be the same to all points, the self-limiting mechanism requires only sufficient – not equal – exposure to all locations to maintain uniformity. This is a notable advantage over ALD's closest cousin, chemical vapor deposition (CVD). This allows for the coating of both large format PV panels and their batchwise processing. This distinct advantage is tempered by the cost of vapor-phase precursors, the excess time required for uniform coatings, and the challenge in recycling excess reactant from each half cycle.

The digital nature of ALD also allows tuning of absorber thickness in order to maximize light absorption without exceeding the charge extraction distance. While subnanometer resolution is overkill for this task, precise thickness control may be utilized in the fabrication of designer optical spacers to enable maximum field strength centered in the solar absorber [3, 4]. In this way, an uncharacteristically large light extinction may be achieved with minimum absorber thickness. In addition to conserving materials, thinner absorber layers minimize series resistance without raising the majority dopant level, which would accelerate recombination.

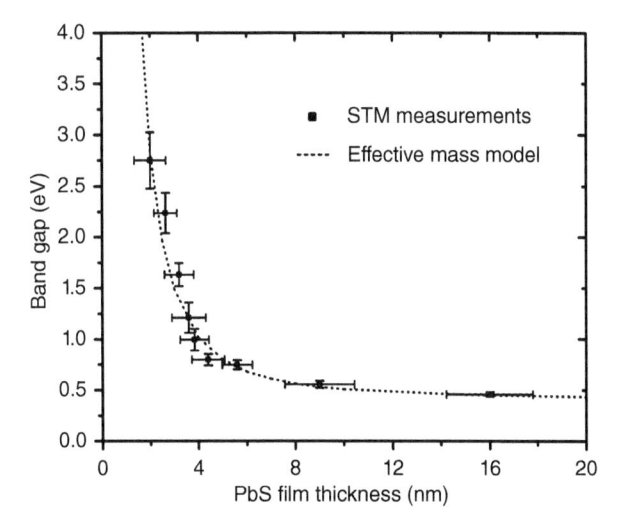

Figure 3.3 Band-gap variation of ALD PbS grown on SiO_2. Band gaps calculated by an effective mass model are shown as a dotted line. By tuning the size of quantum-confined nanoparticles, the wavelength of light absorbed may be precisely tuned. (Dasgupta 2009. Reproduced with permission of American Chemical Society.)

The subnanometer precision of ALD is most keenly leveraged in the vapor-phase synthesis of quantum-confined nanoparticles that enable band-gap tuning of some semiconductors. PbS and NiS are absorbers that take advantage of this effect to shift their prohibitively small band gaps (0.5 and 0.4 eV) into the ideal range for solar absorbers (1.1 to 1.7 eV); see Figure 3.3. The tight control of nanoparticle size distribution is rivaled only by another layer-by-layer method – successive ionic layer adsorption and reaction (SILAR).

3.2.2 Orthogonalizing Light Harvesting and Charge Extraction

The characteristic thickness throughout which photogenerated minority charge carriers can be extracted from a solar absorber is often referred to as the charge extraction length (L_{CE}). L_{CE} is a function of the absorbers' minority carrier lifetime (τ), the mobility of the free charge (μ) through the absorber, and the built-in field (V_{bi}) of the device according to

$$L_{CE} = \sqrt{\tau \mu V_{bi}} \tag{3.3}$$

L_{CE} serves as an upper bound to the *useful* film thickness (which determines the feasible light-harvesting efficiency of a thin film) for a given semiconductor of specific material quality. However, given the *conformal* coating ability bestowed by the self-limiting nature of ALD, this thickness need not be perpendicular to the direction of incident light or charge extraction. That is, ultrathin films arranged in high density parallel to the direction of illumination may enable orders of magnitude greater light harvesting from a film of equal thickness (but much greater volume) (Figure 3.4) (see also Section 8.2.3).

In theory, this allows for ample light absorption from ever-thinner absorber films, thereby dramatically relaxing the requirement for large L_{CE}. This in turn

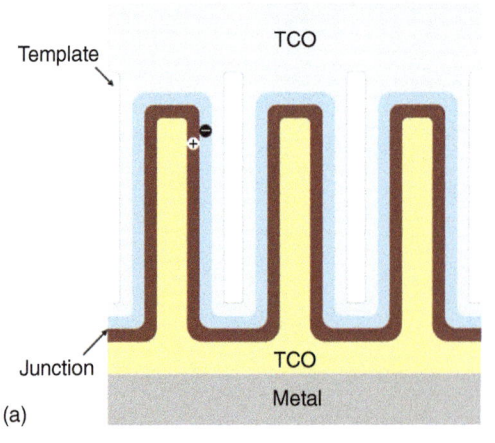

Figure 3.4 The volume of absorber in a projected area may be readily increased without increasing film thickness by performing ALD over a transparent template with large aspect ratio. (a) A "folded junction" geometry may increase the volume of absorber per projected area without increasing film thickness. (b) The same ALD Fe_2O_3 film on flat glass and folded into anodic aluminum oxide membranes (aspect ratio >100).

enables a raft of new materials with small L_{CE} or absorption (or both). Alternatively, the material quality (purity, grain size, defect density) that determines the $\mu\tau$ product can be relaxed. The first strategy has been leveraged in attempts to improve the power efficiency of PVs that utilize earth-abundant materials with modest $\mu\tau$ product including Sb_2S_3 [5] and $CuInS_2$ [6]. Enabling earth-abundant absorbers may prove crucial to the scalability of solar energy conversion. As illustrated in Figure 3.5, many elements utilized in the most efficient PVs to date have limited abundance on our planet. The latter strategy is less often utilized by ALD, but the folded junction approach is well documented both experimentally and theoretically to afford large photocurrent from materials (e.g., Si) with relatively modest material quality [7, 8]. However, this approach is fundamentally limited by the large junction area that results from the orthogonalization of light harvesting and charge extraction. As has been simulated [8, 9] and borne out in experiment, a loss in open-circuit voltage results from the large reverse current densities that scale with junction area. A large junction area may also be further addressed by ALD using a point contact approach [10] (see Chapter 2); however, it will still scale proportionally to a similarly ameliorated planar junction.

Figure 3.5 Worldwide production of some elements used in solar absorbers plotted on a logarithmic scale.

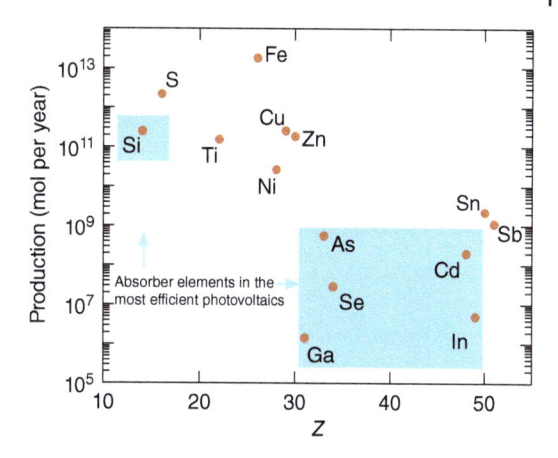

3.2.3 Pinhole-Free Ultrathin Films, ETA Cells

Extremely thin absorber (ETA) cell is one name given to those devices that utilize the orthogonalization of light and charge collection to the greatest extent. In these PVs, the absorber thickness may be 50 nm or less, but the PVs still exhibit ample light absorption. In addition to the challenge of large dark current density and magnified interfacial recombination discussed earlier, these devices are highly susceptible to physical junction shorting via unintentionally connected layers that are caused by an incomplete absorber layer. ALD again intrinsically meets this challenge owing to its non-line-of-sight mechanism. Sputtering and evaporation-based approaches are hopeless in highly convoluted geometries and often prone to failure if even modest roughness or small particles are present. In contrast, ALD has been shown to bury small blemishes to enable pinhole-free thin film with thickness as low as 6 nm [11, 12].

3.2.4 Chemical Control of Stoichiometry and Doping

Control over stoichiometry is paramount to the properties of semiconductors, including solar absorbers. Significant deviations in stoichiometry may result in strikingly different crystalline phases (e.g., FeO, Fe_2O_3, Fe_3O_4) that have distinct optoelectronic properties including disparate band gaps as for the case of iron oxides. Clearly, a different band gap and crystalline phase will have a dramatic effect not only on light absorption but also on the electronic density of states that determine charge transport, relaxation pathways, and excited-state lifetimes. Smaller changes in stoichiometry may also induce phase changes (e.g., Cu_2S, $Cu_{1.97}S$, $Cu_{1.75}S$), but even within the same crystalline phase, the precise stoichiometry may have dramatic effects on electronic properties. In the case of the $Cu_{2-x}S$ family of solar absorbers, each missing Cu acts as a positively charged cation vacancy, which p-dopes the film, Figure 3.6 [13]. This means, for example, that $Cu_{1.99}S$ has a doping density in excess of 10^{20} cm^{-3}. Electronic doping densities of this magnitude result in a degenerate semiconductor, one in which the material begins to act more as a metal than as a semiconductor and is therefore unsuitable for efficient solar energy utilization.

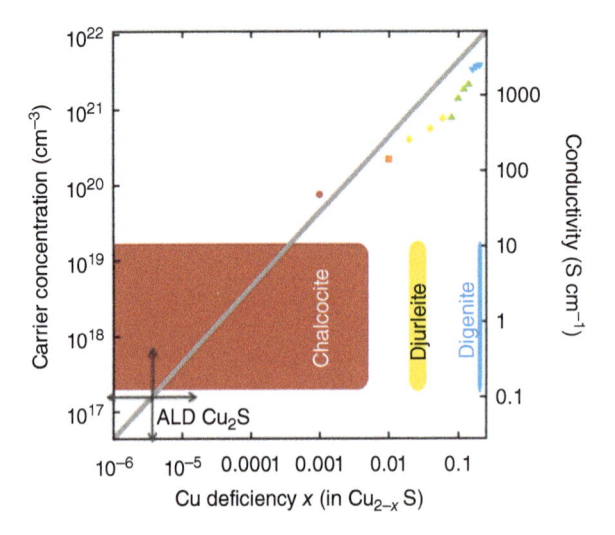

Figure 3.6 Predicted carrier concentration and conductivity of $Cu_{2-x}S$ as a function of Cu deficiency compared to literature data. ALD of Cu_2S using a Cu(I) precursor produces a highly stoichiometric semiconductor as evidenced by its relatively low carrier concentration. (Martinson et al. [13]. Reproduced with permission of Royal Society of Chemistry.)

Atomic layer deposition (ALD) has demonstrated exceptional control over material stoichiometry owing to the uniform surface synthesis reactions that govern it. In contrast to ALD, physical deposition methods utilize targets of known composition but may suffer from uneven evaporation rate/sputter rates and are further susceptible to small fluctuations in the power density that controls the deposition rate of multiple elements. CVD routes, including ALD, instead respond upon the chemical reaction of high-purity chemicals with uniform reactivity – at least when allowed to go to completion as in the case of ALD. The oxidation state of the metal, in particular, especially in the absence of other strong oxidizers or reducing agents has been shown to influence the final stoichiometry. For example, some of the most stoichiometric Cu_2S films – as most sensitively determined by carrier concentration – have been produced by Cu(I) precursor in combination with hydrogen sulfide [13].

Control of the electronic properties of solar absorbers may also be achieved through off-element substitution. For example, the most common solar absorber, crystalline silicon, is often impurity-doped p-type via boron acceptor atoms and n-type via phosphorous donor atoms. Binary semiconductors including In_2O_3 and ZnO respond similarly to incorporation of impurity atoms Sn and Al, respectively, which act as donors to induce carrier concentrations in excess of 10^{20} cm^{-3} for use as transparent conducting oxides [14, 15]. Modest doping levels have also been achieved in solar absorbers. These may be used to switch majority carrier type, thereby enabling a p–n homojunction as in the case of Mg:Fe_2O_3 [16]. Alternatively, ALD impurity doping may simply increase (or decrease) the carrier concentration to larger but still moderate levels, thereby increasing the strength (or length) of the built-in electric field responsible for separating charges. For example, small amounts of Sb significantly increase resistance of ALD-grown SnS for use in the p–i–n design of solar cells [17]. Impurity atoms are simply integrated into ALD films via occasional pulsing of the desired off-atoms in the desired ratio. While a distinct growth rate and surface chemistry for the impurity-atom ALD process results in an elemental composition that is seldom precisely equal to the dose ratio, a highly tunable composition range is often possible. Furthermore, and unlike sputtering, the composition can be refined with ease using the computer-controlled dosing schedule. This also means that a relatively simple ALD recipe can be written to induce graded doping profiles by gradually adjusting the cycle ratios during a single film deposition.

Carefully chosen impurity dopants with sufficient concentrations have also been predicted to form a large but relatively narrow density of states within the band gap of a semiconductor – often referred to as an intermediate band. This intermediate band (IB) is the basis for a novel solar PV device that is theoretically able to exceed the Shockley–Queisser limit (see Section 3.1). The goal is to improve the number of photons harvested with energy lower than the intrinsic gap but larger than the intermediate gaps. The IB creates a stepping stone in intrinsic gap, across which electrons may hop in order to contribute to photocurrent without significant loss of photovoltage. ALD may be a strong choice for these next-generation absorbers due to the spatial and chemical precision with which they may induce local density of states. The first example of an ALD-derived intermediate absorber, V:In_2S_3, was recently described [18].

3.2.5 Low-Temperature Epitaxy

The list of solar absorbers that simultaneously meet the demands of band gap, abundance, and photophysical performance is short enough that any new candidates are highly sought after. One method to expand this list is through epitaxy, the templated and oriented growth of one material on another. In some special cases, new solar absorbers may be enabled by the epitaxial stabilization of metastable crystal structures using ALD, Figure 3.7. For example, epitaxial stabilization of the cubic β-Fe_2O_3 phase on ITO, using the same ALD process that forms α-Fe_2O_3 on other substrates, produced a previously untested solar absorbing phase with a more ideal band gap (1.9 eV) compared to α-Fe_2O_3 (2.1 eV) [19]. Similarly, well-known materials may exhibit improved photophysical properties induced by oriented polycrystalline domains. For example, high-angle grain boundaries have been correlated with recombination in hematite (α-Fe_2O_3). Highly anisotropic materials, including hexagonal α-Fe_2O_3, are also known to exhibit anisotropic charge mobility as a function of crystal orientation. Aligning the c-axis parallel to the direction of charge extraction with atomic layer epitaxy may therefore improve the lifetime–mobility product.

3.3 ALD Processes for Visible and NIR Light Absorbers

The sequential and self-limiting gas-phase reactions that characterize ALD are now known for a plethora of materials that span the periodic table, from lithium

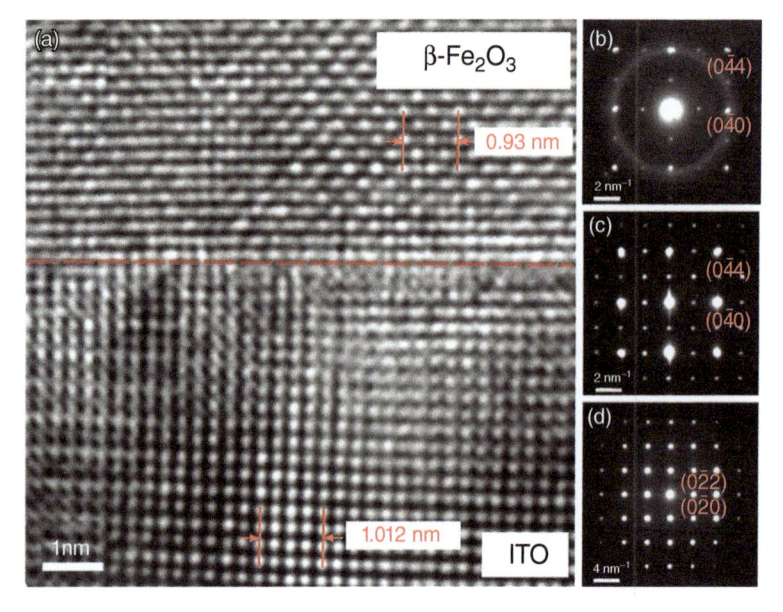

Figure 3.7 (a) High-resolution cross-sectional TEM image showing the β-Fe$_2$O$_3$(001)∥ITO(001) interface produced by the ALD of Fe$_2$O$_3$. (b–d) Electron diffraction patterns along the [100] axes of β-Fe$_2$O$_3$, ITO, and YSZ, respectively. (Emery et al. [19]. Reproduced with permission of American Chemical Society.)

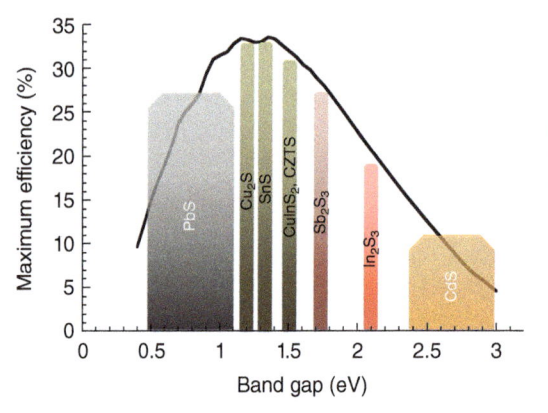

Figure 3.8 Graphical representation of band gaps for several ALD-grown sulfides relevant to solar absorption. The height of each bar corresponds to the theoretical single-gap efficiency limit for photovoltaics.

to bismuth [20]. A small subset of these materials are semiconductors with a band gap suitable for efficiently harvesting power from sunlight (Figure 3.8). The recent proliferation of ALD-grown solar absorbers is largely attributed to the extension of ALD chemistry to metal chalcogenide materials – compounds containing S, Se, or Te. While ZnS was the first ALD process ever demonstrated, the remaining history of ALD is largely rooted in oxides. With a few notable exceptions including Fe$_2$O$_3$ and Cu$_2$O, most oxides exhibit band gaps that are significantly larger than the ideal 0.8–2.0 eV predicted by Shockley and Queisser (Figure 3.1). Aluminum oxide, zirconium oxide, titanium dioxide, and zinc oxide are examples of materials that have received much attention from the ALD community but absorb only the highest-energy solar photons, if any, in the UV-vis-NIR spectrum. In contrast, many chalcogenides including CdTe and $CuInS_2$ are ample solar absorbers with a long history in PV and are now possible by ALD. There are also many lesser known chalcogenide materials that exhibit ample solar energy absorption (e.g., Cu_2S, SnS, and Sb_2S_3).

3.3.1 ALD Metal Oxides for Light Absorption

Very few oxide absorbers, whether grown by ALD or other methods, have been utilized in PV with efficiency greater than 1%. The exception is Cu_2O, for which devices based on oxidized copper foil or electrodeposition have reached an overall power conversion efficiency of 4% [21]. Although ALD has been used to subsequently deposit n-type windows including ZnO [21, 22] or $Zn_{1-x}Mg_xO$ [23], the thick copper oxide layer required for this material is most often and more economically derived from electrodeposition or copper metal oxidation. Very few reports of crystalline Cu_2O through conventional ALD processes have been published to date, and the growth rate per cycle is typically quite low (0.005 nm per cycle) [24]. A fast atmospheric ALD process that utilizes Cu(I)(hfac)(TMVS) (CupraselectTM, Air Products) and water, however, was shown to achieve phase-pure Cu_2O with a growth rate in time of 1 nm min^{-1} [25].

The remaining efficiently absorbing oxides, including Fe_2O_3, have largely been employed in photoelectrochemical (liquid junction) devices that are the topic of Chapter 8.

3.3.2 ALD Metal Chalcogenides for Light Absorption

An increasing number of semiconducting metal sulfides with band gap suitable for solar energy conversions are now synthesized by ALD, with the current status summarized in Table 3.1. Perhaps most strikingly, there is now a technology – SnS – for which ALD-grown absorbers exhibit best-in-class (or nearly so) device performance.

Table 3.1 Semiconducting metal chalcogenide solar absorbers grown by ALD. The highest PV device efficiencies that result from ALD and any deposition method are listed.

	Band gap (eV)	Majority carrier	Power eff. (%)	Record eff. (%)	References
$CuInS_2$	1.5	p-Type	4	12	[26]
CZTS	1.5	p-Type	—	12.6	[27]
Cu_xS	1.2	p-Type	<0.1	10	[28]
SnS	1.3	p-Type	4	4	[29–31]
PbS	0.4 (QD)	p-Type	0.6	6	[32]
Sb_2S_3	1.7	p-Type	3	8	[5, 33, 34]
CdS	2.4 (QD)	n-Type	0.3	3	[35]
In_2S_3	2.1	n-Type	0.4	3	[36]
Bi_2S_3	1.5	n-Type	—	2.5	[37]

3.3.2.1 CIS

The first solar absorber material reported to be successfully grown by ALD was actually a ternary compound – $CuInS_2$ [38]. This material belongs to the well-known family of high extinction, direct-gap solar absorbers that take the chalcopyrite crystal structure. Other members of the family including $CuInSe_2$, $Cu(In,Ga)Se_2$, $Cu(In,Ga)(S,Se)_2$ [CIGS], and $Cu(Zn,Sn)(S,Se)_2$ [CZTS] have been maturing since the early 1970s to achieve power efficiencies that now reach 21%. As many early ALD processes, the surface synthesis [38] constituted a relatively high-temperature (>300 °C) halide precursor process. Although not completely self-limiting, careful alternation of CuCl and $InCl_3$ with H_2S gas was shown to result in a mostly phase-pure $CuInS_2$ thin film. This material was deposited within the pores of a nanoporous TiO_2 framework and further phase-purified with H_2S annealing [39]. Overall solar-to-electricity efficiencies up to 4% are reported, making these devices still some of the most efficient ALD-absorber PV reported to date [26].

3.3.2.2 CZTS

The most complex ALD chalcogenide grown for PV application or otherwise is the quaternary CZTS [27]. This solar absorber is a member of the CIGS family but replaces rare and expensive In and Ga with earth-abundant Zn and Sn. While record efficiencies of CTZS PV are currently limited to 11% (not by ALD), these earth-abundant cousins have the potential to scale to terawatt generation levels at which worldwide energy use is measured. In contrast to the halide-based and high-temperature ALD of $CuInS_2$, the ALD of this quaternary composite was achieved at notably low temperatures (<150 °C) using organometallic precursors. In spite of the relatively low-temperature ALD growth using standard metal–organic precursors, the system requires careful processing owing to at least three possible failure modes: nucleation failure of Cu_2S on SnS_2 underlayers, ion exchange of gas-phase diethylzinc with Cu_2S, and phase stability of SnS_2 (grown by TDMASn and H_2S) on bare quartz. Nevertheless, when deposited in the proper sequence, phase-pure films were obtained after annealing under Ar for 2 h, which exhibit photoactivity in a photoelectrochemical demonstration device.

3.3.2.3 Cu_2S

Cu_2S was one of the first thin film PV technologies seriously explored. It exhibits outstanding earth abundance, and early power efficiencies hit the 10% milestone in 1980 before either CdTe or CIGS. However, the topotaxial cation exchange with CdS, by which Cu_2S was formed, contributed to an unstable junction and suboptimal heavy (p^+) electronic doping of the absorber. A stoichiometric, CdS-free, and stable Cu_2S absorber layer has been elusive via physical vapor deposition routes, Cu metal sulfurization, CVD, and solution methods. In contrast, the low-temperature ALD of Cu_2S performed by alternating bis(N,N'-di-sec-butylacetamidinato)dicopper(I) with H_2S has been shown to produce highly stoichiometric, moderately doped (10^{17} cm^{-3}), and highly oriented crystalline thin films (discussed further in Section 3.2.4) [13, 40].

Furthermore, with ALD-grown oxide overlayers, the electronic properties of these films have been stabilized against the oxidative effects of ambient for at least one month [13, 41].

3.3.2.4 SnS

SnS is a p-type chalcogenide with a nearly ideal 1.3 eV gap that was rediscovered recently for application in solar-to-electricity conversion. The ALD of SnS has been reported via either tin(II) acelylacetonate or N^2,N^3-di-*tert*-butyl-butane-2,3-diamido-tin(II) alternation with H_2S [42–44]. Clean and lightly doped (10^{15} cm^{-3}) thin films with respectable mobility (up to 10 cm^2 V^{-1} s^{-1}) enable thin-film PV external quantum efficiencies over 90% with useful light harvesting as far as 950 nm. The co-optimization of SnS electronic properties alongside Zn(O,S) also grown by ALD recently allowed for a record device efficiency of 4% [31]. Similarly for Cu_2O, a high-speed version of traditional ALD was explored to more quickly reach an absorber thickness commensurate with that required for good light-harvesting efficiency (~1 micron) [30].

3.3.2.5 PbS

Quantum-dot-sensitized solar cells (QDSSCs) are a second class of PVs that have benefited from the ALD of metal chalcogenides. As alternatives to the molecular dyes found in the prototypical dye-sensitized solar cells (DSSCs), PbS [32] QDs have been used to sensitize a porous wide-gap semiconductor framework. Here, ALD is also uniquely suited to the task of uniformly depositing upon and within the high-surface-area oxide frameworks that are common to this class of PVs. Here, ALD may be used to finely tune the QD size, and therefore the energy levels, simply with the number of cycles.

3.3.2.6 Sb_2S_3

Antimony sulfide has formed the foundation of devices with power efficiency in the single digits. Crystalline stibnite, Sb_2S_3, is an intrinsically n-type absorber that collects solar photons out to ~750 nm (~1.7 eV). The most efficient ALD-derived devices utilize a 10-nm-thick film deposited at a low temperature (120 °C) using tris(dimethylamino)antimony, $Sb(NMe_3)$, and H_2S gas. The amorphous film is deposited within the pores of a nanocrystalline TiO_2 framework that serves to increase the volume of absorber within a given projected area, Figure 3.9. Annealing the interdigitated scaffold at 315 °C crystallizes the absorber to the photoactive c-Sb_2S_3 phase.

3.3.2.7 CdS

Cadmium sulfide is a relatively wide-gap (2.4 eV) semiconductor that is most commonly used as an ultrathin buffer layer in PV – a role in which light absorption is to be avoided. Despite its gap and the toxicity of elemental Cd, the forgiving optoelectronic properties of this absorber have led to numerous basic science studies. CdS quantum dots grown by ALD were accurately tuned from 1 to 10 nm by varying the number of dimethyl cadmium + H_2S cycles alone [35].

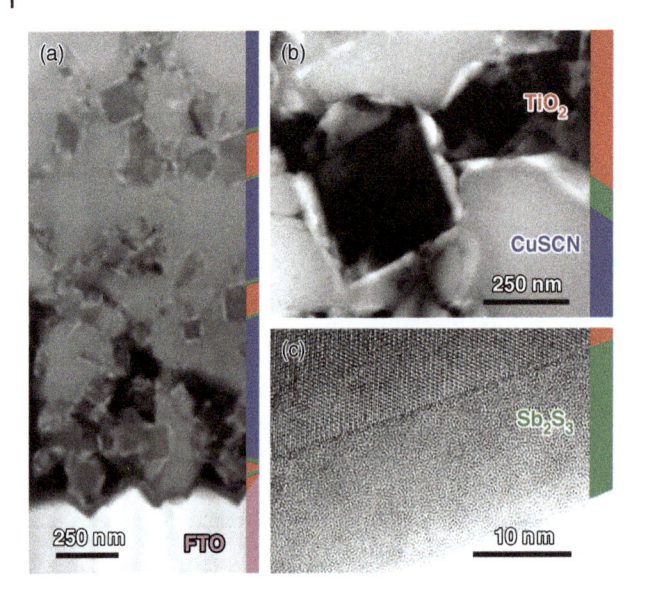

Figure 3.9 Transmission electron micrographs of ALD-grown Sb_2S_3 on nanocrystalline TiO_2 on conductive glass (FTO). The conformal nature of ALD coating is clearly visible. (Wedemeyer et al. [5]. Reproduced with permission of Royal Society of Chemistry.)

3.3.2.8 In_2S_3

Indium sulfide is a 2.1-eV gap absorber that is less commonly considered due to its relatively rare and expensive base metal but has been recently recognized for its promise in buffer layer application and as a substitutionally doped intermediate-band absorber. As an n-type, medium-band-gap semiconductor with relatively low toxicity, it is a contender to replace potentially toxic CdS buffer layers in several PV technologies [45–47]. In its most stable spinel β-In_2S_3 form, powders and single crystals alloyed with transition metals (V [48, 49], Ti [48], Fe [50], Nb [51]) are also being explored for potential use as absorbers in intermediate-band photovoltaics (IBPV), which aim to convert two subband-gap photons into a single excitation of the parent gap via excitation to and from an intermediate band (IB), see Section 3.2.4. The ALD of In_2S_3 was first achieved through relatively high-temperature halide chemistry, then indium(III) trifluoroacetylacetonate, and, most recently, an oxygen-free In(III) N,N'-diisopropylacetamidinate [52].

3.3.2.9 Bi_2S_3

Bismuth sulfide is a 1.3 eV semiconductor based on the relatively heavy Bi metal. While the band gap is nearly ideal for single-gap solar energy conversion, it has elemental abundance comparable to Ag and In. Still the cost of Bi is at least an order of magnitude lower than that of either Ag or In and with none of the toxicity of heavy metals such as lead. As such, it is a potentially useful solar absorber that has been grown by ALD. However, to date, there are no reports of PV devices based on an ALD-grown Bi_2S_3.

3.3.3 Other ALD Materials for Light Absorption

Very few pure elements are semiconductors suitable for solar energy absorption, with Si being the notable exception. Unfortunately, a direct and moderate-temperature route to elemental Si via ALD remains elusive. Nitrides are another class of materials that are often associated with solar energy absorption. Indeed, multijunction devices based on InGaN/GaN are some of the most efficient PVs ever fabricated. However, it is unclear whether there is a significant advantage to be gained from the ALD of these materials. Even more unconventional material classes including mixed metal–organic hybrids have garnered recent attention from the PV community including hybrid metal halide perovskites. While these complex materials have yet to be synthesized by ALD, some of their possible building blocks (including PbS) have been. Subsequent transformations to the hybrid perovskite phase [53] that preserve complex film design demonstrate the applicability and versatility of ALD for future energy materials.

3.4 Prospects and Future Challenges

The ALD of solar absorbers presently includes more than 10 oxides and sulfides, but many more remain. However, those efforts should be targeted at absorbers that will benefit most from the unique control that ALD offers – stoichiometry, conformality, compositional control, and epitaxy.

While thin-film ALD growth of high-quality absorbers is a crucial first step toward more efficient and earth-abundant solar energy conversion devices, interfaces remain the next frontier. ETA designs with exceptionally large junction area remind us of the ever-expanding role that interfaces play in device performance. ALD is equally suited to address these challenges as it produces a homogeneous surface functionality for subsequent amelioration in addition to its potential for chemical selectivity in passivation reactions. As such, ALD will continue to be an essential contributor to the future of solar energy absorption and conversion.

References

1 Liu, X., Coxon, P.R., Peters, M., Hoex, B., Cole, J.M., and Fray, D.J. (2014) *Energy Environ. Sci.*, **7**, 3223.
2 Jewell, A.D., Hennessy, J., Hoenk, M.E., and Nikzad, S. (2013) *Proc. SPIE*, **8820**, 88200Z.
3 Kim, J.Y., Kim, S.H., Lee, H.H., Lee, K., Ma, W., Gong, X., and Heeger, A.J. (2006) *Adv. Mater.*, **18**, 572.
4 Andersen, P.D., Skårhøj, J.C., Andreasen, J.W., and Krebs, F.C. (2009) *Opt. Mater.*, **31**, 1007.
5 Wedemeyer, H., Michels, J., Chmielowski, R., Bourdais, S., Muto, T., Sugiura, M., Dennler, G., and Bachmann, J. (2013) *Energy Environ. Sci.*, **6**, 67.

6 Nanu, M., Reijnen, L., Meester, B., Goossens, A., and Schoonman, J. (2003) *Thin Solid Films*, **431**, 492.
7 Kelzenberg, M.D., Boettcher, S.W., Petykiewicz, J.A., Turner-Evans, D.B., Putnam, M.C., Warren, E.L., Spurgeon, J.M., Briggs, R.M., Lewis, N.S., and Atwater, H.A. (2010) *Nat. Mater.*, **9**, 239.
8 Kayes, B.M., Atwater, H.A., and Lewis, N.S. (2005) *J. Appl. Phys.*, **97**, 114302.
9 Martinson, A.B.F., Hamann, T.W., Pellin, M.J., and Hupp, J.T. (2008) *Chem. Eur. J.*, **14**, 4458.
10 Vermang, B., Fjällström, V., Xindong, G., and Edoff, M. (2014) *IEEE J. Photovoltaics*, **4**, 486.
11 Groner, M.D., Elam, J.W., Fabreguette, F.H., and George, S.M. (2002) *Thin Solid Films*, **413**, 186.
12 Groner, M.D., George, S.M., McLean, R.S., and Carcia, P.F. (2006) *Appl. Phys. Lett.*, **88**, 051907.
13 Martinson, A.B.F., Riha, S.C., Thimsen, E., Elam, J.W., and Pellin, M.J. (2013) *Energy Environ. Sci.*, **6**, 1868.
14 Rit, M., Asikainen, T., Leskelä, M., and Skarp, J. (1996) *MRS Online Proc. Lib. Arch.*, **426**, 513.
15 Elam, J.W., Martinson, A.B.F., Pellin, M.J., and Hupp, J.T. (2006) *Chem. Mater.*, **18**, 3571.
16 Lin, Y., Xu, Y., Mayer, M.T., Simpson, Z.I., McMahon, G., Zhou, S., and Wang, D. (2012) *J. Am. Chem. Soc.*, **134**, 5508.
17 Sinsermsuksakul, P., Chakraborty, R., Kim, S.B., Heald, S.M., Buonassisi, T., and Gordon, R.G. (2012) *Chem. Mater.*, **24**, 4556.
18 McCarthy, R.F., Weimer, M.S., Haasch, R.T., Schaller, R.D., Hock, A.S., and Martinson, A.B.F. (2016) *Chem. Mater.*, **28**, 2033.
19 Emery, J.D., Schlepütz, C.M., Guo, P., Riha, S.C., Chang, R.P.H., and Martinson, A.B.F. (2014) *ACS Appl. Mater. Interfaces*, **6**, 21894.
20 Miikkulainen, V., Leskelä, M., Ritala, M., and Puurunen, R.L. (2013) *J. Appl. Phys.*, **113**, 021301.
21 Lee, Y.S., Chua, D., Brandt, R.E., Siah, S.C., Li, J.V., Mailoa, J.P., Lee, S.W., Gordon, R.G., and Buonassisi, T. (2014) *Adv. Mater.*, **26**, 4704.
22 Brittman, S., Yoo, Y., Dasgupta, N.P., Kim, S.I., Kim, B., and Yang, P.D. (2014) *Nano Lett.*, **14**, 4665.
23 Ievskaya, Y., Hoye, R.L.Z., Sadhanala, A., Musselman, K.P., and MacManus-Driscoll, J.L. (2015) *Sol. Energy Mater. Sol. Cells*, **135**, 43.
24 Dhakal, D., Waechtler, T., Schulz, S.E., Gessner, T., Lang, H., Mothes, R., and Tuchscherer, A. (2014) *J. Vac. Sci. Technol., A*, **32**, 041505.
25 Muñoz-Rojas, D., Jordan, M., Yeoh, C., Marin, A.T., Kursumovic, A., Dunlop, L.A., Iza, D.C., Chen, A., Wang, H., and MacManus Driscoll, J.L. (2012) *AIP Adv.*, **2**, 042179.
26 Nanu, M., Schoonman, J., and Goossens, A. (2005) *Adv. Funct. Mater.*, **15**, 95.
27 Thimsen, E., Riha, S.C., Baryshev, S.V., Martinson, A.B.F., Elam, J.W., and Pellin, M.J. (2012) *Chem. Mater.*, **24**, 3188.

28 Reijnen, L., Meester, B., Goossens, A., and Schoonman, J. (2003) *Chem. Vap. Deposition*, **9**, 15.
29 Park, H.H., Heasley, R., Sun, L., Steinmann, V., Jaramillo, R., Hartman, K., Chakraborty, R., Sinsermsuksakul, P., Chua, D., Buonassisi, T., and Gordon, R.G. (2015) *Prog. Photovoltaics Res. Appl.*, **23**, 901908.
30 Sinsermsuksakul, P., Hartman, K., Kim, S.B., Heo, J., Sun, L.Z., Park, H.H., Chakraborty, R., Buonassisi, T., and Gordon, R.G. (2013) *Appl. Phys. Lett.*, **102**, 053901.
31 Sinsermsuksakul, P., Sun, L., Lee, S.W., Park, H.H., Kim, S.B., Yang, C., and Gordon, R.G. (2014) *Adv. Energy Mater*, **4**, 1400496.
32 Brennan, T.P., Trejo, O., Roelofs, K.E., Xu, J., Prinz, F.B., and Bent, S.F. (2013) *J. Mater. Chem. A*, **1**, 7566.
33 Wu, Y.L., Assaud, L., Kryschi, C., Capon, B., Detavernier, C., Santinacci, L., and Bachmann, J. (2015) *J. Mater. Chem. A*, **3**, 5971.
34 Kim, D.H., Lee, S.J., Park, M.S., Kang, J.K., Heo, J.H., Im, S.H., and Sung, S.J. (2014) *Nanoscale*, **6**, 14549.
35 Brennan, T.P., Ardalan, P., Lee, H.B.R., Bakke, J.R., Ding, I.K., McGehee, M.D., and Bent, S.F. (2011) *Adv. Energy Mater.*, **1**, 1169.
36 Sarkar, S.K., Kim, J.Y., Goldstein, D.N., Neale, N.R., Zhu, K., Elliot, C.M., Frank, A.J., and George, S.M. (2010) *J. Phys. Chem. C*, **114**, 8032.
37 Cao, Y., Bernechea, M., Maclachlan, A., Zardetto, V., Creatore, M., Haque, S.A., and Konstantatos, G. (2015) *Chem. Mater.*, **27**, 3700.
38 Nanu, M., Reijnen, L., Meester, B., Schoonman, J., and Goossens, A. (2004) *Chem. Vap. Deposition*, **10**, 45.
39 Nanu, M., Schoonman, J., and Goossens, A. (2004) *Adv. Mater.*, **16**, 453.
40 Martinson, A.B.F., Elam, J.W., and Pellin, M.J. (2009) *Appl. Phys. Lett.*, **94**, 123017-1 to 123013-3.
41 Riha, S.C., Jin, S., Baryshev, S.V., Thimsen, E., Wiederrecht, G.P., and Martinson, A.B.F. (2013) *ACS Appl. Mater. Interfaces*, **5**, 10302.
42 Kim, J.Y. and George, S.M. (2010) *J. Phys. Chem. C*, **114**, 17597.
43 Kim, S.B., Sinsermsuksakul, P., Hock, A.S., Pike, R.D., and Gordon, R.G. (2014) *Chem. Mater.*, **26**, 3065.
44 Sinsermsuksakul, P., Heo, J., Noh, W., Hock, A.S., and Gordon, R.G. (2011) *Adv. Energy Mater.*, **1**, 1116.
45 Naghavi, N., Abou-Ras, D., Allsop, N., Barreau, N., Bucheler, S., Ennaoui, A., Fischer, C.H., Guillen, C., Hariskos, D., Herrero, J., Klenk, R., Kushiya, K., Lincot, D., Menner, R., Nakada, T., Platzer-Bjorkman, C., Spiering, S., Tiwari, A.N., and Torndahl, T. (2010) *Prog. Photovoltaics Res. Appl.*, **18**, 411.
46 Hariskos, D., Spiering, S., and Powalla, M. (2005) *Thin Solid Films*, **480**, 99.
47 Abou-Ras, D., Rudmann, D., Kostorz, G., Spiering, S., Powalla, M., and Tiwari, A.N. (2005) *J. Appl. Phys.*, **97**, 084908.
48 Lucena, R., Aguilera, I., Palacios, P., Wahnon, P., and Conesa, J. (2008) *Chem. Mater.*, **20**, 5125.
49 Lucena, R., Conesa, J., Aguilera, I., Palacios, P., and Wahnon, P. (2014) *J. Mater. Chem. A*, **2**, 8236.

50 Chen, P., Chen, H., Qin, M., Yang, C., Zhao, W., Liu, Y., Zhang, W., and Huang, F. (2013) *J. Appl. Phys.*, **113**, 213509.
51 Ho, C.H. (2011) *J. Mater. Chem.*, **21**, 10518.
52 McCarthy, R.F., Weimer, M.S., Emery, J.D., Hock, A.S., and Martinson, A.B.F. (2014) *ACS Appl. Mater. Interfaces*, **6**, 12137.
53 Sutherland, B.R., Hoogland, S., Adachi, M.M., Kanjanaboos, P., Wong, C.T.O., McDowell, J.J., Xu, J., Voznyy, O., Ning, Z., Houtepen, A.J., and Sargent, E.H. (2015) *Adv. Mater.*, **27**, 53.
54 Dasgupta, N.P., Lee, W., and Prinz, F.B. (2009) *Chem. Mater.* **21**, 3973, DOI: 10.1021/cm901228x.

4

Atomic Layer Deposition for Surface and Interface Engineering in Nanostructured Photovoltaic Devices

Carlos Guerra-Nuñez[1], Hyung Gyu Park[2], and Ivo Utke[1]

[1] EMPA, Swiss Federal Laboratories for Materials Science and Technology, Laboratory for Mechanics of Materials and Nanostructures, Feuerwerkerstrasse 39, CH-3602 Thun, Switzerland
[2] ETH Zürich, Nanoscience for Energy Technology and Sustainability, Department of Mechanical and Process Engineering, Tannenstrasse 3, CH-8092 Zürich, Switzerland

4.1 Introduction

The past decades have witnessed the rise of new generations of nanostructured devices in nanoelectronics, especially for energy harvesting and storage. Photovoltaic (PV) devices have evolved from the first generation of bulk silicon solar cells to thin films (second generation) and to nanostructured solar cells (third generation) in an attempt to save materials and production costs while improving energy conversion efficiencies. Nanostructured solar cells are currently one of the most desirable technologies that hold the promise to provide sustainable energy at the desired price of 0.03 \$ kWh^{-1} to compete with the grid electricity [1]. For this reason, these types of solar cells have been the focus of research since the invention of a dye-sensitized solar cell (DSSC) in 1991, which achieved a photovoltaic conversion efficiency (PCE) of ∼7.9% using low-purity and inexpensive materials with low-cost processes [2]. Today, the DSSC concept extends to different types of architectures such as nanoparticles, nanotubes, or nanowires and often uses different functional materials such as carbon nanotubes, graphene, quantum dots, and perovskite absorbers.

Nanostructured solar cells are featured by a large surface-to-volume ratio to enhance light absorption and improve charge transport rates on shorter transport distances. This feature is combined with a substantial reduction in materials usage and processing costs. At present, the highest efficiencies published to date for third-generation solar cells are 19.3% and 20.2% using a perovskite absorber [3, 4]. Regardless of the improvements in the efficiencies over the past years, nanostructured solar cells are still subjected to many recombination processes that ultimately lead to a decrease in PCE. One of the major limitations is the photoelectron transport through the numerous interfaces across a network of nanoparticles. The use of nanotube/nanowire metal oxide structures or functional materials as a support is one of the strategies for reducing the number of interfaces and grain boundaries to boost the efficiency [5]. The electrical

Atomic Layer Deposition in Energy Conversion Applications, First Edition. Edited by Julien Bachmann.
© 2017 Wiley-VCH Verlag GmbH & Co. KGaA. Published 2017 by Wiley-VCH Verlag GmbH & Co. KGaA.

performance as well as the mechanical and chemical stability of these devices will be largely influenced by the quality of the interfaces between the different components in a nanostructured solar cell. It is at these interfaces that most of the losses occur via recombination processes or absorber instability. For this reason, ultrathin films deposited at the interfaces, often at a subnanometer scale, can act as recombination barriers and/or protective layers, which can drastically improve the PCE. For this endeavor, atomic layer deposition (ALD) is an appropriate technique, if not the only one, to deposit such films with a subnanometer precision. As detailed in Chapter 1, ALD relies on self-limiting chemical surface reactions of sequentially injected gas molecules (mostly a metal–organic precursor and an oxidizing molecule) to obtain uniform (pinhole-free) films with atomic precision of the film thickness even on highly porous materials. It is this key feature that renders ALD an ideal tool for surface and interface engineering.

ALD has become a widely adopted tool for nanotechnology, which can be devised in two subfields: (i) the fabrication of the essential nanoscale building blocks such as nanotubes/nanowires, nanolaminates, and nanodots and (ii) surface and interface engineering with thin films for a wide range of applications. The work by Knez *et al.* and Kim *et al.* reviews the diverse nanostructures fabricated using ALD [6, 7]. This chapter covers the most relevant research done to engineer the surfaces and interfaces with ultrathin films of different metal oxides deposited by ALD on geometrically complex nanostructures and discusses their application in photovoltaic devices such as in DSSCs, quantum-dot-sensitized solar cells (QDSSC), colloidal quantum dot solar cells (CQDSC), organic solar cells, perovskite solar cells, and photoelectrochemical cells for water splitting. Several review articles are available to complement this chapter on the topic of surface and interface engineering for photovoltaic devices by using ALD [6, 8–16] and other deposition techniques [17, 18].

4.2 ALD for Improved Nanostructured Solar Cells

The DSSC was introduced by Grätzel and coworkers in 1991 as a potential cheap alternative to conventional silicon solar cells [2]. Since then, the concept has extended from the basic TiO_2 nanoparticle structure to different metal oxide nanostructures, substrate materials, absorbing films, and hole-transporting materials. The PCE in nanostructured solar cells is governed by four major processes [19]: (i) photoelectron generation in the light-absorbing material (e.g., ruthenium-based dyes, quantum dots, perovskites); (ii) photoelectron injection to the conduction band of the nanostructured metal oxide (e.g., TiO_2, ZnO, SnO_2); (iii) charge transport within the photoanode, from the metal oxide nanostructure to the transparent conductive oxide (TCO) (e.g., FTO, ITO) and subsequently to the charge collector at the counter electrode through the external circuit; and (iv) regeneration of the photosensitizer (light-absorbing material) by electron donation from the hole transporter material (HTM), solid, or electrolyte such as triiodide electrolyte or Spiro-OMeTAD, respectively. Almost all of these processes occur at different interfaces and thus are limited

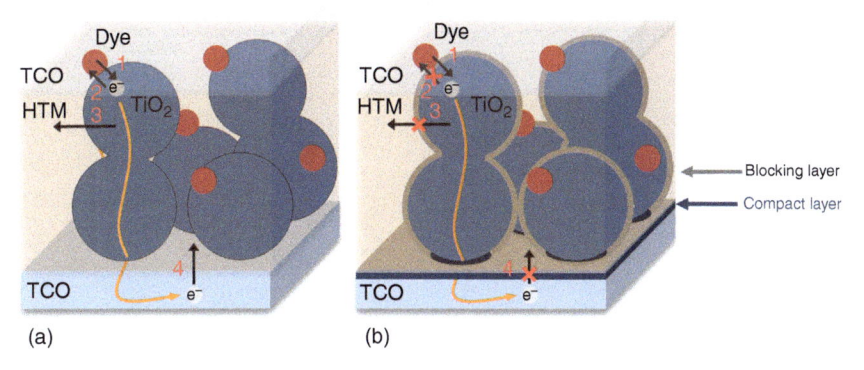

Figure 4.1 Electron transport and possible recombination processes in a DSSC before (a) and after (b) ALD. (1) Injection of a photoelectron to the TiO_2 from the excited dye. (2) The recombination of an injected electron in TiO_2 with an oxidized dye molecule. (3) The recombination of an electron in TiO_2 with the HTM. (4) The recombination of a collected electron in the transparent conductive oxide (TCO) with a hole in the HTM. The deposition of ultrathin compact and blocking layers via ALD can suppress these types of recombination processes.

to interfacial recombination losses. Figure 4.1a illustrates the electron transport and recombination pathways in a nanostructured solar cell based on the classic architecture of the DSSC. For detailed reviews of the different recombination processes in liquid- or solid-state DSSCs and photoelectrochemical cells, the reader can refer to [20, 21]. Here, we only shortly introduce them to show how ALD can be used to reduce them.

The main recombination losses, on the one hand, occur when the injected electrons in the metal oxide (i) recombine with trap states at the surface of the metal oxide structure, with the oxidized absorbing layer (ii), or the HTM (iii). On the other hand, the back transfer of electrons from the TCO to the HTM at the interface is the predominant recombination loss in a DSSC (iv) [22], especially at low light intensities and under open-circuit conditions [23].

Therefore, ultrathin recombination barrier layers are used between the different interfaces to reduce these recombination processes (Figure 4.1b). Often, these layers are tunneling layers of wide-band-gap metal oxides, which can be deposited with superb thickness control by ALD, such as TiO_2, Al_2O_3, ZnO, HfO, or SnO_2. Table 4.1 shows a summary of the various versions of DSSCs where ALD has been used to deposit such films.

4.2.1 Compact Layer: The TCO/Metal Oxide Interface

A compact layer is a thin metal oxide film deposited onto a TCO, to allow the transfer of electrons from the metal oxide nanostructure to the current collector while blocking the back transfer of collected electrons to the HTM, for example, the triiodide electrolyte or Spiro-OMeTAD. If the TCO is incompletely covered by the compact metal oxide layer, recombination losses will be observed at the TCO surfaces directly exposed to the electrolyte. This type of recombination process occurs predominantly under open-circuit conditions and at low light

Table 4.1 Barrier recombination (compact and blocking) layers deposited by ALD on different third-generation solar cells.

Support	Support thickness (μm)	Absorbing material	Hole transporting material	ALD compact layer	Compact layer thickness (nm)	Efficiency (%)	Control cell efficiency (%)	References
Compact layers								
TiO_2 NP	8–9	N719	Pyridine-based electrolyte	TiO_2	50	3.27	2.53	[24]
TiO_2 NP	0.9	D102	Spiro-OMeTAD	TiO_2	20	1.3	<0.1	[25]
TiO_2 NP	15	N719	I_3 electrolyte	TiO_2	4	4.6	4.6	[26]
TiO_2 NP	6	N719	I_3 electrolyte	HfO_2	1.65	6	3.6	[27]
TiO_2 NP	18	N719	I_3 electrolyte	TiO_2	10	8.5	7	[28]
Planar	—	Perovskite	Spiro-OMeTAD	TiO_2	50	12.56	0.32	[29]
TiO_2 NP	18	CdS QDs and N719	Spiro-OMeTAD	TiO_2	22	2.36	1.67	[30]
SnO_2 NP	4	N719	Pyridine-based electrolyte	SnO_2	20	3.7	0.76	[31]
Planar	—	Perovskite (CH_3NH_3I)	P3HT	TiO_2	10	13.6	8.7	[32]
TiO_2 NP	0.25	Perovskite ($CH_3NH_3PbI_{3-x}Cl_x$)	Spiro-OMeTAD	TiO_2	11	7.1	0.01	[33]
Blocking layers								
ZnO NW	15	N719	Pyridine-based electrolyte	TiO_2	21	2.1	0.85	[24]
TiO_2 NP	18	CdS QDs and N719	Spiro-OMeTAD	TiO_2	2.2	2.36	1.67	[30]
SnO_2 NP	4	N719	Pyridine-based electrolyte	Al_2O_3	0.11	3.7	0.76	[31]
TiO_2 NP	16	N719	Pyridine-based electrolyte	Al_2O_3	0.1	6.5	5.75	[34]

TiO$_2$ NP	9	N3	I$_3$ electrolyte	Al$_2$O$_3$	2.2	8.4	6.2	[35]
TiO$_2$ NT	40	N719	I$_3$ electrolyte	Al$_2$O$_3$	0.12	5.75	4.65	[36]
TiO$_2$ NP	1.8	N719	Spiro-OMeTAD	ZrO2	0.22	0.27	1.08	[37]
TiO$_2$ NP	2.7	Y123	Pyridine-based electrolyte	Ga$_2$O$_3$	0.4	4	1.4	[38]
TiO$_2$ NP	2.7	Y123	Pyridine-based electrolyte	ZrO$_2$	0.13	2.65	2.5	[39]
TiO$_2$ NP	2.7	Y123	Pyridine-based electrolyte	Ga$_2$O$_3$	0.3	3.5	2.5	[39]
TiO$_2$ NP	2.7	Y123	Pyridine-based electrolyte	Nb$_2$O$_5$	0.05	3.2	2.5	[39]
TiO$_2$ NP	2.7	Y123	Pyridine-based electrolyte	Ta$_2$O$_5$	0.13	3	2.5	[39]
TiO$_2$ NP	0.3	Perovskite (CH$_3$NH$_3$PbI$_3$)	Spiro-OMeTAD	TiO$_2$	2	11.5	7.2	[40]
TiO$_2$ NR	1.8	Perovskite (CH$_3$NH$_3$PbI$_3$)	Spiro-OMeTAD	TiO$_2$	4.8	13.45	5.03	[41]
ZnO NR	0.35	Perovskite (CH$_3$NH$_3$PbI$_3$)	Spiro-OMeTAD	TiO$_2$	5 cycles	13.4	11.9	[42]
ITO NW/TiO$_2$ NP	30	N719	I$_3$ electrolyte	TiO$_2$	30	6.1	0.1	[43]
ITO NW/TiO$_2$ NP	0.9	Perovskite (CH$_3$NH$_3$PbI$_3$)	Spiro-OMeTAD	TiO$_2$	10	7.5	7.1	[44]
ITO NW/TiO$_2$ NP	20	N719	I$_3$ electrolyte	HfO$_2$	1.3	4.83	2.82	[45]

NP, nanoparticle; NW, nanowire; NT, nanotube; NR, nanorod.

intensities [23]. Initially, compact layers at the TCO/TiO$_2$ nanoparticle interface have been deposited via sol–gel, spray pyrolysis, or sputtering [23, 46]. These initial studies have shown that very thin compact layers in the nanometer range are needed with great uniformity [47]. For this reason, ALD came into the picture to deposit pinhole-free uniform layers and understand the ideal thickness of different metal oxides for effective recombination blocking. The use of ALD to deposit compact layers has been demonstrated in the early work of Law *et al.*, in which they showed that a compact TiO$_2$ layer of 10–50 nm thickness on the FTO and before the TiO$_2$ nanoparticle addition to the substrate already increased the overall PCE by 25% to a final value of 3.27% from 2.53% [24]. Shortly after, Hamman *et al.* showed that a ~14-nm-thick compact film of ALD TiO$_2$ on the FTO substrate prior to the nanoparticle addition can mitigate the redox shunting of the FTO with the electrolyte [48]. In another study, a compact layer of 20 nm of TiO$_2$ using TiCl$_4$ as the precursor was deposited at 150 °C as a low-temperature alternative for flexible substrates. They observed that the use of this compact layer substantially increased the J_{sc} and the V_{oc} of the cells by 460% and 870%, respectively, compared to a cell without this compact layer [25]. Miettunen *et al.* have observed a similar behavior in both FTO and ITO substrates with only 4 nm of ALD TiO$_2$ and reported higher open-circuit voltages at low light intensities with the use of the compact layer [26]. Soon after, Bills *et al.* have reported a 66% increase in the efficiency (from 3.6% to 6%) by depositing only 30 cycles of HfO$_2$ (~1.65 nm) before the nanoparticle addition [27]. These studies show the potential of ALD for studies of compact layers and their effects on DSCC efficiencies compared to other techniques, as ALD can deposit ultrathin films of precisely controlled thickness. In fact, Kim *et al.* reviewed and compared different TiO$_2$ compact layers deposited by different techniques and concluded that ALD-based TiO$_2$ compact layers performed better than other compact layers deposited by hydrolysis or spin casting [28]. They have also concluded that the optimal thickness for an ALD TiO$_2$ compact layer would be between 5 and 10 nm, since thicker compact layers can reduce cell performance due to lower light transmittance. The light transmittance remained relatively unchanged for ALD TiO$_2$ films from 5 to 15 nm; however, a 20-nm-thick ADL TiO$_2$ layer already reduced the light transmittance from ~80% for a bare FTO substrate to ~63%. Hence, there is a trade-off between the blocking properties of the pinhole-free film and the light transmittance at a given thickness. Furthermore, the ALD TiO$_2$ films presented lower light transmittance compared to films with the same thickness prepared by other techniques (electrodeposition, spin casting, and sputtering), which suggests a truly pinhole-free characteristic of ALD films.

Recently, Wu *et al.* have also compared different compact layers of TiO$_2$ deposited by spray pyrolysis, spin coating, and ALD with the same thickness, all used in perovskite solar cells [29]. The films deposited by ALD showed a considerable increase in the PCE compared to the other techniques. The three different films showed exactly the same light transmittance, and, therefore, it was concluded that the pinhole-free characteristic of the ALD films surpassed those deposited by the other techniques even at a thickness of 50 nm. To measure the pinhole structure of each film, they infiltrated Ag nanoparticles (2–10 nm)

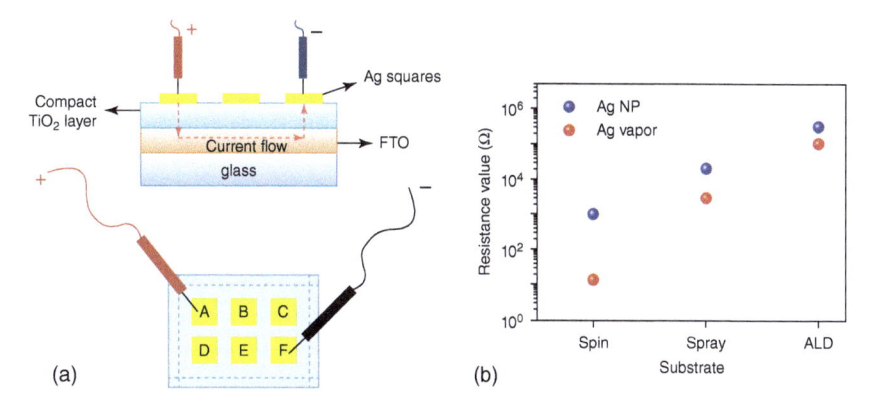

Figure 4.2 (a) Schematic illustration of the resistance measurements of compact TiO_2 layers (deposited by different techniques) after coating with Ag paste and vacuum-evaporated Ag. (b) The average resistance between Ag contacts on the compact TiO_2 layers prepared by different methods (the thickness is reported to be ~50 nm for each sample). (Wu et al. 2014 [29]. Reproduced with permission of American Chemical Society.)

through the potential pinholes and measured the resistance of the films as shown in Figure 4.2.

The lower resistance of the films is caused by the infiltration of the Ag nanoparticles toward the FTO substrate; the high resistivity of the ALD films demonstrated the pinhole-free characteristic. Similarly, Kavan et al. have shown the influence of a pinhole-free compact layer as well as the influence of the crystal structure [49]. In this study, they measured, by cyclic voltammetry, the percentage of pinhole areas as a function of film thickness (1–6 nm) of ALD TiO_2. They have reported that a ~5 to 6-nm-thick TiO_2 compact layer has already less than 1% of pinhole areas, which, however, upon annealing at 500 °C (a typical process after the nanoparticle addition) left uncovered FTO regions and the pinhole area percentage increased to 56%. Not surprisingly, the thickness of ~5 nm is the thickness that previous groups have reported as optimal. In the same study, they also demonstrated that as-deposited crystalline TiO_2 ALD films required slightly larger thickness (6–7 nm) to reach less than 1% of pinhole areas. Similarly, Lin et al. have shown that low-temperature amorphous ALD TiO_2 compact layers performed better than polycrystalline anatase deposited at high temperatures with the same thickness (~15 nm) [30]. They reported lower light transmission for the polycrystalline anatase film and also suggested that this film does not fully block the recombination of holes, due to the presence of grain boundaries. Moreover, they also reported that the TiO_2 film resistivity can be substantially reduced after light soaking for 5 min under simulated illumination.

The compact layer will therefore have a substantial impact on the cell performance. Studies so far have shown that a very thin compact layer is sufficient to block the back transfer of electrons as long as the TCO is covered uniformly with a pinhole-free film. Thicker films will reduce the light transmittance, and thus, it can reduce the PCE. A 5–10-nm-thick amorphous TiO_2 film appears to be the optimal thickness without compromising the light transmittance. Polycrystalline films seem to be less efficient in blocking properties. Typically, TiO_2 has been the material of choice for a compact layer in most of the literature; however, there are reports using SnO_2 [31] and ZnO [50, 51] with similar blocking results. The deposition of a thin compact layer is at present a standard practice to enhance the PV performance in different device architectures, especially in the uprising perovskite solar cells in both planar and nanostructured configurations. For instance, a recent publication has also reported an optimal 10-nm-thick film of ALD TiO_2 as compact layer for a planar perovskite solar cell, reaching 13.6% compared to 8.7% from a spin-coated TiO_2 film of the same thickness [32]. Similarly, an 11-nm-thick film of ALD TiO_2 deposited onto a flexible PET/ITO substrate as a compact layer and using a perovskite absorber displayed a PCE of 7.1% compared to 0.01% without the compact layer [33].

The studies presented in this section reinforce the superior capabilities of ALD compared to other deposition techniques and the importance of a pinhole-free characteristic of ultrathin films for efficient recombination blocking properties.

4.2.2 Blocking Layer: The Metal Oxide/Absorber Interface

As discussed previously, the large surface area is perhaps the main advantage of nanostructured solar cells, but it can also be the major limitation for solar cell performance. Surface defects and trap states in the metal oxide semiconductor act as recombination centers, and injected electrons can recombine with a hole at the metal oxide surface; with a hole in the HTM, or with a hole in an oxidized dye molecule. Therefore, to reduce or eliminate these recombination losses at the interface, a thin coating can be deposited to the metal oxide nanostructure. This coating is known as the blocking layer. The blocking layer material has to have a wider band gap compared to the absorbing film and a more positive conduction band potential compared to the metal oxide nanostructure to create an energy barrier at the interface. The chosen blocking layer material will have a large influence on the J_{sc} and V_{oc}. It has been shown that blocking layer materials with higher isoelectric points (EIP) and lower electron affinities with respect to the metal oxide nanostructure can generate a surface dipole [52]. High conduction band edges and wide-band-gap materials have a better "blocking effect" that can offset the conduction band of the core metal oxide [53]. This results in a shift of the conduction edges and, as a consequence, improves the V_{oc}. Blocking layers with higher EIP resulted in a higher dye absorption, which increased the J_{sc} [53, 54]. Therefore, for optimal cell performance, the material of choice for the blocking layer will depend on the parent metal oxide nanostructure and the light-absorbing material. The ideal blocking layer has to be engineered in such a way to fulfill the following requirements to improve the solar cell efficiency:

i) Complete surface coverage of the metal oxide nanostructure with a uniform and homogeneous coating. This will ensure passivation of the surface defect sites of the metal oxide and avoid electrons to leak through the uncoated surfaces.
ii) It has to be thin enough to allow for efficient electron injection from the dye to the metal oxide in addition to avoid blocking the pores (especially in nanoparticle-type DSCC) for efficient absorber film and HTM infiltration.
iii) It has to be thick enough to avoid the back transfer of electrons to the dye or the HTM.

Semiconducting and insulating metal oxides deposited by different techniques have been studied for these layers. A review of such layers from different deposition techniques can be found in [17]. Unlike other deposition techniques, ALD offers the advantage to study precisely the optimal thickness for each of the metal oxide films used as blocking layer and therefore understand the relation between the properties of the core–shell metal oxides. In this section, we review the different blocking layers deposited by ALD in different types of third-generation solar cells. Table 4.1 presents the metal oxides deposited by ALD that have been used as blocking layers.

Early studies of different metal oxides as blocking layers on TiO_2 nanoparticle DSCC have concluded that a ~1-nm-thick Al_2O_3 deposited by sol–gel coating had the best blocking layer performance [54]. Later, Lin et al. studied the thickness of the Al_2O_3 blocking layers with an atomic precision using ALD, by varying the coating thickness from 1 cycle (~0.12 nm) to 10 cycles (~1.2 nm) [34]. They demonstrated that with only 1 cycle of Al_2O_3 (~0.1 nm), the PCE improved 14% (from 5.75% to 6.50%) and determined that thicker Al_2O_3 layers raised the Fermi level and blocked the dye electron injection to the TiO_2. Compared to the high-temperature sol–gel technique, they concluded that ALD reflects a more realistic value of the optimal thickness of the Al_2O_3 blocking layer thickness due to the better infiltration capabilities of ALD and thickness control. The blocking layer capability of a single cycle of Al_2O_3 was confirmed in a following study using different dyes and electrolytes in a DSSC. In all of the cells used with different electrolytes, the 1 cycle of Al_2O_3 increased the electron lifetimes by ~sixfold compared to the bare TiO_2 electrode [55]. The subnanometer Al_2O_3 blocking layer was also deposited on SnO_2 nanoparticle photoanodes with the same results [31]. Further studies on the subnanometer Al_2O_3 blocking layers continued. In their work, Antila et al. have reported that 1 cycle of Al_2O_3 only increased the V_{oc} values but had no substantial change in the PCE [56]. However, they used different deposition temperatures, heat treatments, and type of dye that makes a straightforward comparison difficult. On another study, Ganapathy et al. have reported that Al_2O_3 thicknesses of 1 and 2 nm yielded a better performance compared to an uncoated cell, and the 2-nm-thick Al_2O_3 blocking layer showed a 35% improvement [35]. The effect of the Al_2O_3 blocking layer has also been studied on a TiO_2 nanotube nanostructure [36]: 1 cycle (0.12 nm) of Al_2O_3 was reported as the optimal thickness for the blocking layer, the PCE increased by 23% (from 4.65% to 5.75%). They also reported that a $TiCl_4$ treatment prior to ALD of Al_2O_3 increased the PCE even further to 8.62% compared to the 6.77% with $TiCl_4$ treatment but no Al_2O_3.

The differences of PCE values obtained in the literature so far using a specific Al_2O_3 blocking layer thickness may be attributed to the different ALD processes applied, and/or types of dyes used in these studies, apart from the ALD temperature and oxidizing agent (water, ozone, plasma) that will determine the hydrogen content in the Al_2O_3 film [57], and the precursor pulse (exposure) time and amount must be adapted to the surface and geometry of porous structures. If this was not taken into account, it could have resulted in insufficient infiltration of the gas precursors. High-aspect-ratio structures or, in this case, tortuous systems would require longer exposure times and doses compared to flat substrates to conformally coat the entire structure [58, 59]. This problem was addressed recently by modifying the TMA exposure times to coat SnO_2 nanoparticle photoanodes [60]. In this study, Dong et al. have demonstrated that longer exposure times of both TMA and H_2O had a positive effect on the dye absorption and blocking properties, which increased the PV efficiency by 125% with only 2 ALD cycles. This was attributed to a better Al_2O_3 infiltration and coverage of the nanoparticle network as a result of the long exposure times.

For sure, the completeness of monolayer film coverage with only 1 or 2 ALD cycles still remains to be elucidated in terms of nucleation (or preferred chemisorption sites) for any material system, that is, the blocking layer material and the surface material on which it is deposited. That being said, the true mechanisms or exact contributions to the PV performance from the ultrathin Al_2O_3 blocking layers are yet to be established. An ultrathin Al_2O_3 layer can passivate the defect surface sites of the metal oxide, increase the dye absorption due to the high EIP, and block the back transfer of electrons from both FTO and metal oxide nanostructure to the electrolyte, and none of these phenomena are exclusive to each other. In this regard, Pascoe et al. have measured the electron transport rates in a DSSC with a TiO_2 nanoparticle photoanode coated with 1–3 cycles of Al_2O_3 and concluded that the role of the alumina blocking layer is purely as a tunneling barrier between conductive electrons and the redox mediator [61]. Notwithstanding, these studies have demonstrated the capabilities of subnanometer control of ALD to deposit ultrathin films and their positive effect as barrier recombination layers in DSSC. With regard to this matter, ALD may be the only technique that possesses this subnanometer control.

Further subnanometer metal oxides that have been studied include the following: Li et al. have coated TiO_2 nanoparticle photoanodes with 2 cycles of ZrO_2 and reported an increase of 4 times the PCE. They have observed that for thicker coatings, the photocurrent decreased due to a reduced charge injection rate from the dye to the TiO_2 [37]. Similarly, the deposition of 4 cycles of ALD Ga_2O_3 demonstrated to increase the J_{sc} and V_{oc} after only 1 cycle (~1 Å), and the maximum V_{oc} (1.11 V) was obtained after 6 cycles, combined with an increase in the fill factor. Nevertheless, as the thickness of the Ga_2O_3 increased, the J_{sc} decreased [38]. Their most efficient cell increased 285% in efficiency with only 4 cycles of ALD Ga_2O_3 (1.4% to 4%). Furthermore, the use of a 5-nm-thick TiO_2 compact layer at the FTO/TiO_2 NP interface raised the efficiency even further to 4.6% using only 2 ALD cycles of Ga_2O_3. This result showed the role of the metal oxide deposited in the nanostructured photoanode, which can act as both compact and blocking layer.

These studies have shown that different metal oxides will perform differently in relation to their film thickness. Recently, Grätzel's group has systematically studied the influence of the thickness of different metal oxides as blocking layers. In one study, they compared four different insulating oxides (Ga_2O_3, ZrO_2, Nb_2O_5, and Ta_2O_5) deposited by ALD. They analyzed in detail the recombination and transport rates as well as the contribution to the V_{oc}, J_{sc}, FF, and PCE as a function of the number of cycles. They have concluded that the position of the conduction band and the oxidation state of the blocking layer metal oxide are of primary importance [39]. The same group has also studied different thicknesses of amorphous TiO_2 ultrathin layers as both compact layer and blocking layer on crystalline TiO_2 nanoparticle photoanodes for perovskite solar cells. The deposition of ultrathin TiO_2 layers tuned the particle size and blocked the back transfer of electrons from both the FTO and the metal oxide nanostructure back to the perovskite absorber. They reported that after 2–3-nm-thick TiO_2 layer, the dye loading decreased without compromising the J_{sc}, but substantially increased V_{oc} and FF. They found that 2–3 nm was the optimal thickness of amorphous TiO_2 to serve as both compact layer and blocking layer; the PCE increased from 7.2% to 11.5%. Thicker films closed the pore network, which resulted in poor filling of the perovskite and HTM [40]. Therefore, for the particular case of using TiO_2 as blocking layer in nanoparticle architectures, there is a limitation to the permissible thickness before closing the pore network. As an alternative, the use of TiO_2 nanorods allowed a better infiltration of the active layers while using >3-nm-thick TiO_2 blocking layers. Mali *et al.* reported that a 4.8-nm-thick amorphous TiO_2 blocking layer substantially increased the PCE from 5.03% to 13.45% in a perovskite-based solar cell [41]. In this case, the nanorods, as seen in Figure 4.3a, possess an inherent TiO_2 compact layer at the FTO interface as shown schematically in Figure 4.3b, and, therefore, the deposited ALD film contributed only to the passivation of the nanorod surface and blocking the backflow of electrons. There is more spacing between the nanorods for efficient

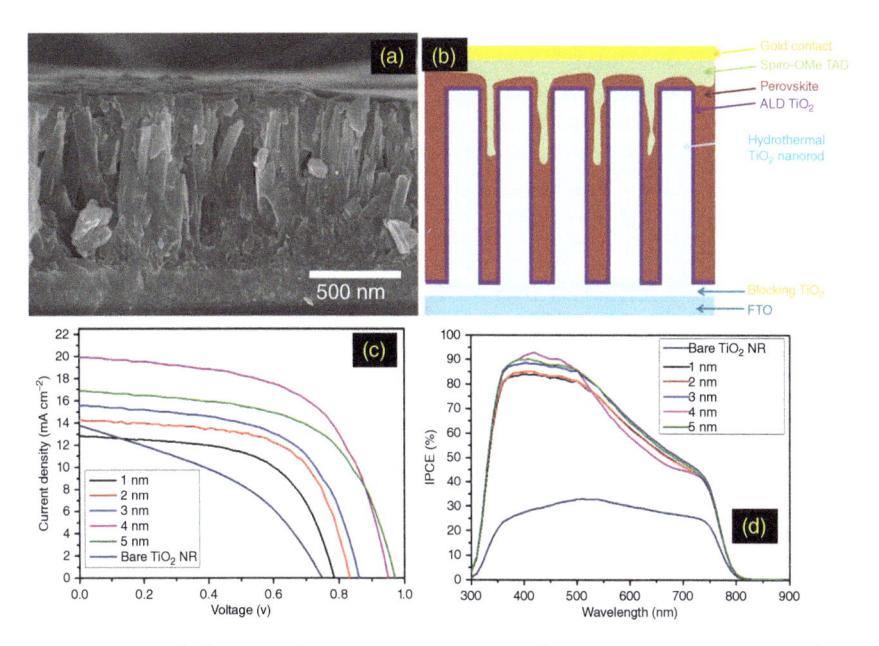

Figure 4.3 Morphology and photovoltaic properties. (a) Cross-sectional FESEM image of the ALD TiO_2 at TiO_2 nanorods. (b) Schematic illustration of the device in (a) representing the pore-filling mechanism. (c) J–V measurements of the devices fabricated with different ALD TiO_2 thicknesses on TiO_2 nanorods measured under 100 mW cm^{-2} of illumination. (d) Respective IPCE spectra recorded in the 300–900 nm wavelength range. (Mali *et al.* 2015 [41]. Reproduced with permission of American Chemical Society.)

infiltration of the perovskite absorber and the hole-transporting material, in this case Spiro-OMeTAD, even after increasing the thickness of the nanorods by ~8 nm with ALD TiO_2 blocking layers. Not surprisingly, the reported optimal thickness of the blocking layer in the work by Mali *et al.* corresponds to ~4-nm-thick compact layer of amorphous TiO_2 (Figure 4.3c,d), the same thickness for blocking the back transfer of electrons from the FTO to the HTM, since this thickness already displayed less than 1% of pinholes in the film as was found by Kavan *et al.* [49].

A similar PCE of 13.4% was obtained with a perovskite absorber using ZnO nanorods as the scaffold coated with 5 cycles of ALD TiO_2 [42].

The deposition of blocking layers by ALD can drastically improve the electrical performance of nanostructured solar cells; especially in solid-state solar cells (e.g., perovskite), the increments are outstanding. The chosen material can in principle passivate the surface defect sites and trap states of the parent metal oxide, increase the dye loading, band offset for improved electron injection, and block the back transfer of electrons. The deposition of a blocking layer by ALD can also simultaneously deposit a compact layer, which will only grow on the naked FTO surfaces without increasing the contact resistance between the metal oxide nanostructure and the FTO as shown in Figure 4.1b.

4.2.3 Surface Passivation and Absorber Stabilization: The Absorber/HTM Interface

One of the main challenges for nanostructured solar cells is the thermal and chemical instability over time. Inorganic absorbers such as ruthenium-based dyes and quantum dots are prone to oxidation, thermal degradation, and detachment from the metal oxide nanostructure, leading to long-term stability problems of the device. The mechanisms of the instability of Ru-dye-based DSSC have been studied in detail to shed some light on how to improve their stability [62–64]: the dye detachment from the TiO_2 nanostructure and the degradation by the electrolyte were the main cause of instability. Adding a controlled amount of dye to the electrolyte solution proved to be a solution to suppress dye detachment for 16 days [65]. Another stabilization alternative is using ALD. ALD can be used to deposit ultrathin films to cover the dye/MO_x structure and completely isolate the photoanode surface from the electrolyte and stabilize the dye to the metal oxide nanostructure. Son *et al.* have reported that ~1 nm of SiO_2 ALD was enough to provide a blocking layer, which passivated the naked TiO_2 surfaces but did not insulate the TiO_2/dye interface [66]. Soon after, Hanson *et al.* found out that the deposition of ALD Al_2O_3 after dye loading not only passivated the naked TiO_2 surfaces but also stabilized the Ru dye to the TiO_2 nanostructure. They observed a decrease in the detachment rate of the dye and a decrease in the electron back transfer by increasing the number of cycles, however, at the expense of the electron injection [67]. Simultaneously, Son *et al.* [68] and Hanson *et al.* [69] have demonstrated the stabilization of absorbing dyes to the TiO_2 nanoparticle structure by the deposition of subnanometer films of ALD TiO_2 on top of the dye/nanostructure. In their research, Son *et al.* reported the stabilization of OrgD-dye molecules with 8 cycles of ALD TiO_2 (using TTIP), which slowed down

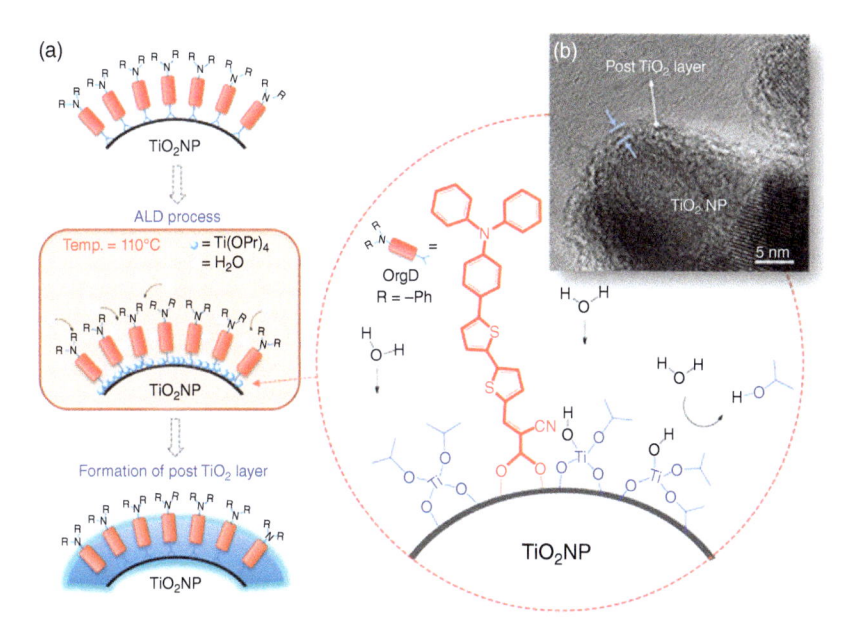

Figure 4.4 (a) Schematic illustration of ALD TiO$_2$ layers postassembly and (b) HR-TEM image of the posttreated TiO$_2$:OrgD with 10 cycles of ALD TiO$_2$. (Son *et al.* 2013 [68]. Reproduced with permission of American Chemical Society.)

the detachment of dye molecules by 50 times. Figure 4.4 shows the schematic diagram of the OrgD-dye stabilization with TiO$_2$. They also showed how the TiO$_2$ post dye loading enhanced the hydrophilicity of the photoanode for its more efficient wettability with aqueous electrolytes [68]. Hanson *et al.* reported a similar behavior using Ru-based dye and stabilizing the dye with up to ~1-nm-thick TiO$_2$ (from TiCl$_4$) or Al$_2$O$_3$ [69]. Later, the same group investigated the combination of TiO$_2$ and Al$_2$O$_3$ layers to stabilize the Ru dye while maintaining efficient electron injections [70].

Over the past years, there has been a transition from Ru-based dye absorbers and liquid electrolytes to perovskite absorbers (e.g., CH$_3$NH$_3$PbI$_3$) and solid hole-transporting materials (e.g., Spiro-OMeTAD), and these materials have demonstrated an impressive increase in conversion efficiencies in a short period of time of about 5 years [71, 72].

This new absorber material with electron and hole diffusion lengths of ~1 µm has generated a debate about which architecture – planar or nanostructured – possesses the best performance. However, the long-term stability of perovskite solar cells is today's big concern and intensively investigated, for instance, the CH$_3$NH$_3$PbI$_3$ films easily degrade in the presence of moisture. To this extent, ALD can be very helpful for uniform encapsulation to protect against oxygen and moisture.

In their study, Dong *et al.* have deposited ALD Al$_2$O$_3$, using TMA and O$_3$ (to avoid the use of water), on top of the perovskite absorber (CH$_3$NH$_3$PbI$_3$) or after the addition of the Spiro-OMeTAD to study the PCE over a period of time. They found out that already with 3 cycles of ALD Al$_2$O$_3$ deposited on top of the

Spiro-OMeTAD layer, the solar cells presented a considerably improved stability compared to those solar cells without a coating. They achieved a PCE of 12.9%, which was ~90% of the initial value after storage in air for 24 days. [73] The Al_2O_3 thin coating protected the cells against air and moisture permeability. However, if the film is too thick, quantum tunneling of the holes would be reduced and therefore affect the overall performance. For this reason, ALD with its immanent thickness control at atomistic level is a promising route, if not the only, to improve the stability of these solar cells.

Another method to improve the long-term stability employing ALD has been devised by Change et al., taken from known encapsulation methods for other types of solar cells with ALD Al_2O_3. They showed that by depositing a 50-nm-thick Al_2O_3 encapsulation coating, that is, after the device is completed, the long-term stability of the device was significantly improved. The initial perovskite cell with a PCE of 10.55% showed negligible degradation after being exposed to ambience (30 °C, 65% relative humidity) for more than 40 days [74]. They demonstrated the gas barrier performance of a 50-nm-thick ALD Al_2O_3 with an oxygen transmission rate (OTR) of 1.9×10^{-3} $cm^3 m^{-2} day^{-1}$ and a water vapor transmission rate (WVTR) of 9.0×10^{-4} $g m^{-2} day^{-1}$. Additionally, they used a 40-nm-thick ALD ZnO film deposited on top of the perovskite, which yielded a PCE of 16.5%.

4.2.4 Atomic Layer Deposition on Quantum Dots

The deposition of ALD ultrathin films on quantum dot (QD) nanostructures has been studied for the last 5 years. QD films have become a promising material for a wide range of optoelectronic devices due to their tunable optical (absorption) properties and inexpensive wet chemical synthesis. Nanocrystals of many different materials and sizes ranging from 2 to 10 nm can be assembled into a film. Their large Bohr exciton radius allows tuning the band gap of the film by increasing or decreasing the nanoparticle size. Additionally, they exhibit multiple exciton generation, or carrier multiplication, which means that they can generate more than one electron–hole pair per incident photon. These features make them ideal for light-emitting diodes and photovoltaics. However, QDs are metastable nanocrystals susceptible to oxidation, photothermal degradation, ripening, and sintering caused by the large surface-to-volume ratio, high surface energy, and meager protection from the organic ligands surrounding the QDs [75].

Pourret et al. have demonstrated that the open structure of QD films allows the diffusion of the gas precursors to fill the pore network. They drop-casted CdSe nanocrystal films onto a quartz sensor for quartz crystal microbalance measurements and reported a 13-fold larger mass uptake than for a flat substrate. The infiltration of ALD ZnO on closed-packed CdSe QD films also showed a substantial increase in the film conductance while retaining the same initial QD crystal size [76]. Ihly et al. have demonstrated that PbS QD films that were infiltrated and overcoated with ALD Al_2O_3 remained stable upon exposure to air for at least 30 days, by monitoring the absorption spectra of the films. Furthermore, the infiltrated alumina provided a nanoscale matrix to inhibit solid-state diffusion, or sintering, between the QD nanocrystals upon exposure at 80 °C [77].

The same group showed that filling and overcoating PbS QD films with ALD Al_2O_3 maintained an efficiency of ~2.3% for a period of 30 days upon exposure to air, compared to an efficiency of <0.5% of an uncoated sample just after a few hours of being exposed to air [78]. They also demonstrated a PbSe QD FET transistor with air-stable electron mobilities above 7 cm^2 V^{-1} s^{-1} [79]. They suggest that trap passivation by ALD is responsible for such high mobilities, by ruling out the possibility of tunnel barrier lowering between QDs that originates from the interparticle distance. Cate et al. later showed that infiltrating QD films with Al_2O_3 or Al_2O_3/ZnO activated carrier multiplication more efficiently compared to uncoated films [80].

Lambert et al. have coated CdSe QDs with a monolayer of Al_2O_3 using thermal and plasma-enhanced ALD. They observed via TEM images and absorption spectra that thermal ALD retained the crystal size of the original QDs, whereas the PE-enhanced ALD did not [81].

Sargent's group has also studied the effect of ALD layers on QD films. Ip et al. have shown that the encapsulation of a completed PbS colloidal QD solar cell (CQDSC) with a 70-nm-thick nanolaminate of Al_2O_3/ZrO_2 deposited via ALD at 80 °C kept the same efficiency of ~6% after being tested in air for a period of 13 days. The 70-nm-thick nanolaminate was able to isolate the device from the atmospheric conditions without compromising the efficiency [82]. The same group also studied different metal oxides as barrier recombination layers between the TiO_2 and PbS QDs to reduce the interfacial recombination. They observed that ZnO deposited by ALD had the best performance by improving the open-circuit voltage and the photocurrent collection. They reported almost one-half reduction in carrier recombination using ALD ZnO buffer layers [83].

Lin et al. fabricated a QD and dye-cosensitized solar cell with ALD TiO_2 barrier recombination layers. First, they deposited a compact layer of TiO_2 between the FTO and TiO_2 mesoporous structure, where the CdS QDs were deposited and coated further with ALD TiO_2 up to 150 cycles; after this blocking layer, they deposited the Ru absorber molecules (N719) [84]. They report a 41.3% of improvement in the PCE compared to those samples without an interfacial layer.

Roelofs et al. systematically studied the role of ALD Al_2O_3 films as blocking layers on solid-state QD-sensitized solar cells (QDSSC). They have compared up to 5 cycles of Al_2O_3 before (TiO_2/Al_2O_3/CdS QD) or after the addition of the CdS QDs (TiO_2/CdS QD/Al_2O_3) as shown schematically in Figure 4.5 and the combination of both (TiO_2/Al_2O_3/QD/Al_2O_3) [85].

Similar to previous reports, they observed that only 1 cycle of Al_2O_3 gave the highest efficiency in any of the aforementioned configurations. The electron lifetimes increased with increasing the Al_2O_3 thickness, which supports the conclusion that Al_2O_3 is undoubtedly a recombination barrier for injected electrons into the TiO_2. The TiO_2/Al_2O_3/QD configuration presented improved electron lifetimes compared to the other configurations because it blocked both the electron recombination to the oxidized CdS QDs and the Spiro-OMeTAD.

Besides the use of ALD on QD films to protect them against oxidation and photothermal degradation, and to enhance the conductivity, ALD has been used to deposit the QD nanocrystals. Such is the case of PbS QDs using

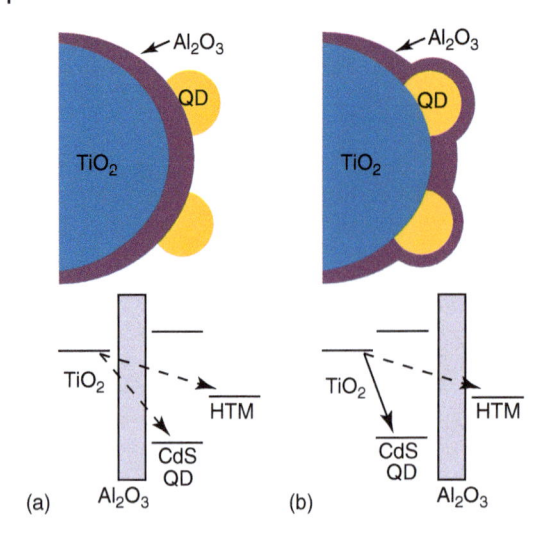

Figure 4.5 Schematic illustration of the different recombination barrier layer configurations in the QDSSC: (a) $TiO_2/Al_2O_3/QD$ and (b) $TiO_2/QD/Al_2O_3$, from the deposition of the ALD Al_2O_3 layer before and after the addition of CdS QDs. Spiro-OMeTAD is the hole-transporting material. Arrows indicate the undesirable recombination pathways. Pathways that may be blocked by the Al_2O_3 barrier layer are shown by dashed arrows. (Roelofs et al. 2013 [85]. Reproduced with permission of American Chemical Society.)

bis(2,2,6,6-tetramethyl-3,5-heptanedionato)lead(II) (Pb(tmhd)$_2$) and a gas mixture of 3.5% H_2S in N_2 [86]. Another example is the deposition of CdS using dimethyl cadmium (DMCd) and *in situ* generated hydrogen sulfide (H_2S), which had been produced by decomposition of thioacetamide (C_2H_5NS) [87, 88].

4.2.5 ALD on Large-Surface-Area Current Collectors: Compact Blocking Layers

So far in this chapter, we have discussed nanostructured solar cells that employ a flat TCO with different metal oxide nanostructures as the scaffold that increases the surface area. However, as mentioned previously, the classic architecture of DSSCs is limited by the tortuous photoelectron path through the nanoparticle network toward the current collector, which increases the recombination rates throughout the numerous interfaces. One of the strategies proposed to minimize the recombination processes of the nanoparticle structure was the use of nanostructured supports with less grain boundaries: TiO$_2$ and ZnO nanotube/nanowires were shown to have several advantages compared to the nanoparticle structure, namely they have a larger pore size for efficient infiltration of the absorbing and hole-transporting materials and a substantially lower amount of grain boundaries. However, they do not possess the high conductivities and transparency as the most common current collectors, FTO and ITO.

To make use of the high electrical properties of FTO or ITO as TCOs, they were used as the nano-scaffold themselves to increase the contact surface area for the absorbing film and thus the current collection area, which reduces the transport distances compared to the nanoparticle film as it is depicted in Figure 4.6.

However, increasing the surface area of the current collector could substantially increase the electron recombination from the TCO to the active materials (e.g., process 4 in Figure 4.1a). Therefore, analogously to a planar TCO, the introduction of a compact blocking layer at the entire TCO nanostructure interface can mitigate the backflow of electrons toward the active materials.

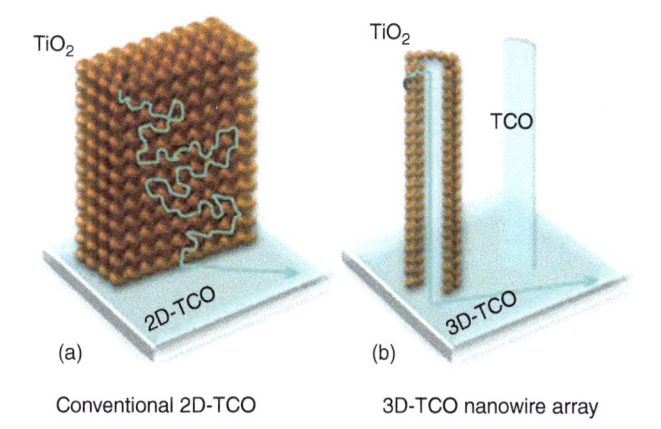

Figure 4.6 Schematic illustrations of a DSSC photoanode consisting of (a) a conventional planar TCO and a TiO_2 nanoparticle film. (b) The proposed 3-D TCO-TiO_2 core–shell nanowire array. (Noh et al. 2011 [89]. Reproduced with permission of John Wiley and Sons.)

This was idealized by Chappel et al. in 2005 [90], but the first experiment using a nanostructured TCO was demonstrated by Martinson et al. some years later [91]. In their report, they used anodic aluminum oxide (AAO) as a template and deposited indium tin oxide (ITO) within and upon the ordered nanoporous array of AAO. They used cyclopentadienyl indium (InCp) and O_3 at a temperature of 220 °C and doped with 1 cycle of tetrakisdimethylamido tin (TDMASn) and H_2O_2 every 10th cycle. They subsequently deposited amorphous TiO_2 as a compact blocking layer within the resulting hollow ITO cylinder array, and on top, they deposited a Ru dye by immersing the films in a dye solution. In this way, the electrons would ideally flow radially through a few nanometers of TiO_2 to be then collected by the underlying ITO NW cylinders. These cells yielded a modest 1.1% in PCE but opened a new possible architecture in nanostructured photovoltaics.

Noh et al. later refined this architecture and used ITO NW arrays, which were deposited directly on an ITO substrate using a vapor transport method, and subsequently coated them with an ALD TiO_2 film and then infiltrated a N719 Ru dye [89]. They pointed out that this configuration yields a considerably reduced surface area compared to the conventional nanoparticle array and, therefore, used a $TiCl_4$ treatment to induce the formation of TiO_2 nanocrystals with a particle size of 5–7 nm. They compared a 25-μm-thick nanowire ITO array coated with ALD TiO_2 and TiO_2 nanoparticles against a 14-μm-thick TiO_2 nanoparticle configuration on a planar ITO. Even though the planar TCO nanoparticle cell displayed almost twice as much dye loading as the NW-NP array, both presented comparable efficiencies, 4.3% and 3.8%, respectively. In addition, the NW-NP array showed 4–10 times faster charge collection in DSSCs compared to the nanoparticle cells. They have also transported this architecture to a stainless steel mesh for flexible photoelectrochemical cells [92]. The same group has achieved a PCE of 6.1% using an array of ITO nanowires of 30 μm in length and a so-called TiO_2 double-shell layer. This double layer is composed of a compact layer of a

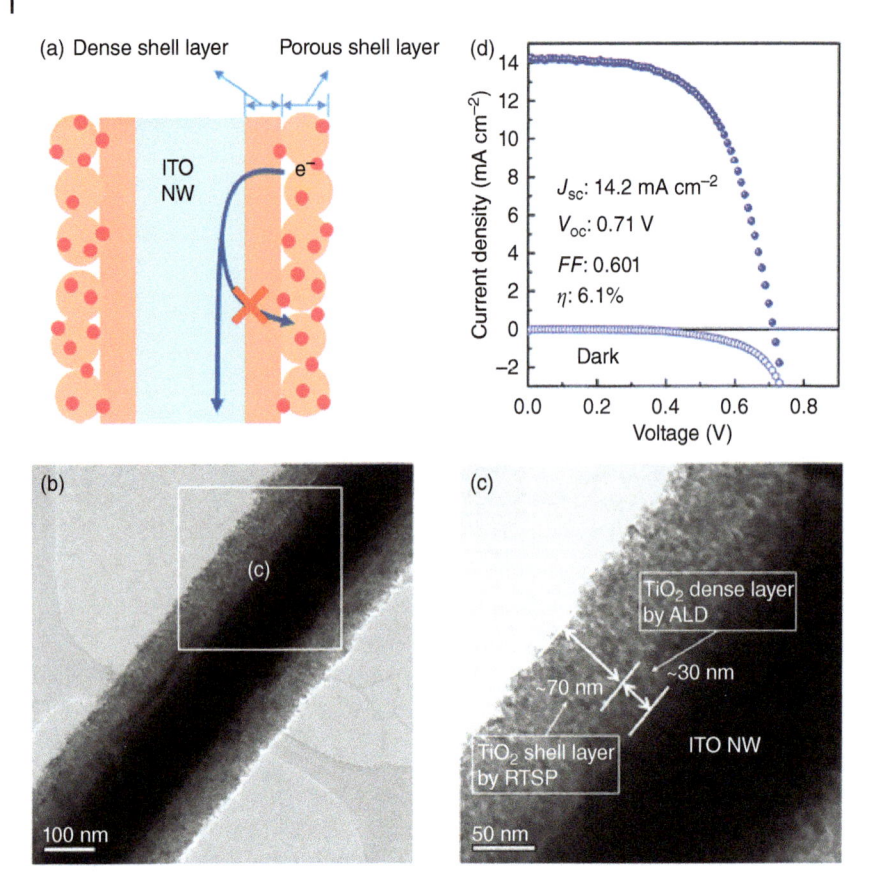

Figure 4.7 (a) Schematic of the TiO_2 double-shell layer. (b and c) TEM images of the TiO_2 double-shell layer on ITO nanowire. (d) J–V measurements of a device with the TiO_2 double-shell layer. (Han et al. 2013 [43]. Reproduced with permission of Royal Society of Chemistry.)

30-nm-thick ALD TiO_2 (TTIP and H_2O) followed by a nanoparticle film with a thickness of 70 nm with nanocrystals of ~2 nm (Figure 4.7) in diameter [43].

In a recent publication, the same group has replaced the Ru dye with a perovskite absorber to obtain a 7.5% in conversion efficiency, using the aforementioned double-shell structure of ITO nanowires coated with compact TiO_2 and TiO_2 nanoparticles [44]. In this study, they report a 40% higher photocurrent density for the perovskite solar cell than using only a nanoparticle network of TiO_2.

In another recent study, Li et al. have investigated HfO_2 and TiO_2 deposited via ALD as blocking layers on ITO NW photoanodes for DSSCs, on which they have also deposited a 70-nm-thick porous film of TiO_2 [45]. They reported that an ALD TiO_2 compact layer of only 1.3 nm performed the best against an ALD HfO_2 film of the same thickness, reaching 4.83% of PCE.

Other alternative materials that can offer a high surface area, excellent electrical and thermal conductivity, and chemical and mechanical stability are functional carbon materials such as carbon nanotubes (CNT), graphene, and carbon foam. These materials in combination with metal oxides have been increasingly used in novel nanoarchitectures. Recently, carbon nanotubes have been employed to improve the transport of the injected photoelectrons from the metal oxide material to the current collector [93–100]. These studies have found that the combination of a certain amount of CNTs, namely ~0.2 to 0.3 wt%, mixed in the TiO_2 nanoparticle structure increased the PCE of DSSCs, often improving by ~54% and reaching absolute values of higher than 8% in PCE [99, 100]. More than ~0.2 to 0.3 wt% addition of CNTs to the nanoparticle network, however, decreased the overall conversion efficiency. This is mainly because of the short circuit and high recombination rates caused by the uncovered CNT surfaces to the electrolyte; the poor CNT surface coverage by the metal oxide is a result of the sol–gel deposition technique used for coating the CNTs. The uncoated CNT surfaces act as leakage channels for the photoelectrons traveling through the CNTs toward the HTM, leading to a decrease in the efficiency [101]. The chemical inertness of the CNTs provides a challenge itself to ensure a complete surface coverage. Therefore, chemical functionalization, by acid or plasma treatments, is a frequently used strategy to introduce functional groups or defects at the surface to improve the deposition of metal oxides onto the CNTs. Several publications that address the coating of carbon nanotubes via ALD with different metal oxides for different applications can be found in the following papers [102–108]. In this chapter, we discuss those that are directly related to solar cells.

Jin et al. have used CNTs coated with TiO_x to fabricate organic solar cells, as shown schematically in Figure 4.8 [109]. They functionalized the CNTs and then transferred them to an FTO substrate by a stamping process with poly(dimethylsiloxane) (PDMS) and coated them with TiO_x via ALD or with a sol–gel procedure, onto which they spin-coated the active layer and then evaporated the Au contact. They compared the difference in PCE between the cells that used functionalized CNTs coated with 10-nm-thick TiO_2 deposited by ALD or sol–gel. Not surprisingly, the CNTs coated with ALD displayed a better surface coverage and conformality compared to those coated with sol–gel, which presented portions of uncoated CNTs. After illumination under 1 sun, the cells with ALD TiO_2-coated CNTs exhibited the highest PCE of 2.54% compared to 0.90% for the sol–gel TiO_2-coated CNTs. The CNTs coated with TiO_x from a sol–gel solution manifested ohmic behaviors similar to those CNT networks without a TiO_x coating; this is due to the inhomogeneous coverage of TiO_x.

Yazdani et al. have demonstrated that the inclusion of a fast transport intermediary in the photoanode such as CNTs can enhance the effective electron diffusion lengths and reduce charge recombination [110]. In this study, they transferred multiwalled CNT arrays to FTO substrates via a PMMA-assisted stamping procedure to coat them with ALD TiO_2, without any functionalization treatment of the CNTs. They were able to show that the use of CNTs significantly suppresses charge recombination, which leads to diffusion lengths two orders of magnitude greater than that in standard mesoporous TiO_2 photoanodes. However, it was mentioned that the dense multiwall CNT array exhibited 99.6–99.9% absorption for wavelengths between 200 and 1200 nm, which was a factor limiting the performance of the photoanode in the completed device. Therefore, there is a threshold on the extent to which a certain amount of CNTs is more beneficial than detrimental, and this still needs to be elucidated in detail using CNTs coated with ALD.

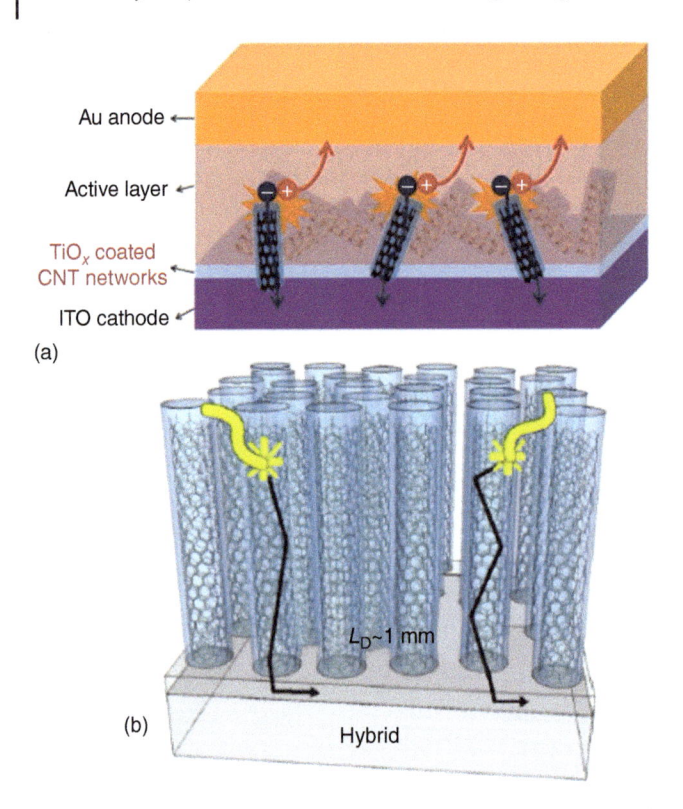

Figure 4.8 Schematic illustration of conformally coated CNT networks with ALD TiO_2 as electron transport layer. (a) In the inverted organic solar cells, the active materials are PCBM as the light absorber and P3HT as the HTM. (Jin *et al.* 2012 [109]. Reproduced with permission of Elsevier.) In (b), the CNTs are transferred to an FTO substrate and then conformally coated with ALD TiO_2 to use as a stratified photoanode in DSSC. (Yazdani *et al.* 2014 [110]. Reproduced with permission of American Chemical Society.)

We recall that the main purpose of replacing the mesoporous TiO_2 with CNTs or mix them is to reduce the numerous grain boundaries and, thus, reduce the recombination rates and improve the electron transport toward the current collector. This is similar to the concept of using TiO_2, ZnO, or ITO and FTO nanowires/nanotubes, however, with a more conductive material. ALD layers onto the CNTs serve an analogous function as compact layers deposited onto the TCO, specifically to prevent leakage currents. Therefore, the complete surface coverage of the parent material is crucial.

4.3 ALD for Photoelectrochemical Devices for Water Splitting

The photoelectrochemical (PEC) splitting of water into H_2 and O_2 is a sustainable alternative to convert sunlight into energy in the form of storable fuel [21]. Briefly, the water molecules in contact with the semiconductor photoelectrodes will be reduced to H_2 by the photogenerated electrons and oxidized to O_2 by the holes (see Chapter 8 for more details). Similar to nanostructured solar cells, one strategy to improve the redox reaction is to use a highly conductive nanostructured supports coated with a thin film of the semiconductor material. In this way, the semiconductor thickness is limited to its carrier diffusion length while the large surface area of the nanostructured support improves the light absorption and the surface contact with water to generate a larger volume of O_2 and H_2. Similarly, due to the large surface area, these nanostructures are also susceptible to surface and interface recombinations, in addition to photocorrosion of the semiconductor in contact with water, such as Si, CuO, and CdS. In this case, surface and interface engineering with blocking layers can reduce these recombination processes as well as protect the active material from corrosion. In this section, we present some examples where ALD has been used to deposit blocking/protective layers in water-splitting devices. Chapter 8 contains a detailed review of ALD films used in a wide range of photoelectrochemical cells.

In the case of semiconductors prone to photocorrosion, ultrathin ALD layers can protect them against degradation, while passivating surface defects and reducing the back reaction of electrons to the aqueous media.

Hwang et al. have used Si nanowires of different lengths (5–20 µm) and deposited ~25 to 40 nm of ALD TiO_2 films from $TiCl_4$ and H_2O to protect the Si from corrosion and determined their potential for the photo oxidation of water [111]. They demonstrated that the Si/TiO_2 core/shell structure with a length of 20 µm improves the photocurrent 2.5 times more than the planar configuration with the same ALD TiO_2 film thickness (Figure 4.9). This result is attributed to a lower reflectance of the light and larger surface area of the nanowire structure, which exemplifies the advantages of nanostructuring. Dasgupta et al. have also used Si nanowires as the conductive nanostructure to deposit ~10- to 12-nm-thick TiO_2 via ALD as a protective layer and then deposited ALD Pt nanoparticles with diameters ranging from 0.5 to 3 nm. They found out that the thin TiO_2 interlayer was required to achieve positive photovoltages with the ALD Pt nanoparticles, with as little as 1 ALD cycle [112].

In contrast to this, Chen et al. have used a planar Si anode and coated it with ALD TiO_2 films of different thicknesses using TDMAT and H_2O as precursors and then deposited a 2-nm-thick film of iridium by PVD, a well-known metal catalyst to promote water oxidation [113]. They reported that a 2- to 3-nm-thick TiO_2 layer protected the underlying Si substrate from corrosion, while still being thin enough to enable efficient transport of electrons and holes between the electrolyte and the Si anode. The thicker TiO_2 layers, namely 10 nm, resulted in increased peak-to-peak splitting (610 mV), which bolsters the use of ultrathin TiO_2 layers for efficient transport of electrons. They have also investigated the endurance of the Si anodes with and without the ALD TiO_2 protective coating; the samples without the protective coating failed under illumination in both 1 M acid and 1 M base solutions within 30 min. The samples with the ALD TiO_2 layer remained stable during the electrochemical life test (8 h).

Liu et al. have used ZnO nanowires as the conductive nanostructure and coated them with only 1-nm-thick TiO_2 using TTIP and H_2O [114]. They demonstrated that a 1-nm-thick TiO_2 coating already resulted in a 25% higher photoelectrochemical activity for water splitting compared to the bare ZnO nanowires. They attribute this enhancement to the passivation of the ZnO surface states through

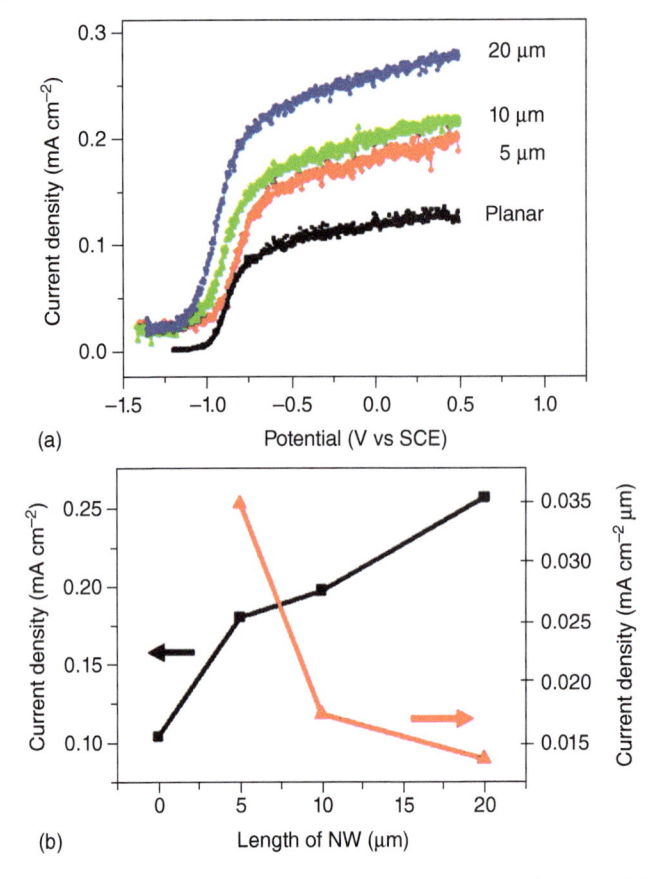

Figure 4.9 (a) Photocurrent density versus potential as a function of the length of n-Si NW/TiO$_2$ arrays: 20 μm (blue line), 10 μm (green line), 5 μm (red line) long NW arrays and n-Si planar/TiO$_2$ (black line). (b) Relationship between photocurrent density and the length of n-Si NW/TiO$_2$, illustrating that longer wire arrays have higher photocurrent. The axis to the right is the current density normalized by the length of the nanowire. (Hwang et al. 2009 [111]. Reproduced with permission of American Chemical Society.)

the partial removal of the deep hole traps, without affecting the minority carrier diffusion due to the ultrathin layer of TiO$_2$. Similarly, Lee et al. have used p-type InP nanopillars and deposited 3–5-nm-thick films of ALD TiO$_2$ and then sputtered a Ru catalyst to investigate the photocathode stability in acidic media for hydrogen production [115]. They have demonstrated the improved wettability and antireflection properties of the InP nanowires compared to the planar substrate. The nanopillars coated with a ∼5-nm-thick ALD TiO$_2$ protective layer remained stable during 4 hours with a constant photocurrent of 37 mA cm^{-2} at 0.23 V (vs NHE). The lowered surface energy resulted in a fast desorption of H$_2$ bubbles, an important characteristic for an efficient water-splitting device. Alternative to TiO$_2$ as a protective layer, Bi$_2$O$_3$ has also been deposited on Si nanowires for water splitting. Weng et al. have coated Si nanowires with Bi$_2$O$_3$ using Bi(thd)$_3$ and H$_2$O [116]. The Bi$_2$O$_3$ improved the chemical stability of the Si nanowires as well as the current density and onset potential, which altogether improved the PEC activity for water splitting. The photocurrent increased by increasing the Bi_2O_3 thickness from 10 to 30 nm due to the increase of surface area of the nanostructure. However, with the increase to 50-nm-thick Bi_2O_3, the current density decreased due to a lower light absorption and larger reflectance of the film at this thickness.

Strandwitz et al. have reported that a 10-nm-thick ALD MnO film from $(Et_2Cp)_2Mn$ and H_2O was sufficient to stabilize planar silicon photoelectrodes in contact with aqueous electrolytes. The MnO-coated n-Si photoanodes were stable for >20 scans at >500 mV and current densities of 28 mA cm^{-2} (vs Ag/AgCl at pH 13.6) [117].

Peng et al. have deposited a 2-μm-thick film of antimony-doped tin oxide nanoparticles (nanoATO) on an FTO substrate [118]. They coated this transparent nanoparticle electrode with different film thicknesses of ALD TiO_2. They reported a peak in the photocurrent density of 0.58 mA cm^{-2} (vs Ag/AgCl, pH = 14) with a ~9-nm-thick TiO_2 film. They also demonstrated that to achieve the highest photocurrent density using a planar FTO, ~50 nm of ALD TiO_2 was needed and only achieved 0.2 mA cm^{-2}. As in a DSSC, the photocurrent density of the nanostructured ATO/TiO_2 can be greatly improved by the larger surface area of nanostructured electrodes.

Similar to DSSCs, large-surface-area current collectors have also been tested for water splitting. Cordova et al. have deposited ALD TiO_2 films ranging from 5 to 20 nm onto an FTO nanoparticle (nanoFTO) scaffold fabricated by solution processing (Figure 4.10). In this way, the nanoFTO scaffold itself is the current collector material, which provides a larger surface area to increase the collection efficiency. In addition, the disorder of the nanoFTO can induce optical scattering

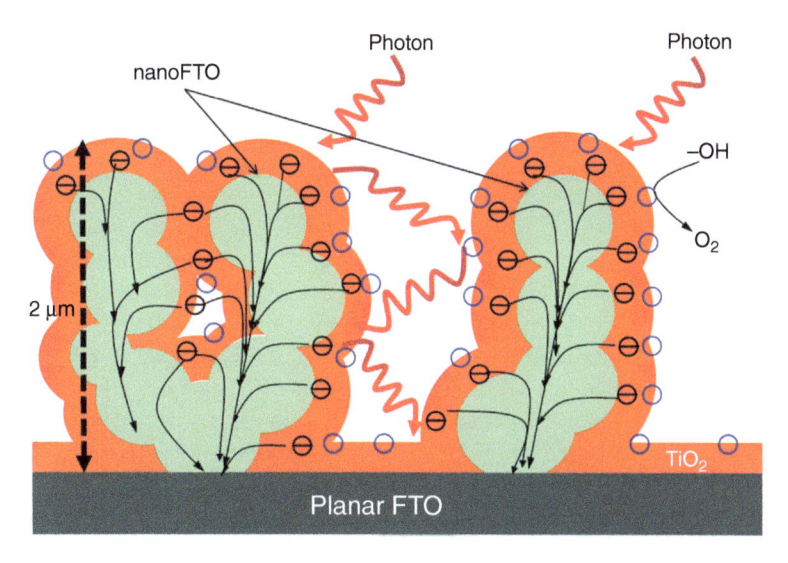

Figure 4.10 Schematic diagram illustrating the light-trapping and charge carrier separation mechanisms for the TiO_2/nanoFTO architecture. A thin film of ALD TiO_2 (orange) is coated over the nanoFTO network (green). On the left, a region is shown where the TiO_2 film causes some pores within the network to become closed, thus preventing an electrochemically active semiconductor–liquid junction from being within reach of photogenerated holes. (Cordova et al. 2015 [119]. Reproduced with permission of Royal Society of Chemistry.)

effects to enhance the light-harvesting efficiency [119]. They have deposited TiO_2 using $TiCl_4$ and water at a temperature of 300 °C and observed a preferential rutile TiO_2 growth, which they attributed to the underlying cassiterite lattice structure of the nanoFTO. They reported that the 10-nm-thick ALD TiO_2 film onto the nanoFTO yielded the highest photocurrent density of 0.7 mA cm^{-2} (vs Ag/AgCl), slightly higher than the previously reported values for nanoATO [118]; however, both reported a similar optimal TiO_2 thickness for water splitting.

In both of these examples, the deposited TiO_2 worked simultaneously as a blocking layer and light absorber, providing valuable information of the charge transport and blocking properties of TiO_2.

4.4 Prospects and Conclusions

ALD has established itself as a standard deposition technique in the solar cell sector, with an increasing number of research groups turning to ALD to further understand and develop more efficient PV devices. We have seen throughout this chapter the superior advantages of using ALD compared to other techniques to deposit a wide range of thin coatings in nanostructured solar cells: barrier recombination layers, active materials, or protective and encapsulation layers. The essential feature of ALD is the precise thickness control of pinhole-free and conformal films from a wide range of materials, as well as doping capabilities. In some ways, because of the use of ALD, third-generation solar cells and water-splitting devices have shown tremendous improvements over the past decade. The introduction of thin and pinhole-free barrier recombination (compact and blocking) layers has enabled the full potential of these devices and shed light on its limitations. The ability to precisely coat geometrically complex structures with high aspect ratios makes ALD ideal for depositing the absorber materials. Additionally, with increasing importance, the pinhole-free characteristic of ALD is desirable to protect the active materials against oxygen and moisture to stabilize the complete device.

The reports shown in this chapter also demonstrate the benefit of nanostructured solar cells compared to its homologous planar configuration. Nanostructuring of solar cells is one strategy to reduce the transport distances of semiconductive materials with short electron diffusion lengths. These nanostructured scaffolds may take several shapes and use diverse functional materials. Although this type of solar cells still need to increase the overall conversion efficiency and solve the long-term stability issues to compete with other solar cell technologies (e.g., Si-based, CdTe or CIGS), the production costs and ease of fabrication make them a promising candidate for future solar cells.

Especially, the perovskite solar cells have displayed remarkable advancements in the span of 3 years, as no other solar cell technology has. Regardless of the still debated optimal configuration (nanostructured or planar) for perovskite absorbers, ALD will remain a major player in the fabrication of these devices for both configurations. The search for new absorbers and HTMs via ALD may find its way to perovskite-based solar cells, as ALD has already been used to deposit

TCOs, compact and blocking layers, and light absorber materials. ALD will continue to be a key asset to the development of modern photovoltaic devices.

Undoubtedly, ALD has been playing an important role not only in increasing the device efficiency and stability but also in understanding the fundamentals of charge transport and recombination mechanisms. As the third-generation solar cells explore new materials and thinner films, the surface and interface engineering will be increasingly important to improve the overall performance and stability of solar cells.

References

1 Beard, M.C., Luther, J.M., and Nozik, A.J. (2014) *Nat. Nanotechnol.*, **9**, 951–954.
2 O'Regan, B. and Grätzel, M. (1991) *Nature*, **353**, 737–740.
3 Zhou, H., Chen, Q., Li, G., Luo, S., Song, T.-b., Duan, H.-S., Hong, Z., You, J., Liu, Y., and Yang, Y. (2014) *Science*, **345**, 542–546.
4 Saliba, M., Orlandi, S., Matsui, T., Aghazada, S., Cavazzini, M., Correa-Baena, J.-P., Gao, P., Scopelliti, R., Mosconi, E., Dahmen, K.-H., De Angelis, F., Abate, A., Hagfeldt, A., Pozzi, G., Graetzel, M., and Nazeeruddin, M.K. (2016) *Nat. Energy*, **1**, 15017.
5 Eder, D. (2010) *Chem. Rev.*, **110**, 1348–1385.
6 Knez, M., Nielsch, K., and Niinistö, L. (2007) *Adv. Mater.*, **19**, 3425–3438.
7 Kim, H., Lee, H.-B.-R., and Maeng, W.J. (2009) *Thin Solid Films*, **517**, 2563–2580.
8 Bakke, J.R., Pickrahn, K.L., Brennan, T.P., and Bent, S.F. (2011) *Nanoscale*, **3**, 3482–3508.
9 Palmstrom, A.F., Santra, P.K., and Bent, S.F. (2015) *Nanoscale*, **7**, 12266–12283.
10 Roelofs, K.E., Brennan, T.P., and Bent, S.F. (2014) *J. Phys. Chem. Lett.*, **5**, 348–360.
11 Liu, M., Li, X., Karuturi, S.K., Tok, A.I., and Fan, H.J. (2012) *Nanoscale*, **4**, 1522–1528.
12 Wang, T., Luo, Z., Li, C., and Gong, J. (2014) *Chem. Soc. Rev.*, **43**, 7469–7484.
13 Bae, C., Shin, H., and Nielsch, K. (2011) *MRS Bull.*, **36**, 887–897.
14 van Delft, J.A., Garcia-Alonso, D., and Kessels, W.M.M. (2012) *Semicond. Sci. Technol.*, **27**, 074002.
15 Niu, W., Li, X., Karuturi, S.K., Fam, D.W., Fan, H., Shrestha, S., Wong, L.H., and Tok, A.I. (2015) *Nanotechnology*, **26**, 064001.
16 Singh, T., Lehnen, T., Leuning, T., and Mathur, S. (2015) *J. Vac. Sci. Technol., A*, **33**, 010801.
17 Saxena, V. and Aswal, D.K. (2015) *Semicond. Sci. Technol.*, **30**, 064005.
18 Graetzel, M., Janssen, R.A., Mitzi, D.B., and Sargent, E.H. (2012) *Nature*, **488**, 304–312.
19 Grätzel, M. (2003) *J. Photochem. Photobiol., C*, **4**, 145–153.
20 Snaith, H.J. and Schmidt-Mende, L. (2007) *Adv. Mater.*, **19**, 3187–3200.

21 Park, H.G. and Holt, J.K. (2010) *Energy Environ. Sci.*, **3**, 1028.
22 Zhu, K., Schiff, E.A., Park, N.G., van de Lagemaat, J., and Frank, A.J. (2002) *Appl. Phys. Lett.*, **80**, 685.
23 Cameron, P.J., Peter, L.M., and Hore, S. (2005) *J. Phys. Chem. B*, **109**, 930–936.
24 Law, M., Greene, L.E., Radenovic, A., Kuykendall, T., Liphardt, J., and Yang, P. (2006) *J. Phys. Chem. B*, **110**, 22652–22663.
25 Jiang, C.Y., Koh, W.L., Leung, M.Y., Chiam, S.Y., Wu, J.S., and Zhang, J. (2012) *Appl. Phys. Lett.*, **100**, 113901.
26 Miettunen, K., Halme, J., Vahermaa, P., Saukkonen, T., Toivola, M., and Lund, P. (2009) *J. Electrochem. Soc.*, **156**, B876.
27 Bills, B., Shanmugam, M., and Baroughi, M.F. (2011) *Thin Solid Films*, **519**, 7803–7808.
28 Kim, D.H., Woodroof, M., Lee, K., and Parsons, G.N. (2013) *ChemSusChem*, **6**, 1014–1020.
29 Wu, Y., Yang, X., Chen, H., Zhang, K., Qin, C., Liu, J., Peng, W., Islam, A., Bi, E., Ye, F., Yin, M., Zhang, P., and Han, L. (2014) *Appl. Phys Express*, **7**, 052301.
30 Lin, Z., Jiang, C., Zhu, C., and Zhang, J. (2013) *ACS Appl. Mater. Interfaces*, **5**, 713–718.
31 Prasittichai, C. and Hupp, J.T. (2010) *J. Phys. Chem. Lett.*, **1**, 1611–1615.
32 Lu, H., Ma, Y., Gu, B., Tian, W., and Li, L. (2015) *J. Mater. Chem. A*, **3**, 16445–16452.
33 Di Giacomo, F., Zardetto, V., D'Epifanio, A., Pescetelli, S., Matteocci, F., Razza, S., Di Carlo, A., Licoccia, S., Kessels, W.M.M., Creatore, M., and Brown, T.M. (2015) *Adv. Energy Mater.*, **5**, 1401808.
34 Lin, C., Tsai, F.-Y., Lee, M.-H., Lee, C.-H., Tien, T.-C., Wang, L.-P., and Tsai, S.-Y. (2009) *J. Mater. Chem.*, **19**, 2999.
35 Ganapathy, V., Karunagaran, B., and Rhee, S.-W. (2010) *J. Power Sources*, **195**, 5138–5143.
36 Gao, X., Guan, D., Huo, J., Chen, J., and Yuan, C. (2013) *Nanoscale*, **5**, 10438–10446.
37 Li, T.C., Góes, M.S., Fabregat-Santiago, F., Bisquert, J., Bueno, P.R., Praslttichal, C., Hupp, J.T., and Marks, T.J. (2009) *J. Phys. Chem. C*, **113**, 18385–18390.
38 Chandiran, A.K., Tetreault, N., Humphry-Baker, R., Kessler, F., Baranoff, E., Yi, C., Nazeeruddin, M.K., and Gratzel, M. (2012) *Nano Lett.*, **12**, 3941–3947.
39 Chandiran, A.K., Nazeeruddin, M.K., and Grätzel, M. (2014) *Adv. Funct. Mater.*, **24**, 1615–1623.
40 Chandiran, A.K., Yella, A., Mayer, M.T., Gao, P., Nazeeruddin, M.K., and Gratzel, M. (2014) *Adv. Mater.*, **26**, 4309–4312.
41 Mali, S.S., Shim, C.S., Park, H.K., Heo, J., Patil, P.S., and Hong, C.K. (2015) *Chem. Mater.*, **27**, 1541–1551.
42 Dong, J., Xu, X., Shi, J.-J., Li, D.-M., Luo, Y.-H., Meng, Q.-B., and Chen, Q. (2015) *Chin. Phys. Lett.*, **32**, 078401.

43 Han, H.S., Kim, J.S., Kim, D.H., Han, G.S., Jung, H.S., Noh, J.H., and Hong, K.S. (2013) *Nanoscale*, **5**, 3520–3526.
44 Han, G.S., Lee, S., Noh, J.H., Chung, H.S., Park, J.H., Swain, B.S., Im, J.H., Park, N.G., and Jung, H.S. (2014) *Nanoscale*, **6**, 6127–6132.
45 Li, L., Xu, C., Zhao, Y., Chen, S., and Ziegler, K.J. (2015) *ACS Appl. Mater. Interfaces*, **7**, 12824–12831.
46 Hore, S. and Kern, R. (2005) *Appl. Phys. Lett.*, **87**, 263504.
47 Xia, J., Masaki, N., Jiang, K., and Yanagida, S. (2007) *J. Phys. Chem. C*, **111**, 8092–8097.
48 Hamann, T.W., Farha, O.K., and Hupp, J.T. (2008) *J. Phys. Chem. C*, **112**, 19756–19764.
49 Kavan, L., Tétreault, N., Moehl, T., and Grätzel, M. (2014) *J. Phys. Chem. C*, **118**, 16408–16418.
50 Ehrler, B., Musselman, K.P., Böhm, M.L., Morgenstern, F.S.F., Vaynzof, Y., Walker, B.J., MacManus-Driscoll, J.L., and Greenham, N.C. (2013) *ACS Nano*, **7**, 4210–4220.
51 Hoye, R.L.Z., Muñoz-Rojas, D., Iza, D.C., Musselman, K.P., and MacManus-Driscoll, J.L. (2013) *Sol. Energy Mater. Sol. Cells*, **116**, 197–202.
52 Diamant, Y., Chappel, S., Chen, S.G., Melamed, O., and Zaban, A. (2004) *Coord. Chem. Rev.*, **248**, 1271–1276.
53 Kim, J.Y., Lee, S., Noh, J.H., Jung, H.S., and Hong, K.S. (2008) *J. Electroceram.*, **23**, 422–425.
54 Palomares, E., Clifford, J.N., Haque, S.A., Lutz, T., and Durrant, J.R. (2003) *J. Am. Chem. Soc.*, **125**, 475–482.
55 Klahr, B.M. and Hamann, T.W. (2009) *J. Phys. Chem. C*, **113**, 14040–14045.
56 Antila, L.J., Heikkilä, M.J., Mäkinen, V., Humalamäki, N., Laitinen, M., Linko, V., Jalkanen, P., Toppari, J., Aumanen, V., Kemell, M., Myllyperkiö, P., Honkala, K., Häkkinen, H., Leskelä, M., and Korppi-Tommola, J.E.I. (2011) *J. Phys. Chem. C*, **115**, 16720–16729.
57 Kozen, A.C., Schroeder, M.A., Osborn, K.D., Lobb, C.J., and Rubloff, G.W. (2013) *Appl. Phys. Lett.*, **102**, 173501.
58 Gordon, R.G., Hausmann, D., Kim, E., and Shepard, J. (2003) *Chem. Vap. Deposition*, **9**, 73–78.
59 Elam, J.W., Routkevitch, D., Mardilovich, P.P., and George, S.M. (2003) *Chem. Mater.*, **15**, 3507–3517.
60 Dong, W., Wang, Z.-D., Yang, L.-Z., Meng, T., and Chen, Q. (2014) *Chin. Phys. Lett.*, **31**, 098401.
61 Pascoe, A.R., Bourgeois, L., Duffy, N.W., Xiang, W., and Cheng, Y.-B. (2013) *J. Phys. Chem. C*, **117**, 25118–25126.
62 Grünwald, R. and Tributsch, H. (1997) *J. Phys. Chem. B*, **101**, 2564–2575.
63 Xue, G., Guo, Y., Yu, T., Guan, J., Yu, X., Zhang, J., Liu, J., and Zou, Z. (2012) *Int. J. Electrochem. Sci.*, **7**, 1496–1511.
64 Sommeling, P.M., Späth, M., Smit, H.J.P., Bakker, N.J., and Kroon, J.M. (2004) *J. Photochem. Photobiol., A*, **164**, 137–144.
65 Heo, N., Jun, Y., and Park, J.H. (2013) *Sci. Rep.*, **3**, 1–6.
66 Son, H.J., Wang, X., Prasittichai, C., Jeong, N.C., Aaltonen, T., Gordon, R.G., and Hupp, J.T. (2012) *J. Am. Chem. Soc.*, **134**, 9537–9540.

67 Hanson, K., Losego, M.D., Kalanyan, B., Ashford, D.L., Parsons, G.N., and Meyer, T.J. (2013) *Chem. Mater.*, **25**, 3–5.
68 Son, H.J., Prasittichai, C., Mondloch, J.E., Luo, L., Wu, J., Kim, D.W., Farha, O.K., and Hupp, J.T. (2013) *J. Am. Chem. Soc.*, **135**, 11529–11532.
69 Hanson, K., Losego, M.D., Kalanyan, B., Parsons, G.N., and Meyer, T.J. (2013) *Nano Lett.*, **13**, 4802–4809.
70 Kim, D.H., Losego, M.D., Hanson, K., Alibabaei, L., Lee, K., Meyer, T.J., and Parsons, G.N. (2014) *Phys. Chem. Chem. Phys.*, **16**, 8615–8622.
71 Park, N.-G. (2013) *J. Phys. Chem. Lett.*, **4**, 2423–2429.
72 Jung, H.S. and Park, N.G. (2015) *Small*, **11**, 10–25.
73 Dong, X., Fang, X., Lv, M., Lin, B., Zhang, S., Ding, J., and Yuan, N. (2015) *J. Mater. Chem. A*, **3**, 5360–5367.
74 Chang, C.-Y., Lee, K.-T., Huang, W.-K., Siao, H.-Y., and Chang, Y.-C. (2015) *Chem. Mater.*, **27**, 5122–5130.
75 Van Huis, M.A., Kunneman, L.T., Overgaag, K., Xu, Q., Pandraud, G., Zandbergen, H.W., and Vanmaekelbergh, D. (2008) *Nano Lett.*, **8**, 3959–3963.
76 Pourret, A., Guyot-Sionnest, P., and Elam, J.W. (2009) *Adv. Mater.*, **21**, 232–235.
77 Ihly, R., Tolentino, J., Liu, Y., Gibbs, M., and Law, M. (2011) *ACS Nano*, **5**, 8175–8186.
78 Liu, Y., Gibbs, M., Perkins, C.L., Tolentino, J., Zarghami, M.H., Bustamante, J. Jr.,, and Law, M. (2011) *Nano Lett.*, **11**, 5349–5355.
79 Liu, Y., Tolentino, J., Gibbs, M., Ihly, R., Perkins, C.L., Liu, Y., Crawford, N., Hemminger, J.C., and Law, M. (2013) *Nano Lett.*, **13**, 1578–1587.
80 ten Cate, S., Liu, Y., Suchand Sandeep, C.S., Kinge, S., Houtepen, A.J., Savenije, T.J., Schins, J.M., Law, M., and Siebbeles, L.D.A. (2013) *J. Phys. Chem. Lett.*, **4**, 1766–1770.
81 Lambert, K., Dendooven, J., Detavernier, C., and Hens, Z. (2011) *Chem. Mater.*, **23**, 126–128.
82 Ip, A.H., Labelle, A.J., and Sargent, E.H. (2013) *Appl. Phys. Lett.*, **103**, 263905.
83 Kemp, K.W., Labelle, A.J., Thon, S.M., Ip, A.H., Kramer, I.J., Hoogland, S., and Sargent, E.H. (2013) *Adv. Energy Mater.*, **3**, 917–922.
84 Lin, X., Yu, K., Lu, G., Chen, J., and Yuan, C. (2013) *J. Phys. D: Appl. Phys.*, **46**, 024004.
85 Roelofs, K.E., Brennan, T.P., Dominguez, J.C., Bailie, C.D., Margulis, G.Y., Hoke, E.T., McGehee, M.D., and Bent, S.F. (2013) *J. Phys. Chem. C*, **117**, 5584–5592.
86 Dasgupta, N.P., Lee, W., and Prinz, F.B. (2009) *Chem. Mater.*, **21**, 3973–3978.
87 Brennan, T.P., Ardalan, P., Lee, H.-B.-R., Bakke, J.R., Ding, I.K., McGehee, M.D., and Bent, S.F. (2011) *Adv. Energy Mater.*, **1**, 1169–1175.
88 Marichy, C., Bechelany, M., and Pinna, N. (2012) *Adv. Mater.*, **24**, 1017–1032.
89 Noh, J.H., Han, H.S., Lee, S., Kim, J.Y., Hong, K.S., Han, G.-S., Shin, H., and Jung, H.S. (2011) *Adv. Energy Mater.*, **1**, 829–835.
90 Chappel, S., Grinis, L., Ofir, A., and Zaban, A. (2005) *J. Phys. Chem. B*, **109**, 1643–1647.

91 Martinson, A.B.F., Elam, J.W., Liu, J., Pellin, M.J., Marks, T.J., and Hupp, J.T. (2008) *Nano Lett.*, **8**, 2862–2866.
92 Hong Noh, J., Ding, B., Soo Han, H., Seong Kim, J., Hoon Park, J., Baek Park, S., Suk Jung, H., Lee, J.-K., and Sun Hong, K. (2012) *Appl. Phys. Lett.*, **100**, 084104.
93 Pint, C.L., Takei, K., Kapadia, R., Zheng, M., Ford, A.C., Zhang, J., Jamshidi, A., Bardhan, R., Urban, J.J., Wu, M., Ager, J.W., Oye, M.M., and Javey, A. (2011) *Adv. Energy Mater.*, **1**, 1040–1045.
94 Lee, T.Y., Alegaonkar, P.S., and Yoo, J.-B. (2007) *Thin Solid Films*, **515**, 5131–5135.
95 Lee, K., Hu, C., Chen, H., and Ho, K. (2008) *Sol. Energy Mater. Sol. Cells*, **92**, 1628–1633.
96 Nath, N.C., Sarker, S., Ahammad, A.J., and Lee, J.J. (2012) *Phys. Chem. Chem. Phys.*, **14**, 4333–4338.
97 Lin, W.J., Hsu, C.T., Lai, Y.C., Wu, W.C., Hsieh, T.Y., and Tsai, Y.C. (2011) *Adv. Mater. Res.*, **410**, 168–171.
98 Yang, Z., Liu, M., Zhang, C., Tjiu, W.W., Liu, T., and Peng, H. (2013) *Angew. Chem. Int. Ed.*, **52**, 3996–3999.
99 Chan, Y.-F., Wang, C.-C., Chen, B.-H., and Chen, C.-Y. (2013) *Prog. Photovoltaics Res. Appl.*, **21**, 47–57.
100 Dembele, K.T., Selopal, G.S., Soldano, C., Nechache, R., Rimada, J.C., Concina, I., Sberveglieri, G., Rosei, F., and Vomiero, A. (2013) *J. Phys. Chem. C*, **117**, 14510–14517.
101 Chen, J., Li, B., Zheng, J., Zhao, J., and Zhu, Z. (2012) *J. Phys. Chem. C*, **116**, 14848–14856.
102 Marichy, C., Donato, N., Latino, M., Willinger, M.G., Tessonnier, J.P., Neri, G., and Pinna, N. (2015) *Nanotechnology*, **26**, 024004.
103 Marichy, C. and Pinna, N. (2013) *Coord. Chem. Rev.*, **257**, 3232–3253.
104 Guerra-Nunez, C., Zhang, Y., Li, M., Chawla, V., Erni, R., Michler, J., Park, H.G., and Utke, I. (2015) *Nanoscale*, **7**, 10622–10633.
105 Zhang, Y., Guerra-Nuñez, C., Utke, I., Michler, J., Rossell, M.D., and Erni, R. (2015) *J. Phys. Chem. C*, **119**, 150203103227001.
106 Deng, S., Verbruggen, S.W., He, Z., Cott, D.J., Vereecken, P.M., Martens, J.A., Bals, S., Lenaerts, S., and Detavernier, C. (2014) *RSC Adv.*, **4**, 11648.
107 Hsu, C.Y., Lien, D.H., Lu, S.Y., Chen, C.Y., Kang, C.F., Chueh, Y.L., Hsu, W.K., and He, J.H. (2012) *ACS Nano*, **6**, 6687–6692.
108 Sun, X., Xie, M., Travis, J.J., Wang, G., Sun, H., Lian, J., and George, S.M. (2013) *J. Phys. Chem. C*, **117**, 22497–22508.
109 Jin, S.H., Jun, G.H., Hong, S.H., and Jeon, S. (2012) *Carbon*, **50**, 4483–4488.
110 Yazdani, N., Bozyigit, D., Utke, I., Buchheim, J., Youn, S.K., Patscheider, J., Wood, V., and Park, H.G. (2014) *ACS Appl. Mater. Interfaces*, **6**, 1389–1393.
111 Hwang, Y.J., Boukai, A., and Yang, P. (2009) *Nano Lett.*, **9**, 410–415.
112 Dasgupta, N.P., Liu, C., Andrews, S., Prinz, F.B., and Yang, P. (2013) *J. Am. Chem. Soc.*, **135**, 12932–12935.
113 Chen, Y.W., Prange, J.D., Dühnen, S., Park, Y., Gunji, M., Chidsey, C.E.D., and McIntyre, P.C. (2011) *Nat. Mater.*, **10**, 539–544.

114 Liu, M., Nam, C.-Y., Black, C.T., Kamcev, J., and Zhang, L. (2013) *J. Phys. Chem. C*, **117**, 13396–13402.
115 Lee, M.H., Takei, K., Zhang, J., Kapadia, R., Zheng, M., Chen, Y.Z., Nah, J., Matthews, T.S., Chueh, Y.L., Ager, J.W., and Javey, A. (2012) *Angew. Chem. Int. Ed.*, **51**, 10760–10764.
116 Weng, B., Xu, F., and Xu, J. (2014) *Nanotechnology*, **25**, 455402.
117 Strandwitz, N.C., Comstock, D.J., Grimm, R.L., Nichols-Nielander, A.C., Elam, J., and Lewis, N.S. (2013) *J. Phys. Chem. C*, **117**, 4931–4936.
118 Peng, Q., Kalanyan, B., Hoertz, P.G., Miller, A., Kim, D.H., Hanson, K., Alibabaei, L., Liu, J., Meyer, T.J., Parsons, G.N., and Glass, J.T. (2013) *Nano Lett.*, **13**, 1481–1488.
119 Cordova, I.A., Peng, Q., Ferrall, I.L., Rieth, A.J., Hoertz, P.G., and Glass, J.T. (2015) *Nanoscale*, **7**, 8584–8592.

Part III

ALD toward Electrochemical Energy Storage

5

Atomic Layer Deposition of Electrocatalysts for Use in Fuel Cells and Electrolyzers

Lifeng Liu

Nanomaterials for Energy Storage and Conversion Research Group, International Iberian Nanotechnology Laboratory (INL), Avenida Mestre Jose Veiga, 4715-330 Braga, Portugal

5.1 Introduction

Fuel cells and electrolyzers are two important electrochemical devices that are commonly touted as core technologies in a clean energy future and have recently received considerable attention from both academia and industry. Particularly, they are proposed to be essential building blocks of future energy infrastructures for a hydrogen economy.

Briefly, a fuel cell is a device converting chemical energy from a fuel into electricity. It consists of three adjacent components: anode, electrolyte, and cathode (Figure 5.1a), which allow positively charged hydrogen ions (i.e., protons) to move between the two sides of the fuel cell. Two chemical reactions occur at the anode and cathode, respectively. At the anode, a catalyst oxidizes the fuel, for example, hydrogen (H_2), transforming it into positively charged protons and negatively charged electrons. The electrolyte is a substance specifically designed such that protons can pass through it, but electrons cannot. A typical electrolyte is made from a polymeric proton-conducting membrane (PEM). Since the PEM is electrically insulating, the free electrons generated at the anode are forced to travel in an external circuit creating an electrical current, while the protons travel through the PEM to the cathode. Once reaching the cathode, the protons are reunited with the electrons coming from the external circuit and the two react with a third chemical, usually oxygen (O_2), to produce water (H_2O). The net result of these two half-cell reactions is that fuel is consumed, water is produced, and an electrical current is created, which can be used to power electrical devices. Fuel cells can produce electricity continuously as long as a source of fuel and O_2 are constantly supplied. They are mainly used as primary and backup power for commercial, industrial, and residential buildings and in remote or inaccessible areas. More attractively, fuel cells can be used to power vehicles including forklifts, automobiles, boats, motorcycles, and submarines. This has become a major driving force for modern fuel cell research.

A water electrolyzer is an electrochemical device working in reverse order of a fuel cell. It also consists of three main components: anode, electrolyte, and

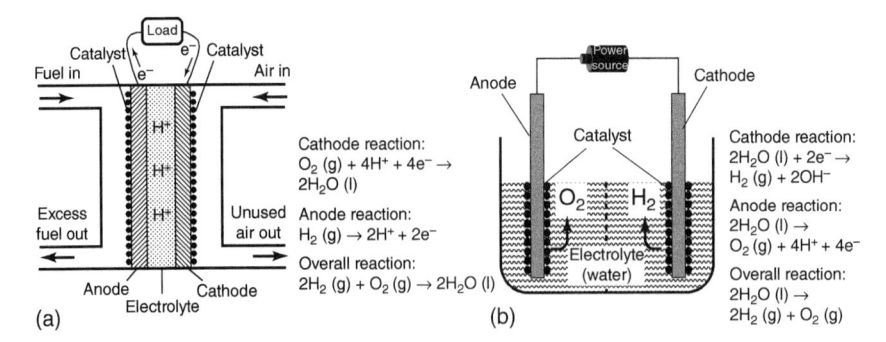

Figure 5.1 Schematic illustration of (a) a proton-exchange membrane fuel cell and (b) a water electrolyzer.

cathode (Figure 5.1b). But different from fuel cells, an electrical power source is needed to connect the anode with the cathode to provide a driving force for H_2O splitting. When the applied external potential exceeds the thermodynamic limit of H_2O decomposition, namely 1.23 V (without considering any loss and nonideality in the electrochemical process), H_2 will principally evolve at the cathode and O_2 simultaneously at the anode. Assuming ideal Faradaic efficiency, the amount of H_2 produced is twice the amount of O_2, and both are proportional to the total electrical charge passed between the two electrodes. At the moment, the H_2 fuel produced by water electrolysis is not economically competitive with that produced by steam reforming of natural gas. However, water electrolysis represents a cleaner and more sustainable approach to H_2 production; moreover, this technology is well suited for coupling with variable sources of renewable energy such as wind and solar and, therefore, can be implemented at a range of scales.

For more technical details about fuel cells and water electrolyzers, readers are recommended to refer to textbooks [1–3] and some review papers published recently [4, 5].

Central to the operation of fuel cells and electrolyzers are catalysts, more specifically electrocatalysts, which mediate charge transfer processes between the electrolyte and device electrodes with minimal losses in efficiency. Unfortunately, many state-of-the-art working electrocatalysts are comprised of expensive platinum-group metals (e.g., Pt, Pd, Ru, and Ir). The high costs and limited availability of these precious metals create prohibitive barriers to market penetration and up-scale production of devices requiring large catalyst loadings for efficient operation. Therefore, a major challenge to be addressed for large-scale commercialization of fuel cells and electrolyzers is to reduce the utilization of Pt-group metals in the electrocatalysts or completely substitute them with nonprecious, earth-abundant alternatives.

Electrocatalysts are usually made of nanoscale particles with a size ranging from tens of nanometers down to subnanometers. In order to make full use of electrocatalysts containing precious metals (e.g., Pt) and prevent them from agglomeration under operation conditions leading to degraded catalytic performance, electrocatalysts are usually dispersed on an electrically conductive support having a high surface area, for example, mesoporous carbon that is typically used in PEM fuel cells [6]. These catalysts can be incorporated onto support materials through conventional methods such as sputtering, electrodeposition, and impregnation followed by chemical reduction [7, 8]. However, these methods do not allow catalyst nanoparticles (NPs) to penetrate into deep regions of porous supports and involve multiple laborious steps and, therefore, cannot render uniform and conformal deposition of catalyst NPs over the whole surface of porous supports.

Atomic layer deposition (ALD), as a powerful thin-film deposition technique, has recently emerged as an effective approach to the fabrication of electrocatalysts. As mentioned in the previous chapters, one of the distinguishing attributes of ALD is its capability to deposit highly uniform and conformal coatings on surfaces with complex topographies and to infiltrate mesoporous materials. Besides its design for growth of ultrathin and continuous thin films, ALD was also widely reported for its capability of depositing discrete NPs. This is due to the deviations from ideal thin-film growth behaviors, which result from, for instance, nucleation delay and/or island growth in the initial ALD cycles. This feature is particularly attractive for the growth of heterogeneous catalysts requiring highly dispersed catalytic species on high-surface-area mesoporous supports. Compared to other synthetic methods, ALD is simpler and allows for large-area, batch production of electrocatalysts with a small particle size and narrow particle distribution. More importantly, ALD offers uniform and conformal dispersion of catalyst NPs over the entire surface of porous substrates, which cannot be readily achieved by any other known methods.

In this chapter, recent advances in fuel cell and water-splitting electrocatalysts fabricated by ALD technique are reviewed, including electrocatalysts of noble metals and metal oxides as well as catalyst supports. The present efforts and future development of ALD for use in synthesis of electrocatalysts are discussed.

5.2 ALD of Pt-Group Metal and Alloy Electrocatalysts

As mentioned previously, Pt-group metal and alloy NPs are still the most efficient and commonly used electrocatalysts in both fuel cells and electrolyzers to catalyze the transformation of chemical energy into electricity or vice versa. Pt-group metals are not only costly but also scarce in the earth's crust. This has, to a large extent, hindered widespread deployment of fuel cells and water electrolyzers. Therefore, reducing the overall loading of Pt-group metal catalysts while retaining sufficiently high catalytic performance has been one of the major challenges for the design of fuel cell and electrolyzer electrodes.

Recent studies have shown that NPs of precious metals such as Pt [9–23], Pd [24–28], and their alloys including PtRu [29–32], PtPd [33, 34], and PtCo [35, 36] can be obtained by ALD and used to catalyze various electrochemical reactions taking place in fuel cells or electrolyzers including H_2 oxidation (HOR), methanol/ethanol/formic acid oxidation (MOR/EOR/FOR), O_2 reduction (ORR), and H_2 evolution (HER). These ALD-derived catalyst NPs can be deposited on high-surface-area carbon supports with an extremely low loading mass (e.g., 0.016 mg cm^{-2}). Nonetheless, they exhibit reasonably good electrocatalytic performance comparable to that of the electrodes fabricated by conventional methods, which have a typical catalyst loading of 0.5 mg cm^{-2} [11]. The fabrication, microstructure, and electrocatalytic performance of these catalysts are discussed in detail in the following sections.

5.2.1 ALD of Pt Electrocatalysts

5.2.1.1 Fabrication and Microstructure

ALD of Pt was first studied in 2000 by Utriainen *et al.* from Helsinki University of Technology, using platinum acetylacetonate [Pt(acac)$_2$] as a precursor [37]. However, Pt(acac)$_2$ was found to be an unsuitable precursor for ALD, because it thermally decomposes at low temperatures, destroying the self-limiting growth mechanism. Later, Leskelä *et al.* reported ALD of Pt using (methylcyclopentadienyl)trimethylplatinum [MeCpPtMe$_3$] as a precursor and found that the deposited Pt films exhibited nanoparticulate feature [38]. MeCpPtMe$_3$ has, nowadays, become the most commonly used precursor for ALD of Pt.

In 2008, King *et al.* reported the ALD of Pt NPs on porous carbon aerogel monoliths having a diameter of 1 cm and thickness of 500 μm [9]. Due to the tortuous porosity of carbon aerogel, very long precursor pulse and purge times (20 min MeCpPtMe$_3$–10 min N$_2$ purge–10 min dry air–10 min N$_2$ purge) were used to maximize the penetration of ALD precursors into and removal of by-products out of the porous carbon aerogel support. Structural characterization reveals that uniformly distributed Pt NPs can form on the inner surface of carbon aerogels at a depth of >10 μm below the monolith surface. However, the average particle size was found to decrease with the increasing depth, exhibiting a nonideal ALD process. This nonideal deposition behavior can be attributed to precursor reactivity and to processes of adsorption–desorption equilibrium of both precursor and reaction product molecules with the pore walls. Remarkably, ALD yielded an ultralow Pt loading of 0.047 mg cm^{-2}, and even at such a low loading level, the Pt-loaded carbon aerogel exhibited high catalytic activity toward carbon monoxide oxidation.

Using ALD, Pt NPs were also successfully deposited on other high-surface-area carbon supports such as carbon cloth [11], carbon nanotubes (CNTs) [10–13, 15, 16], graphene [17], carbon black [19, 20], and graphene oxide [14]. ALD of Pt relies heavily on the surface chemistry of these carbon supports. In many cases, surface functionalization needs to be carried out in order to introduce enough defect sites and oxy-groups on carbon supports for Pt nucleation. Different from the report in [9], the Pt precursor pulse time can be significantly shortened after surface functionalization, which can help to save precious precursors.

Acid treatment is an effective way to functionalize carbon supports [11, 13, 16, 19, 20, 22] for ALD growth. In 2009, Liu *et al.* reported ALD of Pt on functionalized carbon supports obtained by acid treatment [11]. The authors compared the deposition behavior of Pt on both pristine and acid-treated carbon cloth and found that Pt could not be deposited on carbon cloth without acid treatment (Figure 5.2a). In contrast, high-density Pt NPs could be clearly seen on the carbon cloth subjected to 6 h refluxing treatment in concentrated HNO$_3$ at 140 °C (Figure 5.2b). Furthermore, the authors attempted to make fuel cell electrodes by

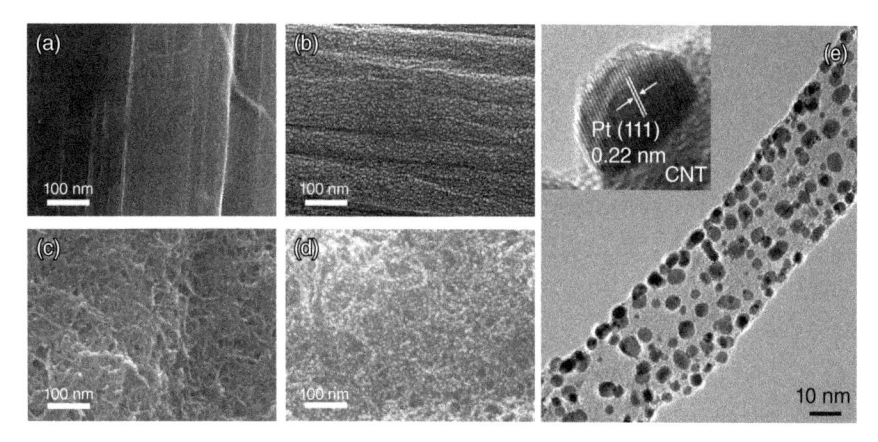

Figure 5.2 Morphology of ALD-derived Pt NPs on different substrates: carbon cloth without acid treatment (a) and with acid treatment for 6 h (b), and CNTs acid-treated for 6 h mixed with 0.06 wt% PTFE (c) and 0.0006 wt% PTFE (d). The cycle numbers were 200 and 300 for carbon cloth and CNTs, respectively. (Liu et al. [11]. Reproduced with permission of John Wiley and Sons.) TEM image of ALD-derived Pt NPs on CNTs (e). The inset is a high-resolution TEM image of a Pt NP. (Hsueh et al. [16]. Reproduced with permission of Elsevier.)

pasting acid-treated CNTs on carbon cloth using polytetrafluoroethylene (PTFE) as a binder. They observed that only when the PTFE concentration was lower than 0.0006 wt%, could Pt NPs be deposited on the carbon cloth/CNT composite (Figure 5.2d). With a higher PTFE concentration (e.g., 0.06 wt%), no Pt deposition was found possibly because PTFE does not react with precursor molecules (Figure 5.2c). These experimental observations indicate that surface functional groups (e.g., C–OH, C–OOH) play a key role in the formation of a Pt monolayer through self-limiting reactions during ALD by providing reactive sites for Pt nucleation [11, 16]. The presence of these functional groups was confirmed by both FTIR and XPS analyses [11, 16]. Besides concentrated HNO_3, citric acid was used to implant oxygen functionalities including carboxyl (O—C=O), carbonyl (C=O), and ether (C—O) groups on CNT surfaces to facilitate the deposition of Pt NPs during ALD, and Pt particle size and distribution similar to those of HNO_3-treated carbon supports were reported [13, 19, 20, 22].

While acid treatment is a time-consuming process, plasma treatment provides a more efficient alternative to surface functionalization of carbon supports [10, 12, 15]. Figure 5.3a schematically shows the difference in CNTs with and without O_2 plasma treatment after a complete ALD cycle [15]. O_2 plasma treatment, even if as short as 5 s, can effectively induce surface oxides on CNTs upon the bombardment of oxygen atoms and positive oxygen ions (O^{2+}, O^+), increasing the number of nucleation sites for Pt growth. As a result, high-density, size-controllable Pt NPs can be uniformly deposited on CNT surfaces after ALD (Figure 5.3b). Moreover, it was demonstrated that the NP size can be adjusted, linearly increasing with the plasma treatment time [15]. However, since O_2 plasma is very aggressive, the treatment should not be performed over 35 s to avoid completely destroying the CNTs. A process of using milder Ar and Ar/O_2 plasma to functionalize CNTs was developed [12]. However, it was found that

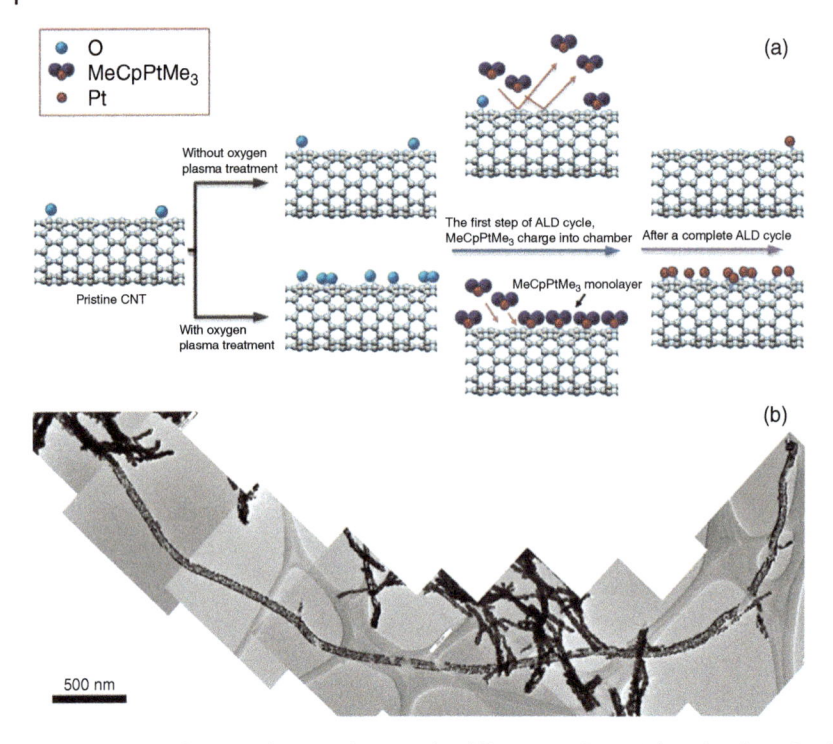

Figure 5.3 (a) Schematic diagram showing the difference in CNTs with and without O_2 plasma treatment after a complete ALD cycle. (Hsueh *et al*. [15]. Reproduced with permission of Institute of Physics.) (b) A stitched TEM image showing uniform deposition of Pt NPs over the entire length of a 25-μm-long O_2 plasma-treated CNT. The deposition was carried out over a 25-μm-thick vertically aligned CNT array for 200 ALD cycles. (Dameron *et al*. [12]. Reproduced with permission of Elsevier.)

no Pt deposition occurred on Ar plasma-treated CNTs, presumably because Ar bombardment removed surface defect sites that should have facilitated Pt nucleation. For Ar/O_2 plasma-treated CNT arrays, Pt deposition took place on the CNT surface, but the deposited NPs were unevenly distributed over the length of CNTs [12].

In addition to acid and plasma treatments, chemical functionalization of CNTs using trimethylaluminum (TMA) was investigated. The TMA functionalization was achieved by exposing the CNT arrays to TMA precursors at 250 °C inside an ALD chamber [12]. TMA functionalization greatly increased the amount of Pt deposited compared to that on unfunctionalized CNTs for the same number of ALD cycles. However, there was a significant variation in Pt coverage from the top to bottom of the CNT array. The size of the deposited Pt NPs was also larger than that of NPs deposited on O_2 plasma-treated CNTs. It is likely that the TMA functionalization only offers chemical sites that are more active with the Pt precursor molecules but does not significantly increase the number of sites available for Pt nucleation on the CNT surface, as the O_2 plasma treatment does [12].

Although surface functionalization facilitates the nucleation of Pt NPs on CNTs, it should be noted that due to the introduction of defects, the chemical and electrical properties of the CNTs are deteriorated. To prevent performance degradation, nitrogen (N)-doped CNTs (N-CNTs) have been used as supports for ALD of Pt and PtRu NPs [31]. It has been reported that N-CNTs with appropriate N-doping concentrations have higher electrical conductivity compared to pristine CNTs [39]. Moreover, N-CNTs can offer better catalyst dispersion, stability, and activity [40, 41]. In fact, N-CNTs themselves are efficient working catalysts for both ORR and OER [42, 43] and can therefore synergize with Pt NPs to catalyze these reactions.

5.2.1.2 Electrochemical Performance

Hydrogen Oxidation Reaction (HOR) The electrochemical performance of ALD-derived Pt/C catalysts toward HOR has been evaluated in full-cell configuration in a membrane electrode assembly (MEA) [11, 13, 16, 22, 23].

In 2012, Shu *et al.* reported the performance of a single H_2-O_2 PEM fuel cell using an anode loaded with ALD-derived Pt catalysts ($0.26\,\mathrm{mg\,cm^{-2}}$) supported on CNTs and a cathode loaded with commercially available Pt catalysts ($0.4\,\mathrm{mg\,cm^{-2}}$, Johnson Matthey), separated by a Nafion® 212 membrane (Dupont Inc.) as the PEM [13]. High-purity H_2 (99.999%) and O_2 (99.999%) were fed at the anode and cathode, respectively, at a flow rate of $200\,\mathrm{cm^3\,min^{-1}}$ at 100% relative humidity. For comparison, a cell consisting of a commercial Pt anode with a Pt loading of $0.4\,\mathrm{mg\,cm^{-2}}$ was made and tested under the same operation conditions.

Figure 5.4a,b show the polarization curves obtained with the cells made of ALD-Pt and commercial Pt anodes at 40, 60, and 80 °C. These curves typically consist of three polarization regions, as described in [1]: (i) activation polarization, (ii) ohmic polarization, and (iii) concentration polarization. As can be seen in Figure 5.4a,b, the polarization curves of the ALD-Pt cells measured at different temperatures are very similar to those of the cells having a commercial Pt anode. However, the maximal power densities of the ALD-Pt cells are much higher than those of commercial Pt cells. Since both the ALD-Pt and commercial Pt cells were operated under the same conditions, it can be concluded that the ALD-derived Pt indeed offers an improved catalytic activity compared to the commercial Pt catalysts. The improved power density can be attributed to the well-dispersed Pt NPs over the entire CNT surfaces achieved by the ALD technique, which serve as active sites for HOR and substantially improve the Pt utilization. The durability of the ALD-Pt and commercial Pt cells was also tested at a constant voltage of 0.7 V for 50 h (Figure 5.4c). The specific current density delivered by the ALD-Pt cells was found to be much higher than that of the commercial Pt cells. Apart from an initial current decay (within 1 h), the current density of the ALD-Pt cells was kept almost constantly at about $2155\,\mathrm{A\,g^{-1}_{Pt}}$, demonstrating that the ALD-derived Pt catalysts can provide excellent durability under operation conditions. This is likely because ALD can help to immobilize the Pt sites on the surface of CNTs via the strong chemical adsorption between Pt atoms and surface oxides (e.g., C–OH, C–OOH).

The electrochemical performance of PEM fuel cells with both anodes and cathodes made by ALD was also investigated. Hsueh *et al.* deposited Pt NP

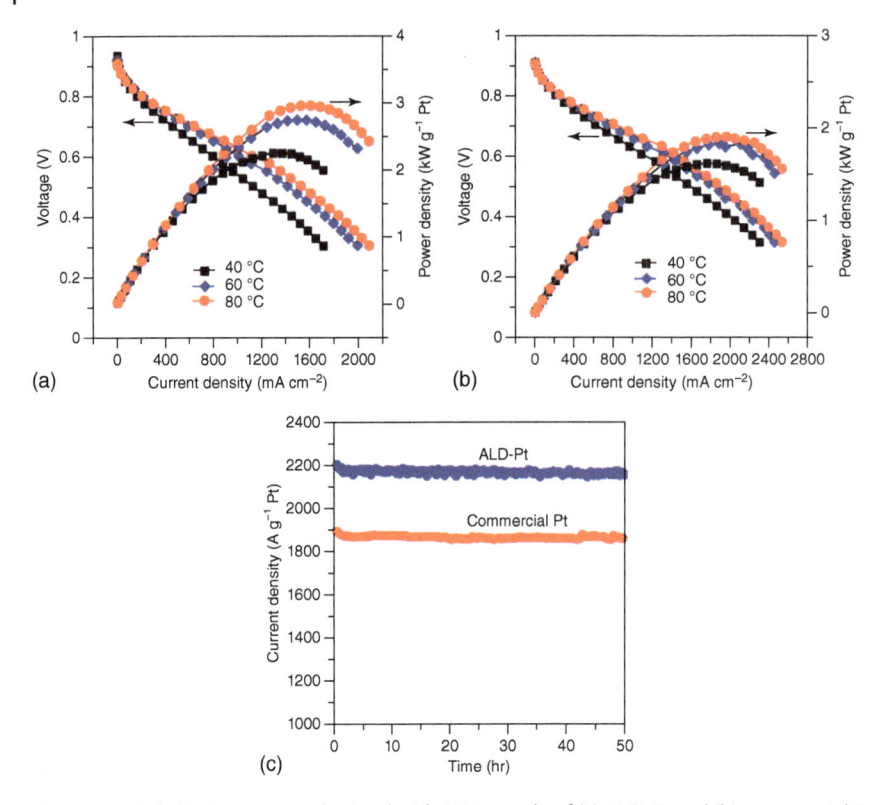

Figure 5.4 Polarization curves obtained with MEAs made of (a) ALD-Pt and (b) commercial Pt (Johnson Matthey) anodes at 40, 60, and 80 °C. (c) Constant-voltage durability test conducted with MEAs made of ALD-Pt and commercial Pt at 0.7 V and 60 °C for 50 h. (Shu et al. [13]. Reproduced with permission of Elsevier.)

catalysts on acid-treated CNTs using ALD for different numbers of cycles to vary the Pt loading [16]. At a given CNT loading (4 mg cm^{-2}), the Pt loadings after ALD for 100, 200, and 400 cycles (hereafter denoted as ALD-100, ALD-200, and ALD-400) were 0.02, 0.05, and 0.11 mg cm^{-2}, respectively. The as-fabricated catalysts were then pasted on a carbon cloth, which was subsequently used as an electrode in a MEA. Figures 5.5 a and b show the polarization curves of MEAs made with Pt/CNT/carbon cloth composite for both anodes and cathodes. The catalyst prepared for the anode was fixed at 100 ALD cycles and that for the cathode changed from 100 to 300 ALD cycles. Among these cells, the cathode prepared with Pt for 200 ALD cycles has the best performance. In order to increase the Pt loading but to keep the constant catalyst particle size, the authors also varied the loading of CNTs at the cathode and fixed the ALD cycle number at 100 and loading of CNTs at the anode at 4 mg cm^{-2}. They found that the cell's performance was enhanced as the loading of CNTs at the cathode increases, and the increase in the power density is nearly proportional to the loading of CNTs (Figure 5.5). The performance of all homemade electrodes is not as good as that of the commercial E-Tek electrodes. However, considering that the Pt loading in all the homemade electrodes is <0.05 mg cm^{-2}, 10 times lower than that for the commercial E-Tek

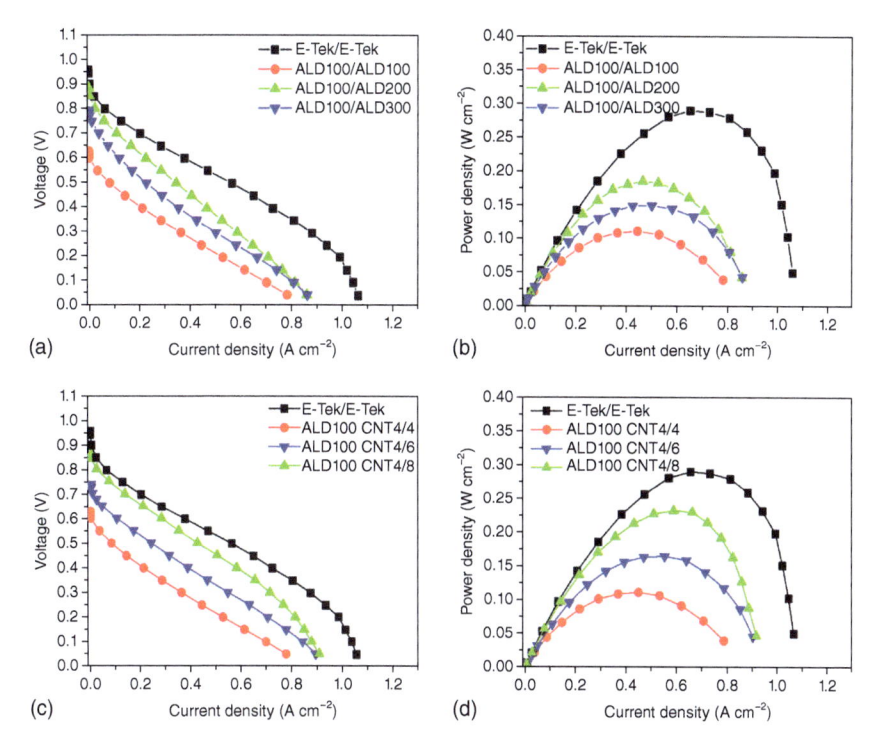

Figure 5.5 Performance of single PEM fuel cells with both anode and cathode made by ALD at various ALD cycle numbers (a and b) and various amount of CNTs (c and d). The cell made with commercial E-Tek electrodes is included for comparison. (Hsueh et al. [16]. Reproduced with permission of Elsevier.)

electrodes (0.5 mg$_{Pt}$ cm^{-2}), the reported performance for homemade electrodes is remarkable. In fact, if comparing Pt-specific power density at a given potential, for example, 0.65 V, the performance of homemade electrodes is significantly better than that of E-Tek electrodes (2.27 kW g^{-1}$_{Pt}$ for homemade electrodes versus 0.18 kW g^{-1}$_{Pt}$ for E-Tek electrodes). Furthermore, if the enhancement factor, defined as the ratio of specific power density of homemade electrodes over that of the E-Tek electrodes, is used to compare the fuel cell's performance, it was found that the enhancement factor of cells made with ALD-Pt catalysts (0.019 mg cm^{-2} at the anode and 0.044 mg cm^{-2} at the cathode) was as high as 11.95, significantly higher than that of many others reported in the literature [16].

Table 5.1 summarizes the performance of several PEM fuel cells made with ALD-derived Pt catalysts at anodes or at both anodes and cathodes. Some cells made with commercial Pt catalysts at both anodes and cathodes are also listed for comparison.

Methanol/Ethanol/Formic Acid Oxidation Reaction (MOR/EOR/FOR) Direct liquid-fed fuel cells (DLFCs) are a class of PEM fuel cells with organic solvents as the feeding fuel [44]. They include direct methanol fuel cells (DMFCs), direct ethanol fuel cells (DEFCs), and direct formic acid fuel cells (DFAFCs), among others. DLFCs offer some advantages over H$_2$–O$_2$ fuel

Table 5.1 Comparison of electrochemical performance of several PEM fuel cells made with ALD-derived Pt catalysts at the anode.

References	Anode Pt loading (mg cm^{-2})	Cathode Pt loading (mg cm^{-2})	Temperature (°C)	Peak power density (kW g$^{-1}_{Pt}$)
[11]	0.5 (E-Tek)	0.5 (E-Tek)	60	0.42
	0.016 (ALD-Pt)	0.5 (E-Tek)	60	10.60
[13]	0.4 (Johnson Matthey)	0.4 (Johnson Matthey)	60	1.83@0.63 V
	0.4 (Johnson Matthey)	0.4 (Johnson Matthey)	80	1.90@0.66 V
	0.26 (ALD-Pt)	0.4 (Johnson Matthey)	60	2.74@0.72 V
	0.26 (ALD-Pt)	0.4 (Johnson Matthey)	80	2.95@0.76 V
[16]	0.019 (ALD-Pt)	0.044 (ALD-Pt)	60	2.69@0.60 V
[22]	0.08 (ALD-Pt)	0.5 (E-Tek)	75	7.60@0.95 V
	0.11 (ALD-Pt)	0.5 (E-Tek)	75	5.10@0.90 V
[23]	0.061 (ALD-Pt)	0.4 (Johnson Matthey)	60	1.10@0.90 V
	0.061 (ALD-Pt)	0.4 (Johnson Matthey)	75	1.15@0.97 V
	0.13 (ALD-Pt)	0.4 (Johnson Matthey)	60	2.16@0.93 V
	0.13 (ALD-Pt)	0.4 (Johnson Matthey)	75	2.32@0.97 V

cells, for example, higher power density, easier storage and transportation, lower operation temperature, and improved safety. So far, ALD-derived Pt catalysts have not been evaluated in full-cell configuration in DLFCs, but they have shown great promise for improving catalytic activity in half-cell reactions toward MOR, EOR, and FOR [14, 17, 19–21].

Sun and coauthors successfully fabricated Pt single atoms, subnanometer clusters, and NPs on graphene nanosheets (GNS) using ALD and investigated the electrocatalytic performance of these ALD-Pt catalysts toward MOR [17]. They found that by fine-tuning the number of ALD cycles (1 s MeCpPtMe$_3$ – 20 s N$_2$ – 5 s O$_2$ – 20 s N$_2$), the size and density of Pt on GNS could be easily modulated. For the Pt/GNS sample subjected to 50 ALD cycles, single atoms of Pt were found to distribute dominantly on the surface of GNS (bright spots on the dark background in Figure 5.6A,A′). After 100 ALD cycles, some larger clusters started to appear, forming NPs with two groups of average size of 1 and 2 nm, respectively (Figure 5.6B,B′). Both single atoms and clusters were observed in this sample, likely because the single atoms appearing in the first 50 cycles had grown in the second 50 cycles, creating bigger clusters and particles, while some new single atoms were formed at the same time. After 150 cycles, the size of preformed clusters and NPs further grew with three groups of sizes at 1, 2, and 4 nm, respectively; similarly, the newly formed single atoms and clusters were also visible (Figure 5.6C,C′). High-resolution scanning transmission electron microscopy (STEM) images show that these Pt clusters and NPs are well crystallized (inset, Figure 5.6B′,C′). The authors performed a detailed X-ray absorption near-edge structure (XANES) analysis and found that the

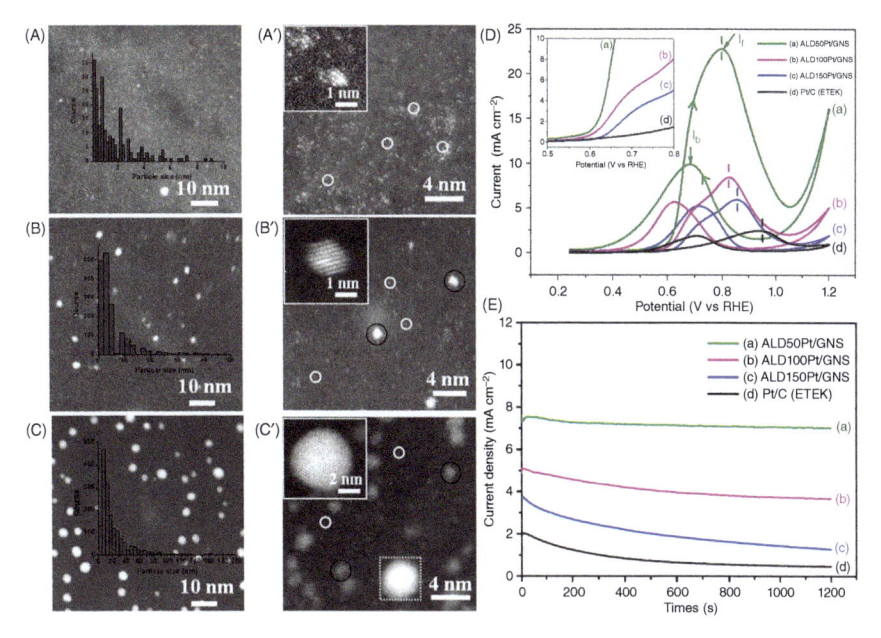

Figure 5.6 HAADF-STEM images of Pt/GNS samples. (A, B, C) present the results with 50, 100, and 150 ALD cycles, respectively, and (A′, B′, C′) show the corresponding magnified images. Inset in each figure shows the corresponding histogram of Pt clusters on GNS. (D) Cyclic voltammograms (CVs) of methanol oxidation on various Pt catalysts. The inset is the enlarged CV curves in the onset potential region of methanol oxidation. (E) Chronoamperometry (CA) curves of various Pt catalysts recorded at a constant potential of 0.6 V versus RHE for 20 min. The electrocatalysis tests were conducted at room temperature in an Ar-saturated aqueous solution containing 1 M MeOH and 0.5 M H_2SO_4. Label of samples: (a) ALD50Pt/GNS, (b) ALD100Pt/GNS, (c) ALD150Pt/GNS, and (d) Pt/C. (Sun et al. [17]. Reproduced with permission of Nature Publishing Group.)

ALD-derived Pt single atoms, clusters, and NPs exhibited a metallic feature. Electrocatalytic tests revealed that the MOR onset potentials of the ALD-Pt/GNS after 50, 100, and 150 cycles (denoted as ALD50Pt/GNS, ALD100Pt/GNS, ALD150Pt/GNS hereafter) as well as commercial Pt/C catalysts were 0.59, 0.60, 0.62, and 0.70 V versus reversible hydrogen electrode (RHE), respectively (inset, Figure 5.6D). The peak potentials for MOR in the forward scan of ALD-Pt/GNS catalysts increases with the number of ALD cycles with the following order: ALD50Pt/GNS (0.79 V) < ALD100Pt/GNS (0.82 V) < ALD150Pt/GNS (0.85 V), all of which show a significant negative shift relative to that of Pt/C (0.96 V). The negative shifts of the onset and peak potentials indicate that ALD-Pt/GNS catalysts, particularly the ALD50Pt/GNS, are able to substantially reduce the overpotential for methanol oxidation, compared to the Pt/C commercial catalysts. Moreover, it was found that the methanol oxidation peak current density of ALD50Pt/GNS catalysts (22.9 mA cm^{-2}) was 2.7, 4.0, and 9.5 times higher than those of ALD100Pt/GNS, ALD150Pt/GNS, and Pt/C, respectively. The authors attributed this enhancement to the intrinsic nature of Pt single atoms and subnanometer clusters, which have more low-coordinated and unoccupied

5d orbitals and stronger Pt–C interaction, according to their XANES analyses. Furthermore, the authors also examined the poisoning tolerance of the ALD-Pt catalysts. They found that the ratio of the forward anodic peak current density (I_f) to the backward anodic peak current density (I_b), I_f/I_b, of ALD50Pt/GNS, is 2.23, much higher than that of other samples, meaning that ALD50Pt/GNS has the best tolerance for carbon monoxide (CO) poisoning (Figure 5.6D). This was also confirmed by the CO stripping experiments performed by the authors [17]. The excellent CO tolerance of ALD-Pt catalysts was further evidenced by chronoamperometric tests, where the ALD-Pt/GNS samples show a slower current decay over time compared to commercial Pt/C catalysts (Figure 5.6E).

In addition, Pt NP catalysts were also deposited on carbon black (CC) [20] and ZnO nanorod (ZnO-NR) arrays supported on carbon cloth [45] using ALD for use to catalyze MOR. Remarkably improved catalytic activity of ALD-derived Pt was observed in both cases. Moreover, the CO poisoning tolerance of both ALD-Pt/CC and ALD-Pt/ZnO-NR was found to be enhanced. This could be attributed to the presence of Pt–O groups and hydroxide species on ZnO surface, respectively, which are favorable for the removal of CO-like intermediates on the Pt surface to release active sites via the so-called bifunctional mechanism. In the case of ALD of Pt on ZnO-NRs, it was also reported that UV light irradiation of the catalyst surface could help to improve the chronoamperometric response by 62%, due to the synergistic effect of photo-oxidation of methanol on ZnO and electro-oxidation of methanol on Pt. In this case, ZnO NRs, on the one hand, can react with CO-like intermediate species on the Pt surface to form CO_2 and release active sites for further methanol oxidation; on the other hand, alter the electronic structure of Pt NPs such that the binding between Pt and CO-like intermediates is remarkably weakened. Consequently, in the presence of UV light, the ALD-Pt/ZnO-NR electrode exhibited 90% higher methanol oxidation activity compared to the commercial Pt/C electrode with the same Pt loading. Similar promotion effect can be expected as well if ALD of Pt is performed on other metal oxide nanostructures [46, 47]. However, whether the ALD-Pt/ZnO-NR composite can find an application in a practical fuel cell is questionable since ZnO is not chemically stable in both acidic and basic solutions.

Apart from MOR the electrocatalytic performance of ALD-derived Pt NPs supported on carbon was tested for EOR and FOR [14, 19–21]. Juang et al. deposited Pt NPs with different sizes and densities on carbon black powders and studied their catalytic activity toward EOR [19]. The size and density modulation was realized by a modified ALD process in which the MeCpPtMe$_3$ pulse time was systematically adjusted from 4 to 20 s. The average size of the deposited Pt NPs was found to monotonously increase with the increasing MeCpPtMe$_3$ pulse time, showing a diffusion-limited growth behavior. When tested in the electrolyte containing 0.5 M H_2SO_4 and 0.5 M ethanol at room temperature (Figure 5.7a), all ALD-Pt catalysts exhibited an ethanol oxidation potential of <0.2 V versus Ag/AgCl, and the forward ethanol oxidation current of ALD-Pt with an average particle size of 2.9 nm (4 s MeCpPtMe$_3$) was as high as 380 A g$^{-1}_{Pt}$ at a scan rate of 50 mV s^{-1}, eight times larger than that of ALD-Pt with a mean size of 10.4 nm (20 s MeCpPtMe$_3$). Furthermore, the small

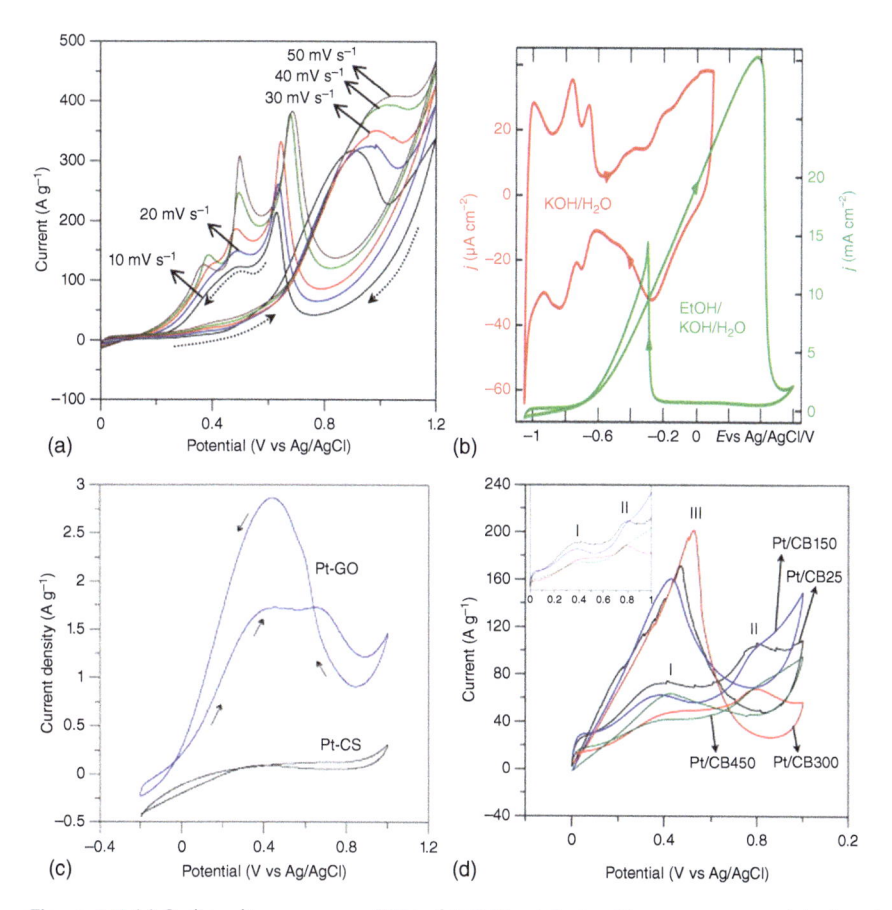

Figure 5.7 (a) Cyclic voltammograms (CVs) of ALD-Pt catalysts with an average particle size of 2.9 nm, recorded in 0.5 M H_2SO_4 + 0.5 M ethanol at various scan rates. (Juang et al. [19]. Reproduced with permission of Elsevier.) (b) CVs of Pt clusters supported on TiO_2 nanotubes carried out in 1 M KOH and 1 M KOH + 1 M ethanol. Scan rate: 50 mV s^{-1}. (Reproduced from Ref. [21] with permission from the Royal Society of Chemistry.) (c) CV profiles of ALD-Pt catalysts measured in 0.5 M H_2SO_4 + 0.5 M HCOOH at 10 mV s^{-1}. (Hsieh et al. [14]. Reproduced with permission of Elsevier.) (d) CVs of different ALD-Pt catalysts measured in 0.5 M H_2SO_4 + 0.5 M HCOOH at 50 mV s^{-1}. The inset shows CV curves within the potential range of 0–1 V versus Ag/AgCl. (Hsieh et al. [20]. Reproduced with permission of Elsevier.)

ALD-Pt NPs were found to exhibit better poisoning tolerance compared to larger particles. The outstanding electrocatalytic performance for EOR of small ALD-Pt NPs was attributed to the remarkably large electrochemically active surface area (ECSA) they delivered (105 m^2 g$^{-1}_{Pt}$), significantly higher than that of other ALD-Pt NPs and many Pt-based catalysts reported in the literature [19]. Very recently, Pt NP catalysts were also deposited on anodized TiO_2 nanotube arrays using ALD, which showed good catalytic activity toward EOR in both acid and base solutions (Figure 5.7b) [21].

DFAFCs can offer higher energy density and much lower fuel crossover compared to DMFCs and therefore attracted much attention in the past decade. So far, there have been only few reports on the electro-oxidation of formic acid using ALD of Pt catalysts. In 2012, Hsieh et al. reported the ALD of Pt NPs on both graphene oxide (GO) and carbon spheres (CS) and studied their electrocatalytic performance toward FOR [14]. Figure 5.7c shows the CV profiles of Pt-GO and Pt-CS measured in 0.5 M H_2SO_4 + 0.5 M HCOOH at 10 mV s^{-1}. Although the average size of Pt NPs on GO and CS is similar and the density of Pt NPs on GO (mainly located at edges of GO) is lower than that on CS, Pt-GO still exhibited much higher mass catalytic activity toward FOR. This was attributed to the high density of oxygenated species at the edges of GO, which offer an antipoisoning effect favoring the oxidation of Pt-CO_{ads} generated from the dissociative adsorption step [14]. The same group of authors also deposited Pt NPs on carbon black powders and studied the annealing effect in reductive atmosphere on the catalytic activity toward FOR [20]. Electrochemical measurements showed that the ALD-Pt catalysts without annealing treatment exhibited the highest activity for FOR (Figure 5.7d), which can be ascribed to the presence of a Pt-O skin layer showing a "bifunctional" effect.

Oxygen Reduction Reaction (ORR) Oxygen reduction reaction is the half-cell reaction occurring at the cathode in a fuel cell. It also plays a vital role in governing the performance in some other electrochemical devices such as metal–air batteries [48]. Since 2011, the use of ALD-derived Pt NP catalysts for ORR has been reported by several groups [12, 18, 49–52]. Dameron et al. deposited Pt NPs on vertically aligned CNT arrays functionalized by TMA and O_2 plasma, respectively, and their electrochemical measurements showed that at 0.9 V versus RHE, the specific activity of ALD-Pt/CNT for ORR (470 µA cm$^{-2}_{Pt}$ for TMA functionalized CNTs and 790 µA cm$^{-2}_{Pt}$ for O_2 plasma-treated CNTs) is much higher than that of commercial Pt/C catalysts, even if Pt/C possesses substantially higher ECSA (i.e., 101 m^2 g$^{-1}_{Pt}$) [12].

The cathode usually operates under harsh conditions, such as low pH (<1), high O_2 concentration, high humidity, and high potentials (e.g., 0.6–1.2 V) [53], in which carbon supports are prone to corrosion. This would result in the agglomeration of Pt NPs and thereby rapid degradation of the cell's performance. In addition, the weak interaction between carbon supports and Pt NPs would lead to sintering of Pt NPs and a consequent decrease in catalytic activity. For these reasons, researchers have been pursuing to replace carbon supports with other corrosion-resistant support materials that can sustain the harsh conditions for ORR. This effort has been made in conjunction with the attempt to reduce Pt utilization using ALD [49–52].

In 2011, Hsu and coworkers reported ALD of Pt on thin films of tungsten carbide (WC) – an inexpensive material whose electronic properties are very similar to those of Pt [49]. The authors attempted to grow a monolayer Pt on WC, which would be an effective ORR catalyst according to their density functional theory (DFT) prediction. However, they found that Pt nucleated and grew on WC in an island growth, rather than the layer-by-layer mode. As a result, they only obtained discrete Pt NPs covering on the WC surface. Nevertheless, the ALD-Pt/WC exhibited improved catalytic activity toward ORR, even if only

20 Pt ALD cycles were used (10 s MeCpPtMe$_3$–30 s N$_2$–2 s O$_2$–30 s N$_2$). Later, Xie *et al.* reported ALD of Pt on the surface of TiSi$_2$ nanonets [50]. Interestingly, they observed that Pt NPs were only selectively deposited on the top and bottom surfaces of the nanonets, namely, *b* crystal planes terminated by Si (Figure 5.8). Moreover, these Pt NPs were multitwinned with Pt {111} surfaces exposed preferentially. After performing control experiments, the authors suggested that the selective deposition of Pt and the unique twinned microstructure of Pt NPs resulted from the interaction between Pt and the TiSi$_2$ nanonet *b* planes. Notwithstanding nonoptimization, the electrocatalytic activity of ALD-Pt/TiSi$_2$ was found to be much higher than that of optimized Pt/C catalysts (160 vs 90 µA cm$^{-2}_{Pt}$ at 0.9 V vs RHE, Figure 5.8b), when measured at the same Pt loading mass (50 µg$_{Pt}$ cm^{-2}).

Very recently, Pt NPs were also deposited on ceramic supports of zirconium carbide (ZrC), a very hard and conducting material with good corrosion resistance [51]. According to the authors' high-resolution TEM and XANES analyses (Figure 5.8), the deposited Pt NPs had strong interaction with the ZrC support at the embedded Pt–ZrC interfaces, which modified the electronic structure of Pt. This not only resulted in enhanced specific and mass activities (Figure 5.8F) about three times higher than those of Pt/C but also remarkably improved the catalytic stability of ALD-Pt/ZrC (fivefold more stable than that of Pt/C). The same group further developed this strategy and successfully encapsulated the ALD Pt NPs in open zirconia (ZrO$_2$) nanocages through a smart area-selective ALD [52]. To do so, Pt NPs were initially deposited on N-CNTs by ALD, followed by the application of a blocking agent (oleylamine) to the Pt NPs surface. ALD ZrO$_2$ selectively grew around the Pt NPs, but not on the Pt surface because of the presence of blocking agents. In this way, a nanocage structure may be formed by precisely controlling the ALD ZrO$_2$ layer. The thus-obtained ALD-ZrO$_2$-Pt/N-CNT catalysts exhibited significantly improved stability toward ORR, losing only about 8% of their initial ECSA after 4000 cycles of accelerated degradation test (ADT). In sharp contrast to this, the ECSA of commercial Pt/C catalyst decreased by 82% after the same cycling test.

In the past decades, supported catalysts have been used in various types of fuel cells to catalyze electrode reactions. As mentioned earlier, for supported catalysts, there are two major drawbacks to be overcome: the corrosion of carbon supports and weak catalyst/support interaction. These problems have been addressed to a certain extent using the strategies discussed in the previous sections; however, there still remain significant obstacles to widespread commercialization of PEM fuel cells. In recent years, the use of supportless nanostructured catalysts such as nanowires and nanotubes [18, 54–57] has captured increasing attention, because this eliminates the need for carbon supports together with the associated corrosion problem. With this consideration in mind, Galbiati *et al.* recently reported supportless Pt nanotube (NT) arrays fabricated by ALD of Pt using porous anodic aluminum oxide (AAO) as a template [18]. Prior to the dissolution of AAO, the membrane was hot-pressed on a proton-exchange Nafion® film, which constitutes the basement of the Pt NT array. The area of Pt NT arrays can be made as large as 47 mm in diameter

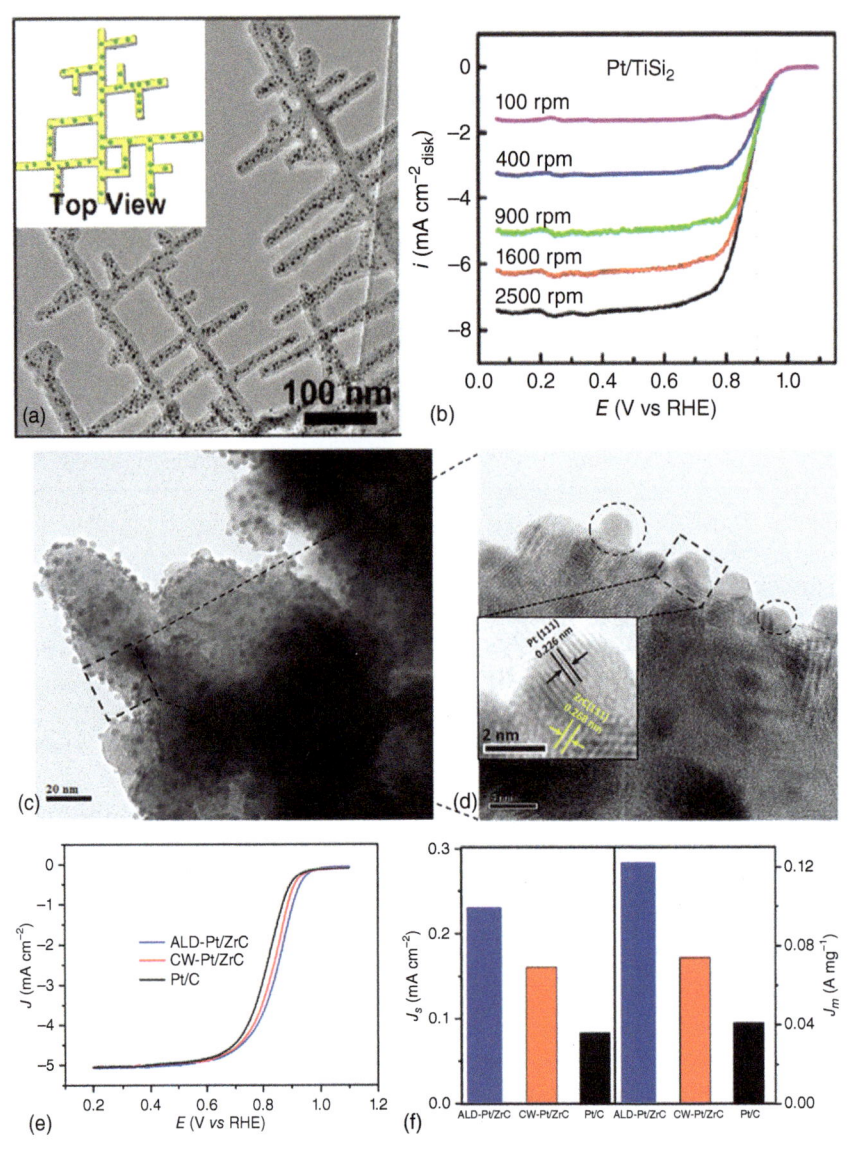

Figure 5.8 (a) Top-view TEM image of the ALD-Pt/TiSi$_2$ heteronanostructures after a typical 50-cycle ALD growth; (b) polarization curves of ALD-Pt/TiSi$_2$ measured in 0.1 M KOH at a scan rate of 10 mV s^{-1} at varying rotation rates. (Xie et al. [50]. Reproduced with permission of American Chemical Society.) (c, d) TEM images of the ALD-Pt/ZrC catalysts. (e) Polarization curves of electrodes made from ALD-Pt/ZrC, chemically reduced (CW)-Pt/ZrC composites and E-Tek Pt/C catalyst in an O$_2$-saturated 0.5 M H$_2$SO$_4$ solution at room temperature (1600 rpm, scan rate: 10 mV s^{-1}). (f) Specific and mass activity at 0.9 V versus RHE for these catalysts. (Cheng et al. [51]. Reproduced with permission of Royal Society of Chemistry.)

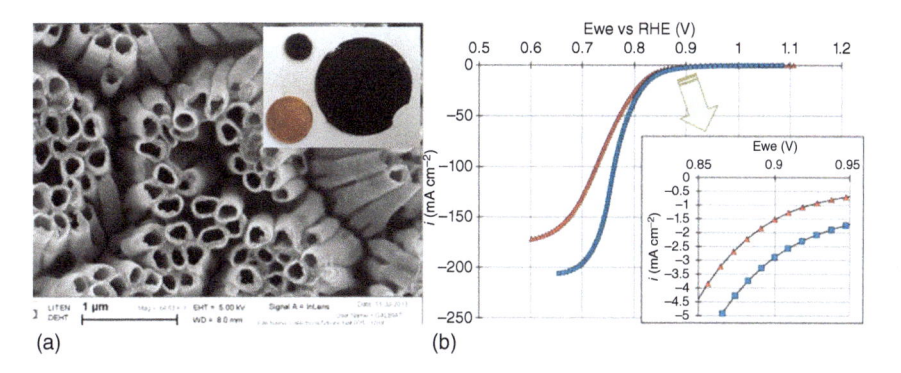

Figure 5.9 (a) SEM image showing ALD-Pt NTs with a thin Nafion® coating layer. Inset: optical photograph of the ALD-Pt NT membrane electrodes. (b) Polarization curves of the ALD-Pt NT (red triangles) and Pt/C (blue squares) electrodes measured in 0.5 M H_2SO_4 in O_2-saturated atmosphere. (Galbiati et al. [18]. Reproduced with permission of Elsevier.)

(inset, Figure 5.9), such that it may be used in a practical fuel cell. The authors directly used the Nafion®-supported Pt NT array as working electrode and tested its electrocatalytic activity toward ORR by simulating the actual fuel cell environment with the Nafion® film in direct contact with electrolyte while the Pt NT array was in contact with the O_2 gas flow. Preliminary results showed that the Pt NT array exhibited a bit higher specific ORR activity compared to Pt/C (37 vs 28 $\mu A\,cm^{-2}_{Pt}$ at 0.9 V vs RHE). Given that the NTs were only 2 μm long and sparsely distributed, there is still much room for improvement in the future by adjusting the structural parameters of the NT arrays.

Hydrogen Evolution Reaction (HER) While considerable efforts have been dedicated to engineering electrocatalysts for use in fuel cells, little attention has been paid to the catalysts for use in water electrolyzers. Especially, the ALD technique has not attracted sufficient interest from researchers working on electrolyzers for the time being. Consequently, there are few reports on ALD of Pt for use in catalyzing HER in the literature [58, 59]. Hsu and coworkers investigated the HER performance of ALD-Pt catalysts deposited on WC particles [58]. According to the authors' DFT calculations, the electronic properties of a monolayer (ML) Pt supported on WC are pretty similar to those of bulk Pt [60], and therefore, ML-Pt/WC should be an efficient catalyst for HER. Although an ML-Pt was not achieved by ALD, the ALD-Pt/WC still exhibited HER activity comparable to that of 10% Pt/C commercial catalysts, but with a nearly 10-fold lower Pt loading. Very recently, Liu et al. reported ALD of Pt NPs and thin films on both p-type silicon and glassy carbon [59]. They compared the HER activity of a 3-nm ALD-Pt layer with that of an e-beam sputtered Pt layer with the same thickness and found that ALD-Pt exhibited slightly higher activity in the high overpotential region. The ALD-Pt needs an overpotential of 33.6 ± 0.4 mV to achieve a current density of −10 mA cm^{-2}, similar to that of the sputtered Pt layer (33.3 (±0.8) mV).

5.2.2 ALD of Pd Electrocatalysts

Similar to Pt, palladium (Pd) is also a noble metal that can find many applications in electrocatalysis, particularly in electro-oxidation of small organic molecules and oxygen reduction reaction [61–63]. ALD of Pd cannot be accomplished as easily as that of Pt. Different precursors such as Pd(keim$_2$)$_2$ (keim$_2$ = CF$_3$C(O)CHC(NBu^n)CF$_3$) and Pd(thd)$_2$ (thd = 2,2,6,6,-tetramethyl-3,5-heptanedionato) have been used, but they all showed unsatisfactory results [64–66]. To date, the best results have been obtained with Pd(hfac)$_2$ (hfac = hexafluoroacetylacetonate), which is currently the most commonly used precursor for Pd ALD [24, 25, 27, 28]. Unlike the ALD processes for other Pt-group metals where an oxidant is usually used as a coreactant, ALD processes based on Pd(hfac)$_2$ requires a true reducing agent, most commonly H$_2$.

Although there are numerous reports concerning ALD of Pd published since the pioneering work done by Senkevich *et al.* in 2003 [67], only few works have been dedicated to electrocatalytic studies using ALD-Pd [24, 28]. Rikkinen and coworkers deposited Pd NPs on Vulcan XC72R carbon support by exposing the support to Pd(thd)$_2$ precursors at 180 °C for a long time (6 h per cycle) [24]. Figure 5.10b shows the morphology of the ALD-Pd/C catalysts, where uniform Pd NPs with a mean particle size of 2.6 nm are evenly distributed on

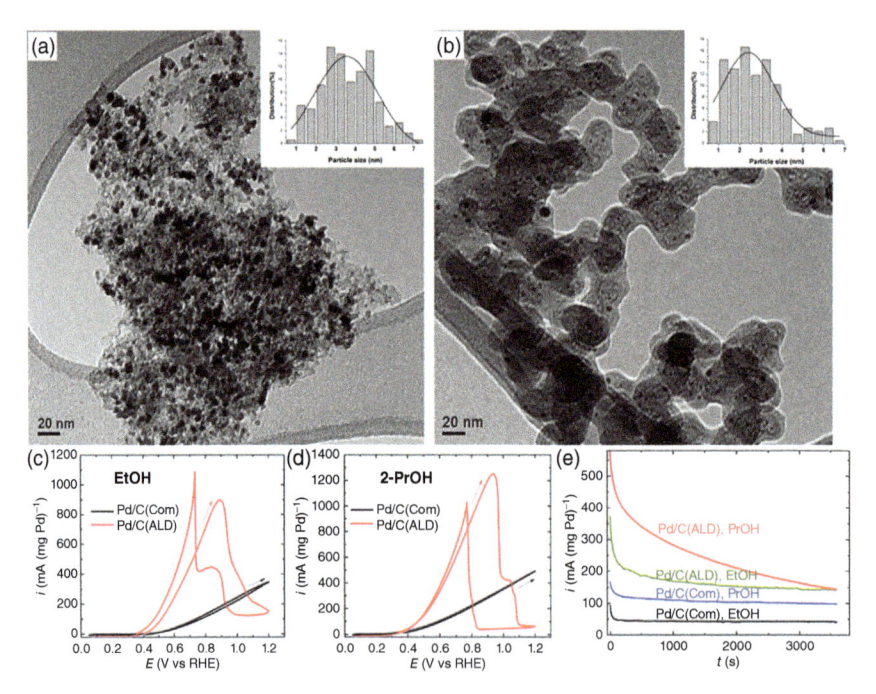

Figure 5.10 TEM images of (a) commercial Pd/C and (b) ALD-Pd/C catalysts. The inset shows the histogram of NP sizes. CV curves of (c) ethanol and (d) isopropanol oxidation measured in the electrolyte containing 0.1 M NaOH and 1 M alcohol at 10 mV s^{-1} and 1800 rpm. The third CV scan is presented. (e) Chronoamperometric curves measured at 0.7 V versus RHE in 0.1 M NaOH + 1 M alcohol. (Rikkinen *et al.* [24]. Reproduced with permission of American Chemical Society.)

the carbon support. Unlike the commercial Pd/C catalysts (Figure 5.10a), only a few aggregates were observed in ALD-Pd/C, suggesting that ALD offers better dispersion. The authors performed CV measurements in 0.1-M NaOH solution and estimated the ECSA of the catalysts based on the charge needed for PdO reduction at the cathodic sweep. It was shown that the ECSA of ALD-Pd/C was >2 times higher than that of commercial Pd/C, although the Pd loading for ALD-Pd/C was significantly lower (3.5 wt% vs 20 wt% for Pd/C). Electrocatalytic tests revealed that the onset potential of ALD-Pd/C is more negative than that of commercial Pd/C for both ethanol and isopropanol electro-oxidation reactions. Compared to commercial Pd/C, ALD-Pd/C also exhibited much higher mass activity for alcohol oxidation (about 2.5 times, Figure 5.10c,d). The stability of both commercial Pd/C and ALD-Pd/C was investigated and compared (Figure 5.10e). Not surprisingly, the ALD-Pd/C catalysts showed much slower current decay over time compared to the commercial Pd/C. It is also worth noting that for ALD-Pd/C catalysts, better catalytic performance was observed when they were used to electro-oxidize isopropanol, indicating that isopropanol would be an interesting fuel for an alkaline direct alcohol fuel cell. However, as shown in Figure 5.10e, surface poisoning is a vital issue for isopropanol oxidation. To overcome this problem, alloying Pd with a metal that can alleviate the poisoning effect is necessary.

Pd has been demonstrated to be more efficient than Pt for catalyzing formic acid oxidation [68, 69]. Recently, Assaud et al. studied electrocatalytic activity of ALD-derived Pd catalysts toward FOR [28]. They performed ALD using $Pd(hfac)_2$ and formalin as precursors (1 s $Pd(hfac)_2$–30 s purge–3 s formalin–30 s purge) on Ni-coated nanochannels of AAO membranes. The Pd/Ni configuration was used to enhance the catalytic performance. Their preliminary results showed that FOR could be accomplished at low potentials with a high oxidation current. Although the authors did not compare the catalytic performance of ALD-Pd catalysts with that of commercial Pd/C, the mass activity of ALD-Pd catalysts was indeed high comparable to the literature values. Specifically, when 40 Pd ALD cycles were used, the mass activity could be as high as 0.83 A mg^{-1}_{Pd}. However, the stability of these catalysts was very poor as Ni can be easily dissolved in acid solutions.

Besides, Pd has proved to be able to efficiently catalyze ORR. However, the relevant investigations were only carried out toward the application in lithium–air batteries [63, 70]. This is beyond the scope of this chapter.

5.2.3 ALD of Pt-Based Alloy and Core/Shell Nanoparticle Electrocatalysts

Nanoparticles of Pt-based alloys and core/shell nanostructures have been widely proposed to be more efficient electrocatalysts compared to Pt NPs. So far, a wide range of Pt-based alloys and core/shell NPs has been successfully fabricated. Readers are recommended to read some review papers published recently [71–75]. Pt alloys and core/shell nanostructures have several advantages over pure Pt: (i) the utilization of precious Pt can be effectively reduced by alloying with a transition metal or by transition metal core/Pt shell structures; (ii) alloying or lattice strain induced at core/shell interfaces modifies electronic structure of Pt promoting electrocatalytic reactions; (iii) the ensemble effect arising from the introduction of a second metal can help to improve tolerance against poisoning. Therefore, Pt-based alloys and core/shell NPs are regarded as promising electrocatalysts for next-generation PEM fuel cells.

5.2.3.1 ALD of Pt Alloy Nanoparticle Electrocatalysts

PtRu Alloys PtRu has been extensively studied as it is known as the best electrocatalyst toward MOR [72]. However, PtRu fabricated by ALD has so far been rarely explored [29–32]. In 2010, Jiang *et al*. reported ALD of PtRu thin films with different compositions using a "supercycle" consisting of "x" cycles of Pt ALD followed by "y" cycles of Ru ALD [30]. Each supercycle was repeated until the total number of individual ALD cycles reached a target number. The composition of PtRu films could be modulated by adjusting the values of x and y. The authors tested the electrocatalytic activity of these films for MOR and concluded that the film having a stoichiometric Pt:Ru ratio of about 1 : 1 had the highest activity.

Compared to thin films, ALD of PtRu alloy NPs is more challenging, because Ru can in principle either be deposited on the pre-existing Pt clusters forming bimetallic PtRu alloys or nucleate itself on the support forming isolated Ru clusters. Christensen and coworkers attempted to codeposit PtRu on alumina particles [29]. To differentiate between separate Ru and Pt NPs and PtRu alloys, the authors performed XANES analyses and confirmed the bimetallic nature of the deposited NPs. Alloying of Ru with Pt was also verified by X-ray diffraction analysis by Johansson *et al*. in a recent work [32]. The formation of PtRu alloys upon ALD is likely because Ru has higher surface energy compared to Pt, and therefore, Ru is expected to nucleate as islands on Pt [31]. Johansson *et al*. recently showed that uniform PtRu NPs can be evenly deposited on N-CNTs (Figure 5.11a,b) [32]. They investigated the electrocatalytic performance of ALD-PtRu NPs with different compositions toward MOR. According to the anodic scans they obtained in the first cycle, ALD-PtRu/N-CNT catalysts show a very pronounced promotional effect at low overpotentials in a region of technological interest for a DMFC (Figure 5.11). However, at higher overpotentials (e.g., >0.6 V), the catalytic current of ALD-PtRu/N-CNT is not as high as that of ALD-Pt/N-CNT because at this potential, Pt alone can dissociate water with a reasonable rate to form OH groups and oxidize the poisoning carbon species. For long-term operation, ALD-PtRu/N-CNT catalysts showed much better performance, evidenced by the fact that all ALD-PtRu/N-CNT catalysts delivered higher current densities compared to ALD-Pt/N-CNT (Figure 5.11d).

PtCo Alloys Recently, it has been reported that PtCo NPs can be successfully deposited on carbon black (Vulcan 72R) in an ALD reactor by exposing the carbon support to $Pt(acac)_2$ and $Co(acac)_2$ in sequence at 180 °C for 6 h [35, 36]. Note that although the authors claimed this process an "ALD," no typical self-limiting reactions were involved. The deposition of NPs essentially resulted from thermal decomposition of precursors. For comparison, the authors also synthesized PtCoPt and PtCoPtPt catalysts by performing one and two more Pt "ALD" cycles, respectively, after the deposition of PtCo. Since all samples were not obtained in

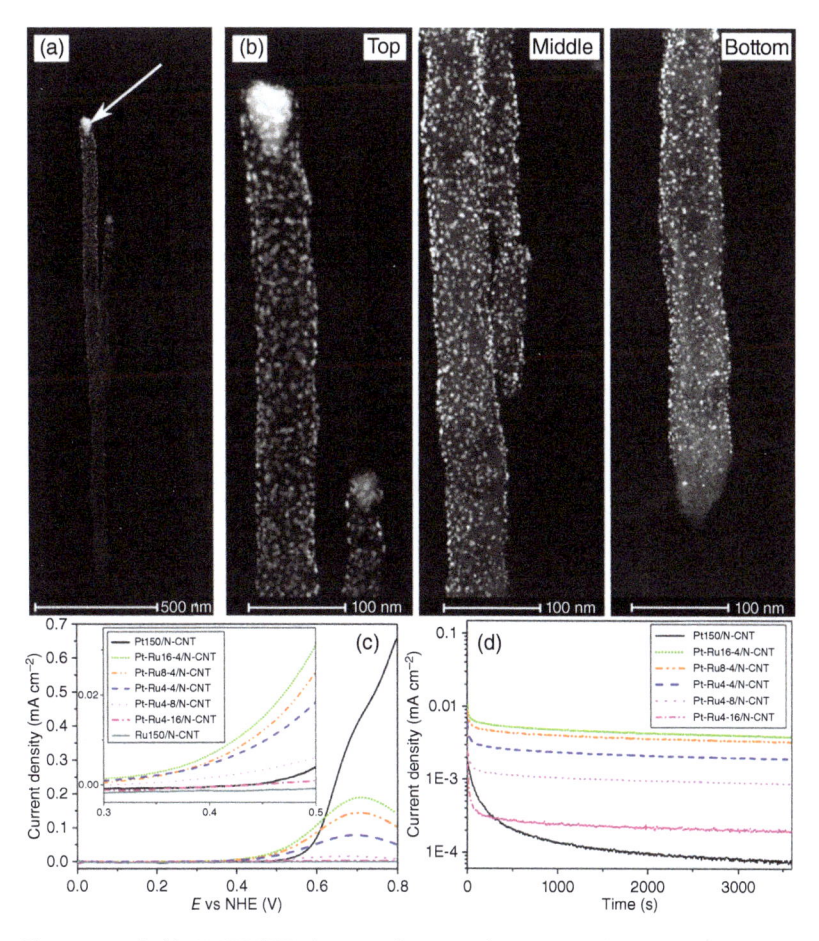

Figure 5.11 (a, b) HAADF STEM images of a 1.8 μm long N-CNT decorated with PtRu catalyst NPs. The arrow points at the CNT growth catalyst (Ni). (c) Anodic scans (1st CV cycle) of the ALD-Pt, -PtRu, and -Ru catalyst measured in 0.5 M H_2SO_4 + 1 M MeOH at 10 mV s^{-1}. (d) Chronoamperograms of the catalysts measured at 0.4 V versus NHE in the same electrolyte. (Johansson et al. [32]. Reproduced with permission of Elsevier.)

the same experimental batch, the size and distribution of NPs of different samples were inconsistent and therefore difficult to correlate with each other [35]. Nevertheless, according to the authors' electrocatalytic tests, PtCo catalysts exhibited remarkably enhanced performance for both MOR and ORR in an acid solution, when compared to other references. The same authors also investigated the EOR performance of the PtCo catalysts at various temperatures and concluded that the introduction of Co by "ALD" can effectively enhance the catalytic performance for EOR, mainly through promoting the reaction via a 12 e$^-$ path [36]. However, poor stability at elevated temperatures is still a challenging issue to address before using these catalysts in a practical fuel cell.

Recently, a general strategy has been reported on selective growth of a secondary metal on the primary metal surface while avoiding growth on the support to exclude monometallic NP formation [34]. To address the challenges of the mismatched deposition temperature and surface chemistry for the ALD processes of two different metals, the authors first established appropriate ALD conditions to accomplish selective deposition of a secondary metal on the primary metal surfaces, but not on oxide support surfaces. They successfully obtained well-mixed bimetallic alloys of PtPd, PdRu, and PtRu by fine-tuning the deposition sequences. The deposition of two metals was performed at the same temperature, making precise control over the size, composition, and structure of alloy NPs practically possible. However, no electrocatalytic tests were carried out using these bimetallic alloy NPs.

5.2.3.2 ALD of Core/Shell Nanoparticle Electrocatalysts

Core/shell NPs often show improved catalytic properties compared to their alloyed counterparts or to mixtures of monometallic NPs, because the lattice strain created at the core/shell interfaces and the heterometallic bonding interactions can modify the surface electronic properties, promoting electrocatalytic reactions [76].

Kessels and coworkers recently demonstrated that Pd/Pt and Pt/Pd core/shell NPs can be fabricated solely by ALD, taking advantage of the island growth of primary metal cores on oxide substrates, followed by subsequent selective deposition of shells of the second metal on the preformed primary metal NPs [33, 77]. For example, to obtain Pd/Pt core/shell NPs, they first performed ALD of Pd on Al_2O_3 substrates at 100 °C using $Pd(hfac)_2$ and H_2 plasma as precursors [33]. This resulted in the formation of Pd NPs with an average diameter of 2.6 ($\pm$0.5) nm and a density of 8.6×10^{11} NPs cm^{-2}. Selective growth of Pt was then accomplished at 300 °C without breaking vacuum using $MeCpPtMe_3$ and O_2 gas as precursors wherein a low O_2 partial pressure of 7.5 mTorr was used during the O_2 exposure. Under this condition, the dissociative chemisorption of O_2, a key step for Pt growth, only takes place on the preformed Pd NPs, but not on the Al_2O_3 substrate. Therefore, Pt will be selectively deposited on the surface of Pd NPs, forming a core/shell structure (Figure 5.12a,b). The Pd/Pt core/shell NPs have an average diameter of 4.1 ($\pm$0.5) nm and a density virtually equal to that of the preformed Pd NPs, suggesting that the Pt ALD process only takes place on the Pd cores without creating new monometallic Pt NPs. In their following work, the authors further demonstrate that both the size of Pd cores and thickness of Pt shells can be precisely tuned at a subnanometer level by adjusting the cycle numbers during ALD of Pd and Pt, respectively [77]. Using the same strategy, they also obtained Pt/Pd core/shell NPs supported on Al_2O_3 substrates. Furthermore, they showed that these Pt/Pd core/shell NPs can be deposited on high-aspect-ratio GaP nanowires (Figure 5.12c). This opens up a new possibility of using Pt/Pd core/shell NPs in semiconductor photoelectrocatalysis.

Although Pd/Pt and Pt/Pd core/shell NPs can be successfully obtained using the aforementioned selective ALD, the use of a plasma renders the method unsuitable for high-surface-area substrates where the radical species in the plasma would recombine before reaching the inner surfaces. Moreover, ALDs

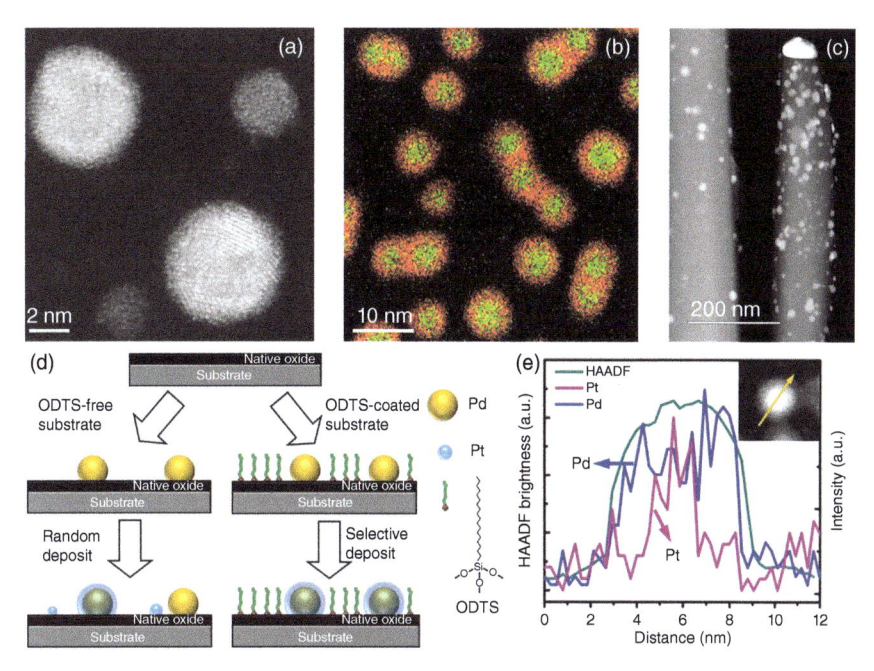

Figure 5.12 (a) HAADF-STEM image and (b) EDX mapping of Pd/Pt core/shell NPs (150 cycle Pd; 50 cycle Pt) grown on an Al_2O_3-covered Si_3N_4 TEM window. (c) HAADF-STEM image showing Pt/Pd core/shell NPs deposited on Al_2O_3-coated GaP nanowires. The large particle at the top of the nanowire is a gold particle, which is used to grow the nanowires in the VLS process. (Weber et al. [77]. Reproduced with permission of Institute of Physics.) (d) Schematic illustration for fabricating the core/shell NPs through area-selective ALD on ODTS modified substrates. (e) HAADF-STEM line scan over a typical Pt/Pd core/shell NP. Inset: the HAADF-STEM image. (Adapted from [78]. http://www.nature.com/articles/srep08470 Used under creative commons license: https://creativecommons.org/licenses/by/4.0/.)

of Pd and Pt were carried out at different temperatures, which makes the growth of core/shell NPs practically inconvenient. By properly selecting the deposition temperature and the appropriate coreactant, Elam et al. recently prepared Pd/Pt and Pt/Pd core/shell NPs at a constant deposition temperature without the need for a plasma [34]. To achieve this, however, many efforts have to be made to fine-tune the ALD parameters. Chen and coworkers reported a general area-selective ALD technique that allows one to fabricate core/shell NPs using only thermal ALD [78]. They used octadecyltrichlorosilane (ODTS) self-assembled monolayers (SAMs) to modify the substrate surface. Since ODTS does not react with ALD precursors, the ALDs of primary and secondary metals only take place in the pinholes of the SAMs, as schematically illustrated in Figure 5.12d. Particularly, new nucleation sites can be effectively blocked by surface ODTS SAMs in the second deposition stage. In this way, uniform Pd/Pt and Pt/Pd core/shell NPs with a narrow size distribution can be produced in a high yield (Figure 5.12e). This method can also be extended to the fabrication of core/shell NPs composed of other metals.

5.3 ALD of Transition Metal Oxide Electrocatalysts

While noble metals and their alloys are the state-of-the-art electrocatalysts being used in fuel cells and electrolyzers, transition metal oxides (TMOs) have recently emerged as promising alternatives to noble metals to catalyze several important reactions including MOR, ORR, and OER [79, 80]. Given the significantly low costs and high natural abundance of TMOs, replacement of noble metal catalysts by TMO counterparts would potentially lower the production costs of fuel cells and electrolyzers and promote widespread deployment of these devices.

ALD represents a powerful technique for TMO deposition. However, the present efforts to depositing TMO catalysts using ALD are mainly devoted to the application in photoelectrochemical devices, which will be described in detail in Chapter 8. TMO catalysts fabricated by ALD for use in fuel cells or electrolyzers have only received little attention so far [81–83].

In 2012, Tong et al. reported the fabrication of CNT-NiO hybrids and studied their electrocatalytic properties toward MOR [81]. Using bis(cyclopentadienyl) nickel (Cp_2Ni) as the precursor and ozone as the coreactant, uniform NiO

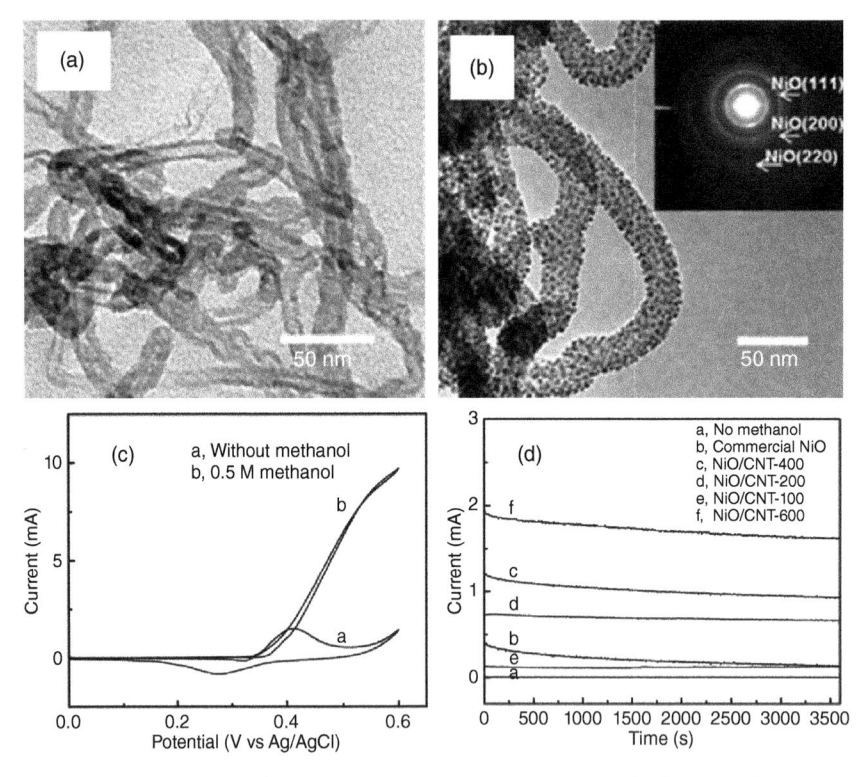

Figure 5.13 TEM images of (A) pristine CNTs and (B) CNT-NiO hybrids obtained after 400 ALD cycles. Inset: SAED pattern. (C) Cyclic voltammograms of CNT-NiO catalysts measured in 0.5 M methanol + 1 M KOH solution at a scan rate of 50 mV s^{-1}. (D) Chronoamperometric curves of CNT-NiO catalysts with different loadings recorded at the potential of 0.45 V versus Ag/AgCl. (Tong et al. [81]. Reproduced with permission of Wiley.)

NPs were deposited on the surface of CNTs (Figure 5.13A) with a high particle density. Unlike the previously reported, the authors did not perform an acid pretreatment to CNTs. They claimed that oxygenated species and defect sites can be introduced on the CNT surface during the ozone half cycles, providing nucleation sites for NiO. The authors demonstrated that the size of NiO NTs monotonously increased with the increasing number of ALD cycles, being 1.5 nm after 100 cycles and 6.3 nm after 600 cycles. Furthermore, they tested the electrocatalytic performance of the CNT-NiO toward MOR and found that the hybrid catalysts exhibited pronounced oxidation current in the presence of methanol and remarkably enhanced long-term stability compared to that of commercially available NiO nanopowders (Figure 5.13). However, they failed to provide further information about the specific and mass activities of the samples so that it is difficult to make comparison with the literature.

Manganese oxide has emerged as a bifunctional electrocatalyst that is active for both ORR and OER [84]. Using ALD, Bent *et al.* deposited MnO_x thin films over glassy carbon substrates and investigated the electrocatalytic performance of the deposited MnO_x toward ORR and OER [82]. The as-deposited MnO_x was found to uniformly cover the whole surface of glassy carbon (Figure 5.14a), having a Mn:O stoichiometric ratio of 1 : 1. After annealing the film at 480 °C in air for 10 h, MnO was transformed into Mn_2O_3. Meanwhile, the film became rough and porous (Figure 5.14b), likely resulting from a reaction of the underlying glassy carbon during thermal annealing. These two samples exhibited different electrocatalytic properties: the MnO film has very poor catalytic activity for ORR but excellent catalytic activity for OER, while the Mn_2O_3 film shows very good catalytic activities for both ORR and OER, comparable to those of noble metal references (Figure 5.14c,d). Apart from MnO_x, Fe_2O_3 is a popular TMO electrocatalyst for OER as well. However, a planar Fe_2O_3 thin film usually exhibits poor catalytic activity because of the low electrical conductivity of Fe_2O_3. Recently, Bachmann *et al.* demonstrated that through proper nanostructuring, the area-specific catalytic activity of Fe_2O_3 electrodes can be dramatically improved [83]. They deposited Fe_2O_3 into a porous AAO template through ALD using ferrocene and ozone as precursors (Figure 5.14e). This resulted in a vertically aligned array of Fe_2O_3 nanotubes (Figure 5.14f). When used to oxidize water, the nanostructured Fe_2O_3 electrode exhibited significantly enhanced oxidation current density, one order of magnitude higher than that of the flat reference sample (Figure 5.14g).

5.4 Summary and Outlook

ALD is a powerful thin-film deposition technique based on sequential and self-limiting surface reactions, and it allows the deposition to occur on substrates with three-dimensional surface topologies or with high aspect ratios. Nonideal ALD often results in the formation of discrete nanoparticles, which provides an attractive approach to the preparation of heterogeneous catalysts where catalytic clusters uniformly dispersed on highly porous substrates are preferred. In this respect, nanoparticles of various noble metals and their alloys

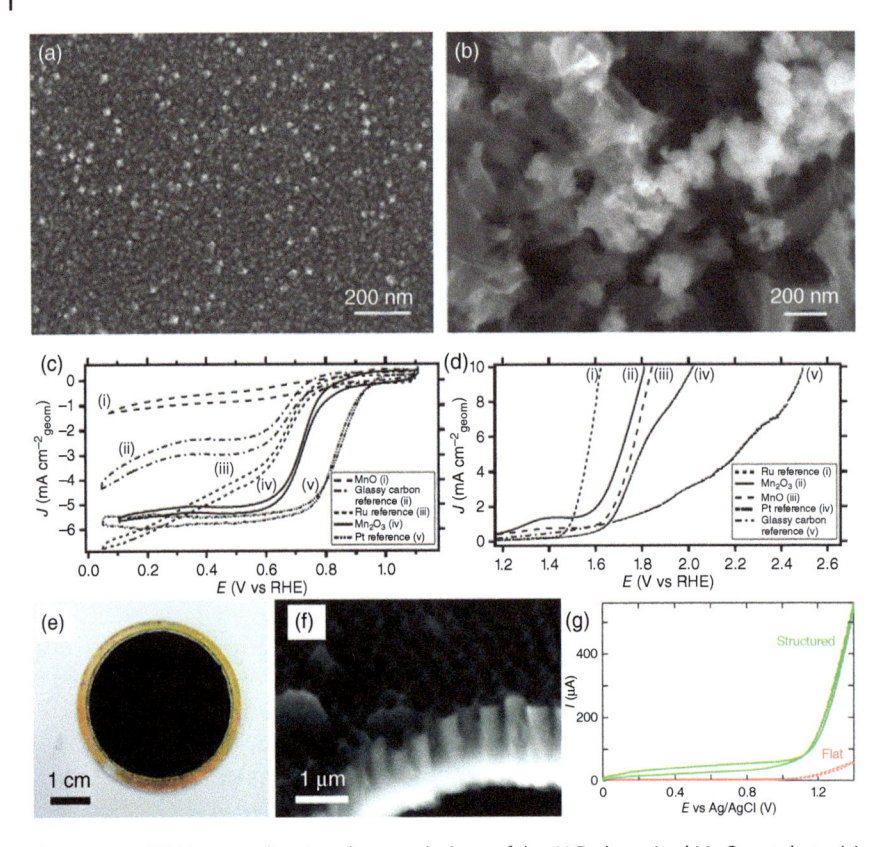

Figure 5.14 SEM images showing the morphology of the ALD-deposited MnO_x catalysts. (a) As-deposited MnO. (b) Mn_2O_3 obtained by annealing MnO at 480 °C in air for 10 h. (c) ORR performance of the MnO_x catalysts tested in O_2-saturated 0.1 M KOH at 1600 rpm. (d) OER performance of the MnO_x catalysts tested in O_2-saturated 0.1 M KOH solution. (Pickrahn et al. [82]. Reproduced with permission of John Wiley and Sons.) (e) Digital photograph of a porous Fe_2O_3 electrode. (f) SEM image of the porous Fe_2O_3 electrode. (g) Cyclic voltammogram of a freshly prepared nanostructured electrode (green curve), compared with a flat electrode of the same macroscopic area (0.30 cm^{-2}) prepared with ALD of Fe_2O_3 on an ITO film sputtered on Si (red dotted curve). Scan rate: 20 mV s^{-1}. pH = 7. (Gemmer et al. [83]. Reproduced with permission of Elsevier.)

have been successfully deposited on high-surface-area carbon supports using ALD. Compared to conventional preparation methods, ALD renders uniform dispersion of size-controllable catalyst nanoparticles over the entire surface of catalyst supports. Moreover, ALD is believed to be able to robustly immobilize the catalytic species on the surface of supports due to the strong chemical interaction between them. Consequently, ALD-derived electrocatalysts have shown improved catalytic activity and stability in comparison with the commercially available catalysts, when used in fuel cells or electrolyzers.

Notwithstanding remarkable progress, full potential of ALD for the fabrication of electrocatalysts to be used in fuel cells and water electrolyzers has by far not

5.4 Summary and Outlook

been reached. Future efforts should be directed, but not limited, to the following aspects:

1) Using new carbon-based catalyst supports without a need for pretreatment. Pretreatment by either acid or plasma is still needed presently in order to deposit noble metal and alloy nanoparticles on carbon supports. However, pretreatment in many cases will deteriorate the chemical and electrical properties of carbon supports, leading to poor catalytic performance. Therefore, new supports that have intrinsically high electrical conductivity and enough surface defect sites for catalyst nucleation, for example, N-doped CNTs/graphene, should be used. N-doped CNTs/graphene by themselves are catalytically active for ORR in alkaline solutions and can offer a synergistic effect on ORR of the ALD-deposited catalysts.
2) More efforts should be made to electrocatalytic study of alloy and core/shell nanoparticle catalysts derived by ALD. ALD has proven to be a powerful technique to fabricate uniform, size- and composition-controllable alloy and core/shell catalysts. However, the electrocatalytic performance of these alloy and core/shell catalysts is not clear yet.
3) Full-cell electrochemical tests need to be performed for use of ALD-derived Pt catalysts in DLFCs. Presently, all full-cell tests using the ALD-derived Pt catalysts were conducted in H_2–O_2 fuel cells. For use in DLFCs, only electrochemical data from half-reactions (e.g., MOR, EOR, and FOR) are available. To justify the benefit of ALD-derived Pt and Pt-alloy catalysts, full-cell tests under operation conditions must be conducted.
4) Catalysts for use in electrolyzers should be developed vigorously. Current efforts to using ALD technique to fabricate electrocatalysts are primarily devoted to the application in fuel cells, and few researches are dedicated to developing HER and OER catalysts for electrolyzers. This does not match the recent rapid development in this vibrant field.
5) Developing inexpensive Pt-free electrocatalysts. Although Pt is the most efficient catalyst for many electrochemical reactions, it is expensive and has limited availability in the earth's crust and therefore practically cannot be used in fuel cells or electrolyzers on a large scale. There has been a trend of replacing Pt with less expensive materials, for example, Pd and earth-abundant metal oxide catalysts. ALD of Pd was already reported, but the electrocatalytic performance of ALD-derived Pd has not been investigated under operation conditions. Metal oxides are widely believed to be a good alternative to Pt for ORR. However, few metal oxide ORR catalysts have been synthesized using ALD, even if ALD is an effective approach to the deposition of metal oxides.
6) Development of supportless catalysts. Supportless catalysts have been proposed to be able to alleviate catalyst agglomeration, Ostwald ripening, and catalyst loss that the supported catalysts often suffer from under operation conditions. There are some reports published on ALD-derived Pt nanotubes for use as supportless electrocatalysts. However, further investigation is still needed.

Acknowledgment

The author acknowledges the support of the FCT Investigator Grant (IF/01595/2014).

References

1 Vielstich, W., Lamm, A., and Gasteiger, H.A. (eds) (2003) *Handbook of Fuel Cells: Fundamentals, Technology, and Applications*, John Wiley & Sons, Inc..
2 Sorensen, B. (2012) *Hydrogen and Fuel Cells*, 2nd edn, Elsevier Ltd.
3 Bessarabov, D., Wang, H.J., Li, H., and Zhao, N.N. (2015) *PEM Electrolysis for Hydrogen Production: Principles and Applications*, Taylor & Francis.
4 Zhang, H.W. and Shen, P.K. (2012) *Chem. Soc. Rev.*, **41**, 2382–2394.
5 Carmo, M., Fritz, D.L., Merge, J., and Stolten, D. (2013) *Int. J. Hydrogen Energy*, **38**, 4901–4934.
6 Litster, S. and McLean, G. (2004) *J. Power Sources*, **130**, 61–76.
7 Chan, K.Y., Ding, J., Ren, J.W., Cheng, S.A., and Tsang, K.Y. (2004) *J. Mater. Chem.*, **14**, 505–516.
8 Wee, J.H., Lee, K.Y., and Kim, S.H. (2007) *J. Power Sources*, **165**, 667–677.
9 King, J.S., Wittstock, A., Biener, J., Kucheyev, S.O., Wang, Y.M., Baumann, T.F., Giri, S.K., Hamza, A.V., Baeumer, M., and Bent, S.F. (2008) *Nano Lett.*, **8**, 2405–2409.
10 Hsueh, Y.C., Hu, C.T., Wang, C.C., Liu, C., and Perng, T.P. (2008) *ECS Trans.*, **16**, 855–862.
11 Liu, C., Wang, C.C., Kei, C.C., Hsueh, Y.C., and Perng, T.P. (2009) *Small*, **5**, 1535–1538.
12 Dameron, A.A., Pylypenko, S., Bult, J.B., Neyerlin, K.C., Engtrakul, C., Bochert, C., Leong, G.J., Frisco, S.L., Simpson, L., Dinh, H.N., and Pivovar, B. (2012) *Appl. Surf. Sci.*, **258**, 5212–5221.
13 Shu, T., Liao, S.J., Hsieh, C.T., Roy, A.K., Liu, Y.Y., Tzou, D.Y., and Chen, W.Y. (2012) *Electrochim. Acta*, **75**, 101–107.
14 Hsieh, C.T., Chen, W.Y., Tzou, D.Y., Roy, A.K., and Hsiao, H.T. (2012) *Int. J. Hydrogen Energy*, **37**, 17837–17843.
15 Hsueh, Y.C., Wang, C.C., Liu, C., Kei, C.C., and Perng, T.P. (2012) *Nanotechnology*, **23**, 405603.
16 Hsueh, Y.C., Wang, C.C., Kei, C.C., Lin, Y.H., Liu, C., and Perng, T.P. (2012) *J. Catal.*, **294**, 63–68.
17 Sun, S.H., Zhang, G.X., Gauquelin, N., Chen, N., Zhou, J.G., Yang, S.L., Chen, W.F., Meng, X.B., Geng, D.S., Banis, M.N., Li, R.Y., Ye, S.Y., Knights, S., Botton, G.A., Sham, T.K., and Sun, X.L. (2013) *Sci. Rep.*, **3**, 1775.
18 Galbiati, S., Morin, A., and Pauc, N. (2014) *Electrochim. Acta*, **125**, 107–116.
19 Juang, R.S., Hsieh, C.T., Hsiao, J.Q., Hsiao, H.T., Tzou, D.Y., and Huq, M.M. (2015) *J. Power Sources*, **275**, 845–851.
20 Hsieh, C.T., Hsiao, H.T., Tzou, D.Y., Yu, P.Y., Chen, P.Y., and Jang, B.S. (2015) *Mater. Chem. Phys.*, **149–150**, 359–367.

21 Assaud, L., Schumacher, J., Tafel, A., Bochmann, S., Christiansen, S., and Bachmann, J. (2015) *J. Mater. Chem. A*, **3**, 8450–8458.
22 Hsieh, C.T., Liu, Y.Y., Tzou, D.Y., and Chen, Y.C. (2014) *J. Taiwan Inst. Chem. Eng.*, **45**, 186–191.
23 Hsieh, C.T., Liu, Y.Y., Tzou, D.Y., and Chen, W.Y. (2012) *J. Phys. Chem. C*, **116**, 26735–26743.
24 Rikkinen, E., Santasalo-Aarnio, A., Airaksinen, S., Borghei, M., Viitanen, V., Sainio, J., Kauppinen, E.I., Kallio, T., and Krause, A.O.I. (2011) *J. Phys. Chem. C*, **115**, 23067–23073.
25 Feng, H., Libera, J.A., Stair, P.C., Miller, J.T., and Elam, J.W. (2011) *ACS Catal.*, **1**, 665–673.
26 Liang, X.H., Lyon, L.B., Jiang, Y.B., and Weimer, A.W. (2012) *J. Nanopart. Res.*, **14**, 943.
27 Weber, M.J., Mackus, A.J.M., Verheijen, M.A., Longo, V., Bol, A.A., and Kessels, W.M.M. (2014) *J. Phys. Chem. C*, **118**, 8702–8711.
28 Assaud, L., Monyoncho, E., Pitzschel, K., Allagui, A., Petit, M., Hanbucken, M., Baranova, E.A., and Santinacci, L. (2014) *Beilstein J. Nanotechnol.*, **5**, 162–172.
29 Christensen, S.T., Feng, H., Libera, J.L., Guo, N., Miller, J.T., Stair, P.C., and Elam, J.W. (2010) *Nano Lett.*, **10**, 3047–3051.
30 Jiang, X.R., Gur, T.M., Prinz, F.B., and Bent, S.F. (2010) *Chem. Mater.*, **22**, 3024–3032.
31 Johansson, A.C., Yang, R.B., Haugshoj, K.B., Larsen, J.V., Christensen, L.H., and Thomsen, E.V. (2013) *Int. J. Hydrogen Energy*, **38**, 11406–11414.
32 Johansson, A.C., Larsen, J.V., Verheijen, M.A., Haugshoj, K.B., Clausen, H.F., Kessels, W.M.M., Christensen, L.H., and Thomsen, E.V. (2014) *J. Catal.*, **311**, 481–486.
33 Weber, M.J., Mackus, A.J.M., Verheijen, M.A., van der Marel, C., and Kessels, W.M.M. (2012) *Chem. Mater.*, **24**, 2973–2977.
34 Lu, J.L., Low, K.B., Lei, Y., Libera, J.A., Nicholls, A., Stair, P.C., and Elam, J.W. (2014) *Nat. Commun.*, **5**, 3264.
35 Sairanen, E., Figueiredo, M.C., Karinen, R., Santasalo-Aarnio, A., Jiang, H., Sainio, J., Kallio, T., and Lehtonen, J. (2014) *Appl. Catal., B*, **148–149**, 11–21.
36 Santasalo-Aarnio, A., Sairanen, E., Aran-Ais, R.M., Figueiredo, M.C., Hua, J., Feliu, J.M., Lehtonen, J., Karinen, R., and Kallio, T. (2014) *J. Catal.*, **309**, 38–48.
37 Utriainen, M., Kroger-Laukkanen, M., Johansson, L.S., and Niinisto, L. (2000) *Appl. Surf. Sci.*, **157**, 151–158.
38 Aaltonen, T., Ritala, M., Sajavaara, T., Keinonen, J., and Leskelä, M. (2003) *Chem. Mater.*, **15**, 1924–1928.
39 Mabena, L.F., Ray, S.S., Mhlanga, S.D., and Coville, N.J. (2011) *Appl. Nanosci.*, **1**, 67–77.
40 Chen, Y.G., Wang, J.J., Liu, H., Li, R.Y., Sun, X.L., Ye, S.Y., and Knights, S. (2009) *Electrochem. Commun.*, **11**, 2071–2076.
41 Chen, Y.G., Wang, J.J., Liu, H., Banis, M.N., Li, R.Y., Sun, X.L., Sham, T.K., Ye, S.Y., and Knights, S. (2011) *J. Phys. Chem. C*, **115**, 3769–3776.

42 Gong, K.P., Du, F., Xia, Z.H., Durstock, M., and Dai, L.M. (2009) *Science*, **323**, 760–764.
43 Tian, G.L., Zhao, M.Q., Yu, D.S., Kong, X.Y., Huang, J.Q., Zhang, Q., and Wei, F. (2014) *Small*, **10**, 2251–2259.
44 Soloveichik, G.L. (2014) *Beilstein J. Nanotechnol.*, **5**, 1399–1418.
45 Su, C.Y., Hsueh, Y.C., Kei, C.C., Lin, C.T., and Perng, T.P. (2013) *J. Phys. Chem. C*, **117**, 11610–11618.
46 Zhang, H., Zhou, W., Du, Y., Yang, P., Wang, C., and Xu, J. (2010) *Int. J. Hydrogen Energy*, **35**, 13290–13297.
47 Song, H., Qiu, X., Li, X., Li, F., Zhu, W., and Chen, L. (2007) *J. Power Sources*, **170**, 50–54.
48 Cheng, F.Y. and Chen, J. (2012) *Chem. Soc. Rev.*, **41**, 2172–2192.
49 Hsu, I.J., Hansgen, D.A., McCandless, B.E., Willis, B.G., and Chen, J.G. (2011) *J. Phys. Chem. C*, **115**, 3709–3715.
50 Xie, J., Yang, X.G., Han, B.H., Shao-Horn, Y., and Wang, D.W. (2013) *ACS Nano*, **7**, 6337–6345.
51 Cheng, N.C., Banis, M.N., Liu, J., Riese, A., Mu, S.C., Li, R.Y., Sham, T.K., and Sun, X.L. (2015) *Energy Environ. Sci.*, **8**, 1450–1455.
52 Cheng, N.C., Banis, M.N., Liu, J., Riese, A., Li, X., Li, R.Y., Ye, S.Y., Knights, S., and Sun, X.L. (2015) *Adv. Mater.*, **27**, 277–281.
53 Ferreita, P.J., Ia O', G.J., Shao-Horn, Y., Morgan, D., Makharia, R., Kocha, S., and Gasteiger, H.A. (2005) *J. Electrochem. Soc.*, **152**, A2256.
54 Debe, M.K. (2012) *Nature*, **486**, 43–51.
55 Chen, Z.W., Waje, M., Li, W.Z., and Yan, Y.S. (2007) *Angew. Chem. Int. Ed.*, **46**, 4060–4063.
56 Liu, L.F., Pippel, E., Scholz, R., and Gosele, U. (2009) *Nano Lett.*, **9**, 4352–4358.
57 Liu, L.F. and Pippel, E. (2011) *Angew. Chem. Int. Ed.*, **50**, 2729–2733.
58 Hsu, I.J., Kimmel, Y.C., Jiang, X.Q., Willis, B.G., and Chen, J.G. (2012) *Chem. Commun.*, **48**, 1063–1065.
59 Liu, R., Han, L.H., Huang, Z.Q., Ferrer, I.M., Smets, A.H.M., Zeman, M., Brunschwig, B.S., and Lewis, N.S. (2015) *Thin Solid Films*, **586**, 28–34.
60 Esposito, D.V., Hunt, S.T., Stottlemyer, A.L., Dobson, K.D., McCandless, B.E., Birkmire, R.W., and Chen, J.G.G. (2010) *Angew. Chem. Int. Ed.*, **49**, 9859–9862.
61 Yin, Z., Lin, L.L., and Ma, D. (2014) *Catal. Sci. Technol.*, **4**, 4116–4128.
62 Long, N.V., Thi, C.M., Yong, Y., Nogami, M., and Ohtaki, M. (2013) *J. Nanosci. Nanotechnol.*, **13**, 4799–4824.
63 Lei, Y., Lu, J., Luo, X.Y., Wu, T.P., Du, P., Zhang, X.Y., Ren, Y., Wen, J.G., Miller, D.J., Miller, J.T., Sun, Y.K., Elam, J.W., and Amine, K. (2013) *Nano Lett.*, **13**, 4182–4189.
64 Aaltonen, T., Ritala, M., Tung, Y.L., Chi, Y., Arstila, K., Meinander, K., and Leskela, M. (2004) *J. Mater. Res.*, **19**, 3353–3358.
65 Hamalainen, J., Puukilainen, E., Sajavaara, T., Ritala, M., and Leskela, M. (2013) *Thin Solid Films*, **531**, 243–250.
66 Lashdaf, M., Hatanpaa, T., Krause, A.O.I., Lahtinen, J., Lindblad, M., and Tiitta, M. (2003) *Appl. Catal., A*, **241**, 51–63.

67 Senkevich, J.J., Tang, F., Rogers, D., Drotar, J.T., Jezewski, C., Lanford, W.A., Wang, G.C., and Lu, T.M. (2003) *Chem. Vap. Deposition*, **9**, 258–264.
68 Zhu, Y.M., Khan, Z., and Masel, R.I. (2005) *J. Power Sources*, **139**, 15–20.
69 Ha, S., Larsen, R., and Masel, R.I. (2005) *J. Power Sources*, **144**, 28–34.
70 Lu, J., Lei, Y., Lau, K.C., Luo, X.Y., Du, P., Wen, J.G., Assary, R.S., Das, J., Miller, D.J., Elam, J.W., Albishri, H.M., El-Hady, D.A., Sun, Y.K., Curtiss, L.A., and Amine, K. (2014) *Nat. Commun.*, **4**, 2383.
71 Bing, Y.H., Liu, H.S., Zhang, L., Ghosh, D., and Zhang, J.J. (2010) *Chem. Soc. Rev.*, **39**, 2184–2202.
72 Zhao, X., Yin, M., Ma, L., Liang, L., Liu, C.P., Liao, J.H., Lu, T.H., and Xing, W. (2011) *Energy Environ. Sci.*, **4**, 2736–2753.
73 Long, N.V., Yang, Y., Thi, C.M., Minh, N.V., Cao, Y.Q., and Nogami, M. (2013) *Nano Energy*, **2**, 636–676.
74 Liu, B., Liao, S.J., and Liang, Z.X. (2011) *Prog. Chem.*, **23**, 852–859.
75 Strasser, P. (2009) *Rev. Chem. Eng.*, **25**, 255–295.
76 Lei, Y., Liu, B., Lu, J.L., Lobo-Lapidus, R.J., Wu, T.P., Feng, H., Xia, X.X., Mane, A.U., Libera, J.A., Greeley, J.P., Miller, J.T., and Elam, J.W. (2012) *Chem. Mater.*, **24**, 3525–3533.
77 Weber, M.J., Verheijen, M.A., Bol, A.A., and Kessels, W.M.M. (2015) *Nanotechnology*, **26**, 094002.
78 Cao, K., Zhu, Q.Q., Shan, B., and Chen, R. (2015) *Sci. Rep.*, **5**, 8470.
79 Doyle, R.L., Godwin, I.J., Brandon, M.P., and Lyons, M.E.G. (2013) *Phys. Chem. Chem. Phys.*, **15**, 13737–13783.
80 Hong, W.T., Risch, M., Stoerzinger, K.A., Grimaud, A., Suntivich, J., and Shao-Horn, Y. (2015) *Energy Environ. Sci.*, **8**, 1404–1427.
81 Tong, X.L., Qin, Y., Guo, X.Y., Moutanabbir, O., Ao, X.Y., Pippel, E., Zhang, L.B., and Knez, M. (2012) *Small*, **8**, 3390–3395.
82 Pickrahn, K.L., Park, S.W., Gorlin, Y., Lee, H.B.R., Jaramillo, T.F., and Bent, S.F. (2012) *Adv. Energy Mater.*, **2**, 1269–1277.
83 Gemmer, J., Hinrichsen, Y., Abel, A., and Bachmann, J. (2012) *J. Catal.*, **290**, 220–224.
84 Gorlin, Y. and Jaramillo, T.F. (2010) *J. Am. Chem. Soc.*, **132**, 13612–13614.

6

Atomic Layer Deposition for Thin-Film Lithium-Ion Batteries

Ola Nilsen, Knut B. Gandrud, Amund Ruud, and Helmer Fjellvåg

University of Oslo, Department of Chemistry, Centre for Materials Science and Nanotechnology (SMN), P.O. Box 1033 Blindern, 0315 Oslo, Norway

6.1 Introduction

Thin-film coatings become increasingly important in electrochemical systems such as for the lithium-ion batteries. This does not only apply to the emerging all solid-state thin-film battery designs, but is also equally true for the more traditional powder-based batteries using liquid electrolytes. In this perspective, atomic layer deposition (ALD) has a special position since it can provide pinhole-free films with utmost control of coating thickness, even on geometrically complex surfaces [1] (see also Chapter 1). It should be emphasized that these fields, both from the battery design principle and from relevant process development by ALD, are still relatively immature. There is currently no definite design or solution for large-scale implementation of coating of battery materials yet. This makes the field particularly interesting for fundamental research and development. This chapter aims at giving an overview of the status and possibilities in application of ALD for advancement of lithium-ion batteries.

Reversible electrochemical systems are challenging from an interface point of view. This is especially true for the lithium-based batteries where large differences in electrochemical potential are typically sought. An electrochemical battery consists, in rough terms, of a cathode, an electrolyte, and an anode, with current collectors on both sides. In its charged state, the only part that prevents a reaction from taking place between the often extreme electrochemical potential difference between the cathode and anode is the poor ability of the electrolyte to conduct electrons. However, as the difference in electrochemical potential between the cathode and anode is increased, the number of possible alternative, and mostly undesired, reactions increases with it. The most typically known examples are decomposition of the electrolyte and formation of solid electrolyte interface (SEI) layers [2]. However, one frequently also observes dissolution of electroactive elements from the cathode with subsequent plating on the anode, as well as interface reactions between the electrodes and their respective current collectors, among others [3, 4]. These undesired side-reactions leads to loss of electroactive material and typically an overall increase in the impedance of the cell. There is a balance

Atomic Layer Deposition in Energy Conversion Applications, First Edition. Edited by Julien Bachmann.
© 2017 Wiley-VCH Verlag GmbH & Co. KGaA. Published 2017 by Wiley-VCH Verlag GmbH & Co. KGaA.

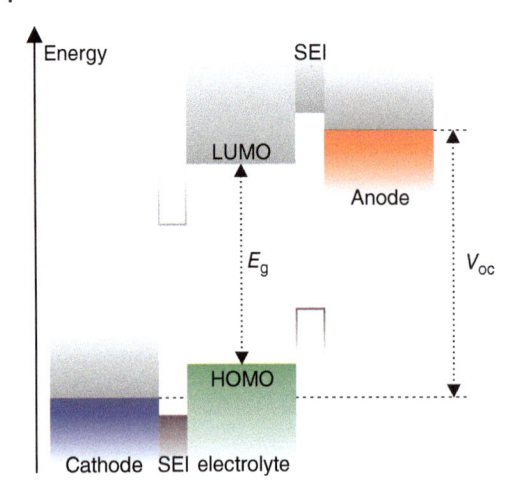

Figure 6.1 Schematic energy diagram of a battery with SEI layers toward both the cathode and anode. Gray and colored areas represent vacant and occupied energy levels, respectively. The positions of the LUMO for the SEI layer toward the cathode and the HOMO for the SEI layer toward the anode is largely arbitrary. V_{oc} is the open-circuit potential and E_g is the band gap of the electrolyte.

between the electrochemical potential and overall durability of a battery. However, through continued optimization of design, the balance is constantly shifted toward increasing volumetric and specific capacity and power.

One design principle to avoid or limit undesired side-reactions is to add interface layers that are sustainable toward the degradation mechanism of the particular interface. The overall idea behind this principle has been described earlier by Goodenough and Kim [2]. The often observed SEI layers are in fact such interface layers leading to increased stability of the battery. However, these are typically not long-term stable and slowly increase in thickness, leading to higher impedance and loss of active material. The overall principle is reproduced in Figure 6.1 and relates to choosing an interface material with HOMO and LUMO levels (or valence and conduction band, respectively) at such positions that electrochemical decomposition of the electrolyte is prevented. If the HOMO level of the SEI layer toward the cathode is lower than the Fermi level of the cathode, then redox reactions between the electrolyte and the cathode are prevented. Similarly, the LUMO of the SEI layer toward the anode must be at a higher level compared to its Fermi level in order to prevent transport of electrons to the electrolyte.

6.2 Coated Powder Battery Materials by ALD

Stabilization or prevention of SEI layers through ALD coating has been demonstrated with great success in many cases. Rather good overviews of coating of powder-based battery materials are given in [5–7]. A selection of the types of materials tested so far are as follows: Al_2O_3 [8–16], $LiAlO_2$ [3], TiO_2 [17–20], TiN [21, 22], HfO_2 [23, 24], ZrO_2 [25], $LiTaO_3$ [26], LiF and AlW_xF_y [27], $FePO_4$ [28]; however, their function should also be viewed in relation to the type of electrolyte and electrode materials. Application of such additional interface layers is a balance between achieving good protection while not deteriorating the ionic conductivity and reducing the overall capacity. The optimal thickness of the interface layer is typically a few cycles, and it is evident that many of the layers applied are so thin that they should be unable to form a complete pinhole-free coating.

Since the main functionality of some of these layers is to prevent dissolution of electroactive elements into the electrolyte, it can be speculated that the ALD coating primarily passivates those sites on the particle surfaces that would also be most prone to be dissolved.

There are also additional or alternative benefits from ALD coatings on both cathode and anode materials. One of the main limitations in the application of $LiCoO_2$ as cathode material has been that only half of the Li is accessible before the lattice oxygen atoms also become oxidized during charging, with possible detrimental outcomes. It has been proven that the extent of active Li can be increased from 0.5 to 0.7 by the application of a layer of Al_2O_3 on the particles of $LiCoO_2$. The mechanisms are unclear but should relate to an increase in structural stability [29].

Silicon is currently being introduced as a replacement for carbon as anode material. However, its application is challenged by the large volumetric change during cycling, which hampers the stability of most naturally formed SEI layers. An molecular layer deposition (MLD) process employing trimethylaluminum (TMA) and glycerol has been added to silicon anode material to grow a mechanical, robust, while still flexible, barrier [30]. The nano-Si composite electrodes are capable of providing capacities of nearly 900 mAh g^{-1} for over 100 cycles, as compared to only 5 cycles for the bare nano-Si material. In a similar manner, TMA and ethylene glycol have been applied to sulfur cathodes for lithium sulfur batteries, showing a significantly prolonged cycle life compared to uncoated and Al_2O_3-coated cathode material [31].

The majority of the materials listed earlier as tested so far are not known as good ionic or electronic conductors and hence expected to increase the overall impedance of the battery. If these coatings are applied to free-flowing powder materials, the overall electronic conductivity of the electrode will be hampered since the coating will also be present at the grain boundaries between the particles and at the interface between the particles and the carbon additive when electrode tapes are produced. One procedure to overcome this effect is to rather coat the finalized electrode tape than the powder itself [3, 16]. This results in a reduced overpotential compared to powder coating and also improved kinetics [16]. The benefit of coating the electrode tape rather than the powder material is well demonstrated for the properties of passivated natural graphite anodes at elevated temperatures where natural SEI layers rapidly deteriorate [16]. The Al_2O_3-coated natural graphite rapidly decays and behaves more poorly than uncoated material due to reduced electronic conductivity, while coated electrode tape shows a significant improved cyclability (Figure 6.2). The major consequences between these principles are related to design of the ALD reactors and production logistics.

One method to increase the power capability of the passivated layer is to use a solid-state Li-ion conducting material as the passivated layer. This has been demonstrated and discussed in [26] and [3] where $LiTaO_3$ and $LiAlO_2$ have been tested as passivation layers, respectively. These layers may actually function as short-range lithium storage layers providing initial high-power capabilities.

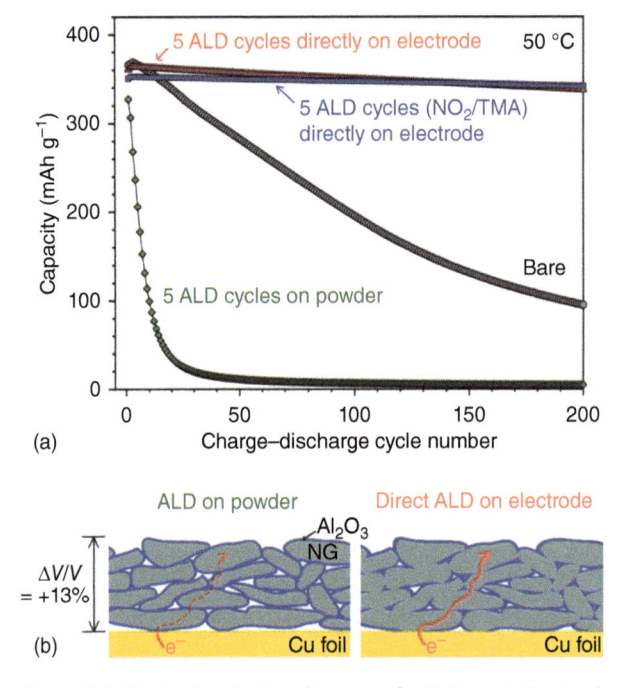

Figure 6.2 Electrochemical performance for ALD-coated natural graphite composite electrodes. (a) Cycle performance at 50 °C, (b) schematic representation of transport in composite electrodes prepared by ALD on powder and by ALD directly on the electrode. (Jung et al. 2010 [16]. Reproduced with permission of John Wiley and Sons.)

This argument is demonstrated in [28] where amorphous $FePO_4$ is used as electrochemical buffer layer and Li storage layer on $LiNi_{0.5}Mn_{1.5}O_4$. An increase in high rate capacity and electrochemical stability is demonstrated as compared to the uncoated powder material; however, the overall capacity is reduced, probably due to the relatively lower electrical conductivity of $FePO_4$.

Based on the observations so far, the ideal passivation layer should show both high Li-ion and electronic conductivities while being stable toward the electrolyte. The simultaneous need for electronic conductivity is not compatible if the electrolyte itself decomposes and forms SEI layers. Then the passivation layer should be electronically insulating and perhaps preferably deposited on the prepared cathode tape rather than on the free-flowing powder [16].

6.3 Li Chemistry for ALD

In order to realize deposition of Li-ion conducting materials by ALD, a suitable Li-containing precursor is required. At first glance, this may seem trivial considering the extent of materials deposited by ALD over time [1]. Deposition of Li-containing materials was first demonstrated in 2009 [32]. It was then also evident that the hygroscopic nature of lithium oxide or hydroxide and, in fact, a number of lithium-containing materials lead to uncontrolled growth through

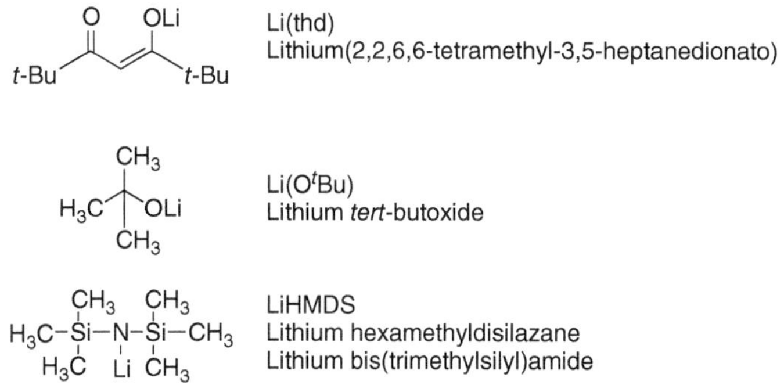

Li(thd)
Lithium(2,2,6,6-tetramethyl-3,5-heptanedionato)

Li(O^{t}Bu)
Lithium *tert*-butoxide

LiHMDS
Lithium hexamethyldisilazane
Lithium bis(trimethylsilyl)amide

Figure 6.3 The three most common precursors for deposition of Li-based materials.

what is known as the reservoir effect [33–37]. It has later been demonstrated that this does not only apply to the mobility of water in the film during growth but also involves bulk transport of lithium ions. A consequence of this is that it has been proven possible to lithiate thicker films of oxide materials as the last step of film growth rather than simultaneously during growth [38]. These effects have made it appear more challenging to deposit lithium-containing materials by ALD than most other materials since the likelihood of obtaining films showing uncontrolled growth is high.

An overview of the current lithium-based chemistries for ALD was recently given in the review by Nilsen *et al.* [39], where a range of 11 different lithium precursors were reported to be tested. There has been little progress in new Li-based chemistries since this review and rather more focus on application of the Li(O^{t}Bu) precursor [3, 26, 37, 40, 41]. This precursor is relatively inexpensive, easy to handle, and is typically applied by sublimation at 130–160 °C. The frequently used alternatives, Li(thd) (Hthd = 2,2,6,6-tetramethyl-3,5-heptadione) and LiHMDS (HMDS = hexamethyldisilazane), are typically applied by sublimation at 175–200 and 60–75 °C, respectively, Figure 6.3.

The hydroscopic nature of lithium-based materials is always a challenge regarding lithium depositions, also when handling precursors before a deposition. Li(thd) is the most stable of the commonly used lithium precursors and can in general be stored and handled under ambient conditions without any problem. This is not the case when handling the more volatile Li(O^{t}Bu) and LiHMDS. Both chemicals will deteriorate outside a glovebox and handling under ambient conditions should be kept to an absolute minimum. It is a paradox that these are the two most commonly used lithium sources, but they are preferred because of their low sublimation temperature and relative good thermal stability during depositions.

6.4 Thin-Film Batteries

The development of consumer electronics has now come so far that further advancement requires specialized energy sources. The form factor of present pouch cells based on powder electrodes and liquid electrolytes does not allow for further radical redesign of mobile devices into wearable products or concealed energy sources for the upcoming Internet of things. A solution to these challenges is to use batteries based on thin-film structures utilizing solid-state electrolytes. Such batteries have already been demonstrated where several designs are commercial available, mainly based on production techniques such as sputtering. The main challenge in the realization of such solid-state thin-film batteries is the limited ionic conductivity of current solid-state electrolytes as compared to liquid-based electrolytes [42]. Liquid-based electrolytes typically have a room-temperature ionic conductivity of 10^{-3} S cm while the presently best commercial solid-state alternative has a conductivity in the range of 10^{-6} S cm [42]. As will be evident through the examples in this chapter, it is common to observe a reduction in specific conductivity when the material is deposited as a thin film, although the causes for this may be varied and are currently unclear.

The main benefits of turning to all solid-state battery designs are increased safety due to removal of the highly flammable liquid electrolyte and increased long-term stability due to prevention of formation of SEI layers, in addition to the new possibilities that arise from its rather thin form factor – such as flexible batteries. One of the current main drawbacks, however, is its limited capacity being typically just above 1 mAh cm^{-2} (THINERGY, Infinite Power Solutions). The currently used solid-state electrolyte is sputtered LiPON with a practical thickness larger than 1 μm, although much thinner films have been demonstrated [43]. This is a significant reduction in thickness when compared to the liquid-based electrolytes. However, recollecting its relatively poor conductivity still renders the overall possible power that can be drawn rather low. This is initially not a problem considering that the typically foreseen applications for such batteries require low power. However, this will also prevent rapid charging of the batteries. The power capabilities will increase if notably thinner electrolyte can be produced while still being pinhole-free – a typical task for ALD. Additionally, the power can be increased even further if the electrodes are converted from the current 2D surface design to complex 3D structures [44].

For thin-film batteries, a thin solid-state electrolyte is beneficial. However, in order to maintain or increase the overall capacity, rather thick cathode and anode are required, which again may hamper the possibility for achieving high power rates. Thick coatings may appear challenging for ALD. However, by applying the same complex 3D structure as mentioned beneficial for the electrolyte, a thin coating on such a structure will appear as a thick coating when measured in terms of capacity per substrate area. Additionally, by constructing the batteries of thinner layers, a larger proportion of the material will remain electrochemically active throughout its use [45–48]. Realization of 3D structured all-solid-state batteries is currently one of the top priorities in thin-film battery research and development.

6.5 ALD for Solid-State Electrolytes

A necessity for realizing 3D all-solid-state batteries is a deposition process for a good solid-state electrolyte. A number of such possible candidates have been deposited by ALD so far. None of these show convincingly good ionic conductivity; however, they may still be usable for low power designs or as barrier materials for powder-based electrodes.

6.5.1 Li_2CO_3

Li_2CO_3 was the first lithium-containing material to be deposited in a controlled manner by ALD [32]. It was deposited using the [Li(thd) + O_3] pair of precursors and is a typical by-product of any processes resulting in the formation of Li_2O or LiOH, when exposed to air.

A similar pulsing scheme has also been applied for the LiHMDS precursor, [LiHMDS + H_2O + CO_2], resulting in uniform films with a growth rate of about 0.35 Å per cycle and a deposition range of 89–380 °C [36]. Even though the LiHMDS precursor contains silicon, the deposited films are almost free from silicon impurities, indicating a simple reaction mechanism probably of ligand exchange type. A similar reaction scheme was tested during QCM investigations using the [Li(O^tBu) + O_3 + CO_2] combination, proving stable growth after an initial reaction sequence of about 25 cycles [33].

Li_2CO_3 is not considered as a good ionic conductor with an ionic conductivity of ca. 10^{-10} S cm^{-1} at room temperature [49] but is typically foreseen as a constituent of SEI layers formed on the anodes in liquid-based electrolytes.

6.5.2 Li–La–O

Lithium lanthanum oxide was deposited using [Li(thd) + O_3] + [La(thd)$_3$ + O_3] as a step toward the development of the lithium lanthanum titanate (LLT) electrolyte material. It was possible to control the lithium content over the whole Li/La range. However, the chosen precursor combination resulted in incorporation of large amounts of carbonates in the film, as shown in [32].

6.5.3 LLT

The perovskite-based LLT material, $Li_{0.32}La_{0.30}TiO_z$, was grown by ALD as a possible lithium-ion conducting material using a combination of [Li(O^tBu) + H_2O] + [TiCl$_4$ + H_2O] + [La(thd)$_3$ + O_3] [50]. During this investigation, it was evident that the pulsing order of the different combinations of precursors was important in order to obtain uniform films. Rather nonuniform and air-sensitive films with high content of chlorine contaminations were obtained when [TiCl$_4$ + H_2O] was pulsed before [Li(O^tBu) + H_2O]. Later experience with similar systems also points at the fact that chlorine-based chemistries are not directly compatible with lithium-based processes [51]. The overall lithium

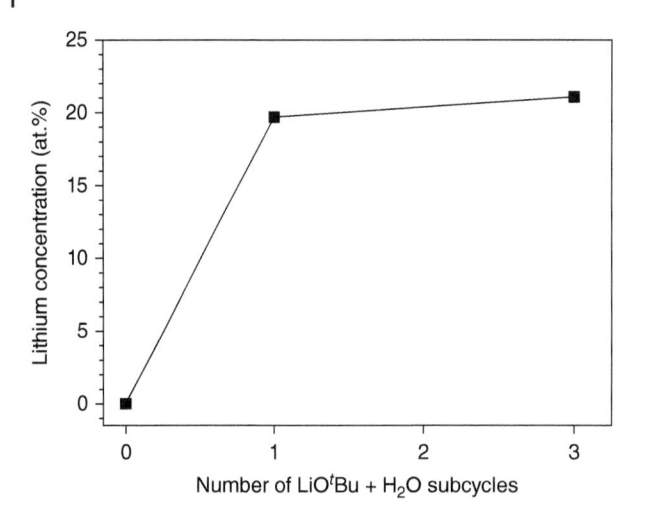

Figure 6.4 Lithium content of the films, as measured by TOF-ERDA, as a function of number of subsequent lithium subcycles n in the pulsing scheme of $400 \times (1 \times TiO_2 + 3 \times La_2O_3 + n \times Li_2O)$. (Aaltonen et al. 2010 [50]. Reproduced with permission of Royal Society of Chemistry.)

content varied with the pulsing scheme in a nonlinear manner. By applying a pulsing scheme of $1 \times [TiCl_4 + H_2O] + 3 \times [La(thd)_3 + O_3] + n \times [Li(O^tBu + H_2O]$, it was only possible to control the Li content up to a level of 20 at.%, see Figure 6.4. Similar limitations in variations in Li content have been observed for several other processes, as will be shown later.

The LLT material has been considered as a promising candidate to replace LIPON in all-solid-state lithium-ion thin-film batteries due to its potentially much higher Li-ion conductivity [52]. However, the practically achieved ionic conductivities when deposited as thin films are still rather low. In addition, it has limited stability toward metallic lithium and would require a barrier layer of, for example, $LiAlO_2$ to be used toward lithium anodes.

6.5.4 Li–Al–O ($LiAlO_2$)

Formation of lithium aluminate should be a highly suitable task for ALD considering how well ALD processes containing aluminum are known. All attempts so far have used the $[LiO^tBu + H_2O]$ process in combination with either $[TMA + H_2O]$ or $[TMA + O_3]$ for the formation of ideally $Li_xAlO_{1.5+x}$ [3, 34, 35, 37]. Surprisingly, high growth rates are observed for both processes, showing rates up to 2.8 Å per cycle. QCM and FTIR analyses of the processes [34, 35] show an enhanced mass increase between TMA and the prior deposition cycles of $[Li(O^tBu + H_2O]$. The mass increase during the next TMA exposure depends on the number of preceding $[Li(O^tBu + H_2O]$ cycles, while the overall growth is less affected by the number of $[TMA + H_2O]$ cycles. This behavior makes it difficult to control the Li content of the Li–Al–O material for higher contents of Li where possibly water is absorbed and released by the bulk material during cycling via a reservoir effect, leading to uncontrolled growth rates, Figure 6.5. A more detailed investigation of

Figure 6.5 Li–Al–O growth rate measured by ellipsometry as a function of % LiOH ALD cycles. Gray-shaded area designates the region of stable growth with constant, linear growth as a function of ALD cycles. (Comstock and Elam 2013 [35]. Reproduced with permission of American Chemical Society.)

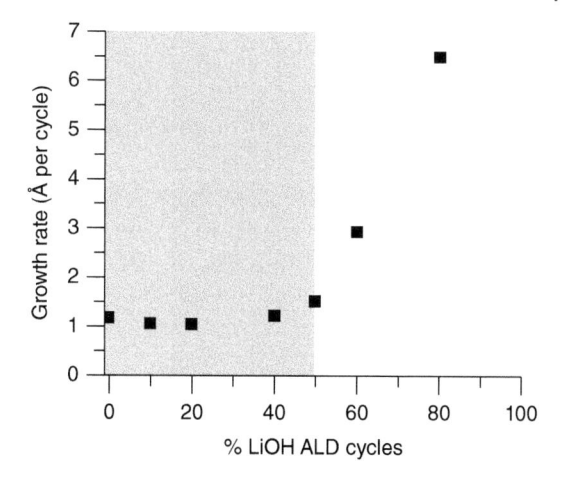

the [LiO^{t}Bu + H$_2$O] + [TMA + O$_3$] process has shown an independence between the pulsing ratio and the obtained composition [37]. Variations in Li/Al pulsing ratio do not affect the composition, microstructure, or electrical properties of the film.

The ionic conductivity of the LiAlO$_2$ material was characterized by impedance spectroscopy to be in the range of 1×10^{-7} S cm^{-1} at 90 °C [34] and 5.6×10^{-8} S cm^{-1} at room temperature [3]. Its ionic conductivity is perhaps too low for an ideal electrolyte material, but it is stable toward current anode materials and is suggested as a barrier layer between anodes and alternative electrolytes [53] and also as a protection layer on cathode materials to prolong their stability under cycling [35]. Its ability to improve the electrochemical stability of LiNi$_{0.5}$Mn$_{1.5}$O$_4$/graphite Li-ion batteries has been demonstrated in [3], which show significantly improved properties compared to similar films of Al$_2$O$_3$, especially at elevated temperatures.

6.5.5 Li$_x$Si$_y$O$_z$

LiHMDS is a Li$_x$Si$_y$O$_z$ silicon and lithium-containing precursor most suitable for the formation of lithium silicates as a single-source precursor when reacted with ozone [54, 55]. The growth rate is strongly temperature-dependent and varies from 0.3 to 1.7 Å per cycle in the deposition range between 150 and 400 °C, although ensuring good uniformity and control over the thickness. The stoichiometry of the deposited film is temperature dependent with a decreasing Li content with increasing deposition temperature. The lithium silicates are potential electrolyte materials although the reported conductivities of Li$_2$SiO$_3$ and Li$_4$SiO$_4$ are in the range of 10^{-8}–10^{-7} S cm^{-1} [56]. There is hope to increase this conductivity to practical values by alloying with other elements such as Al, which has shown to increase the ionic conductivity.

6.5.6 Li–Al–Si–O

Aluminosilicates have been obtained using the [Li(O^{t}Bu) + H$_2$O], [TMA + H$_2$O], and [TEOS + H$_2$O] (TEOS = tetraethylorthosilane) process [41]. The material was amorphous as deposited at 290 °C but crystallized into LiAlSiO$_4$ upon annealing at 900 °C. The room-temperature ionic conductivity of the films has been measured by impedance to be in the range of 10^{-7}–10^{-9} S cm^{-1} with an activation energy between 0.46 and 0.84 eV, depending upon the film composition [41].

6.5.7 LiNbO$_3$

LiNbO$_3$ may be best known as a ferroelectric material; however, it is also a Li-ion conductor with conductivities in the range of 10^{-9}–10^{-5} S cm^{-1} depending on crystallinity, with the largest values for the amorphous material [57–59]. It has been deposited by ALD using a combination of the [LiHMDS + H$_2$O] and [Nb(OEt)$_5$ + H$_2$O] processes at 235 °C [60]. The aim of the work was not to develop a lithium-ion conductor, but it showed that it was possible to control the Li content over large ranges by varying the pulsing ratio of the precursors. The films were uniform and the process was self-limiting apart from the Li-rich compositions, which resulted in large gradients. Even though a silicon-containing precursor was chosen, silicon was not detected in the final product.

6.5.8 LiTaO$_3$

LiTaO$_3$, which is related to LiNbO$_3$ with regard to both ferroelectricity and ionic conductivity, has been deposited by ALD using [LiO^{t}Bu + H$_2$O] and [Ta(OEt)$_5$ + H$_2$O] at 225 °C [26, 40] both as protective layers on the high-voltage-based cathode material LiNi$_{1/3}$Co$_{1/3}$Mn$_{1/3}$O$_2$ [26] and on high-aspect-ratio surfaces to prove its applicability for 3D battery structures. The ionic conductivity of the coating was measured to be 2×10^{-8} S cm^{-1} at room temperature. The coated cathode material shows highly thickness-dependent properties with increased rate capabilities for thin coatings and low potential cut-off (3.0–4.5 V, five-layer coating), while thicker layers (10-layer coating) are best for decreasing the electrode degradation.

6.5.9 Li$_3$PO$_4$

Crystalline Li$_3$PO$_4$ can be formed using the precursor combination [Li(O^{t}Bu) + TMPO] (TMPO = trimethylphosphate) with a growth rate of about 0.7 Å per cycle or [LiHMDS + TMPO] [61]. The growth rate when using LiHMDS increases from about 0.4 Å per cycle up to around 1.3 Å per cycle for depositions in the range of 275–350 °C. Crystalline Li$_3$PO$_4$ has a somewhat too high electronic conductivity to be applied on its own but represents a clear contribution toward deposition of the highly desired electrolyte material, lithium phosphorus oxynitride (LiPON), which also shows good electrochemical stability toward metallic Li.

6.5.10 Li$_3$N

LiHMDS will also react directly with NH$_3$ to form the nitride Li$_3$N, which is a rather air-sensitive compound [36]. The nitride was deposited uniformly at 167 °C with a growth rate of 0.95 Å per cycle and required rather short pulse times. Lithium nitride is a good ionic conductor [62]; however, it suffers from a low breakdown voltage that prevents it from practical use in batteries in its pure state [63]. Formation of this nitride may be regarded as one step toward the development of even more suitable materials for Li-ion batteries.

6.5.11 LiPON

LiPON is the most commonly used solid-state thin-film electrolyte with ionic conductivity in the range of 10^{-6} S cm^{-1} and activation energy of 0.5 eV [64]. Both PVD and CVD have been used to deposit LiPON, and recently, it was reported that it can also be deposited by plasma ALD [65]. Both amorphous and crystalline films were deposited from the pulsing sequence LiO^{t}Bu + H$_2$O + TMP + ^{P}N$_2$ at 250 °C. The N$_2$ content and crystallinity were controlled by varying the ^{P}N$_2$ pulse length from short to long ^{P}N$_2$ pulses, 0–20 s led to an N variation from 0% to 16.3%, and phase transition from amorphous to crystalline LiPON occurs at 7 s ^{P}N$_2$ pulse length.

The ionic conductivity is clearly dependent on the N content of the films, Figure 6.6. The highest bulk ionic conductivity at room temperature was determined to be 1.45×10^{-7} S cm^{-1} by electrochemical impedance spectroscopy; this is somewhat lower than those previously reported for films deposited by PVD.

A seemingly vital part turning LiPON into an ionic conductor is the presence of P—N bonds. In order to achieve this without plasma, Nisula et al. used a precursor already containing this functionality, diethyl phosphoramidate (H$_2$NP(O)(OC$_2$H$_5$)$_2$) [66]. Alternating pulsing of LiHMDS and (H$_2$NP(O)(OC$_2$H$_5$)$_2$) produced amorphous films in the temperature range of 250–330 °C. The advantage of thermal ALD over plasma ALD was shown by the uniform deposition on 3D structured Si substrate.

Both approaches use impedance spectroscopy to determine the ionic conductivity. Kozen et al. reported an ionic conductivity of 1.45×10^{-7} S cm^{-1} [65],

Figure 6.6 Ionic conductivity of ALD LiPON films plotted as a function of N content along with linear fit to the data. (Kozen et al. 2015 [65]. Reproduced with permission of American Chemical Society.)

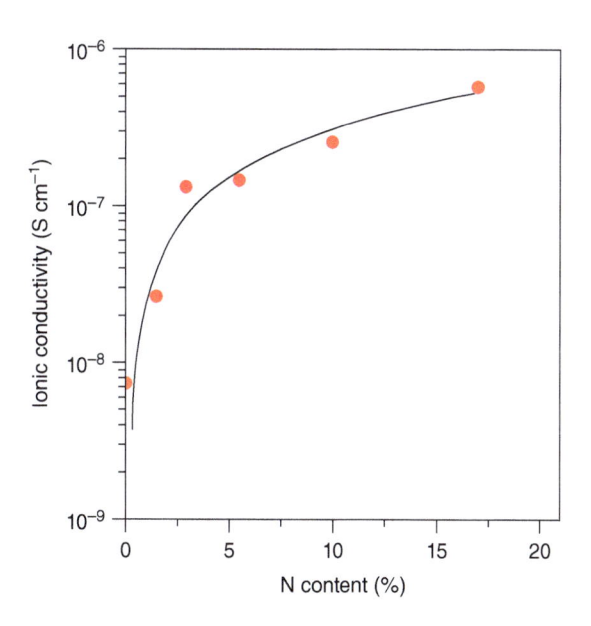

whereas Nisula et al. reported a value of 6.6×10^{-7} S cm^{-1} [66]. These values are somewhat lower than those previously reported for PVD-based LiPON materials [64].

6.5.12 LiF

Deposition of pure LiF is most relevant to optical components; however, it is also an important constituent in Li-ion conducting materials such as Li_3AlF_6 and Li_2NiF_4 that show room-temperature conductivities of the order 10^{-6} S cm^{-1} [67]. ALD deposition of lithium fluoride has been demonstrated using the precursor combination [Li(thd) + TiF_4]. The films are crystalline with a growth rate of 1.5–1.0 Å per cycle in the temperature range of 250–350 °C [68]. This process can be enhanced further by introducing Mg(thd)$_2$ into the deposition sequence without obtaining Mg impurities in the films [69].

6.6 ALD for Cathode Materials

A range of different cathode materials have been deposited by ALD. Several of these have been deposited in the charged unlithiated state, resulting in simpler deposition processes.

6.6.1 V_2O_5

The first report on the deposition of a Li-ion cathode material by ALD was in 2003, where V_2O_5 was used as model material for the investigation of the lithium intercalation mechanism [70]. The vanadium oxide was produced using the [VO(O^{i}Pr)$_3$ + H$_2$O] (VO(O^{i}Pr)$_3$ = vanadyl triisopropoxide) process at a deposition temperature of 105 °C. The result was as-deposited amorphous films, which crystallized to V_2O_5 after annealing in air at 400 °C [70]. The electrochemical properties of similar amorphous V_2O_5 films have been investigated as model material in [45], proving superior properties as compared to crystalline films. The amorphous material was capable of reversible intercalation of Li up to a Li content of $Li_{2.9}V_2O_5$ without becoming limited by nonreversible phase transitions of ω- and γ-V_2O_5. This resulted in a capacity of 455 mAh g^{-1} for a 200-nm film cycled between 4 and 1.5 V; however, its specific capacity decreased for thicker films.

While amorphous materials result in homogeneous films with very low surface roughness, its crystalline counterpart typically develops rough surfaces during growth and thereby also a larger interface area. The processes [VO(O^{i}Pr)$_3$ + H$_2$O] and [VO(O^{i}Pr)$_3$ + O$_3$] result in amorphous and crystalline films, respectively, and their electrochemical properties are compared in [46]. The crystalline films have a higher capacity than the amorphous counterpart for 1Li/V_2O_5 and 2Li/V_2O_5, though their capacities are comparable for 3Li/V_2O_5.

Vanadium oxides can also be grown using the [VO(thd)$_2$ + O$_3$] process, where the surface texture varies significantly with the deposition temperature [47]. A particularly rough surface texture is obtained when deposited at 235 °C, Figure 6.7, where the oxide shows an electrochemical capacity of up to

Figure 6.7 (a) SEM images of V_2O_5 samples obtained by 5000 ALD cycles on silicon substrate, showing plate-like morphology. (b) Simulation of a surface equivalent to the sample deposited using 5000 cycles. (Østreng et al. 2014 [47]. http://pubs.rsc.org/en/content/articlehtml/2014/ta/c4ta00694a Used under creative commons license: CC BY SA 3.0 https://creativecommons.org/licenses/by/3.0/.)

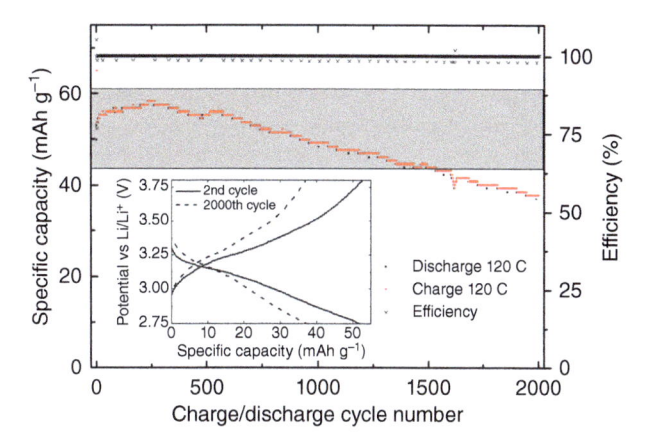

Figure 6.8 Discharge rate cycling stability at 120 C conducted directly after the rate performance test to 960 C. The gray band indicates a window with less than 80% capacity loss relative to the initial capacity (55 mAh g^{-1} at 120 C). Coulombic efficiency is close to 100%. Inset: charge and discharge curves shown for the 2nd and 2000th cycles. (Østreng et al. 2014 [47]. http://pubs.rsc.org/en/content/articlehtml/2014/ta/c4ta00694a Used under creative commons license: CC BY SA 3.0 https://creativecommons.org/licenses/by/3.0/.)

105 mAh g^{-1} for cycling in the range of 2.75–3.80 V at 1 C. The current cycling range corresponds to insertion of 1Li per V_2O_5 and a theoretical capacity of 147 mAh g^{-1}. The capacity at high rate discharge is surprisingly good for this material due to the high surface area. For cycling at 120 C, a capacity of about 55 mAh g^{-1} is maintained for 650 cycles, and the cell lasts for 1530 cycles before reaching 80% of its initial capacity; see Figure 6.8.

6.6.2 LiCoO$_2$

LiCoO$_2$ can be deposited using the Li(O^{t}Bu) precursor in an O$_2$ plasma processes in combination with CoCp$_2$ [71, 72]. The growth rate is reported to be 0.6 Å per cycle at a deposition temperature of 325 °C, showing well saturating pulsing behavior. The films required a subsequent annealing at 700 °C to become electrochemically active and showed good cycling behavior, although at somewhat lower capacity than anticipated, Figure 6.9. The origin of the reduced activity may be

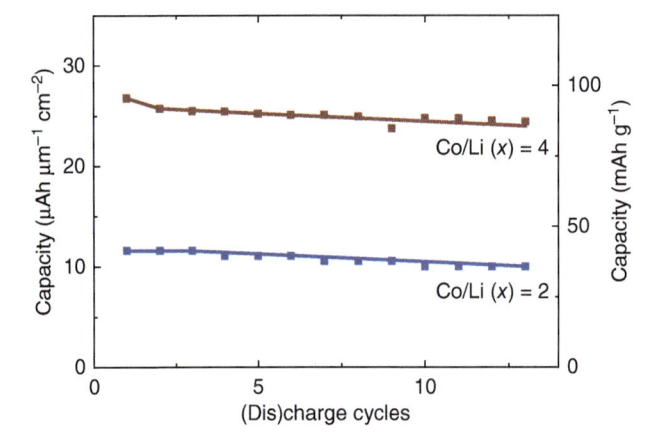

Figure 6.9 Constant current (CC) (dis)charge cycling between 3.0 and 4.1 V (0.35 C-rate) for an ALD-deposited LiCoO$_2$ ($x = 4$) film on Si/TiO$_2$/Pt, using LiClO$_4$ in ethylene carbonate/diethyl carbonate (EC/DEC 1/1) as liquid electrolyte. The electrochemical storage capacity upon cycling showing data for $x = 2$ and $x = 4$, where x denotes the Co/Li pulsing ratio. (Donders et al. 2013 [71]. Reproduced with permission of The Electrochemical Society.)

linked to loss of lithium material by the formation of lithium carbonates during growth or subsequent treatments.

6.6.3 MnO$_x$/Li$_2$Mn$_2$O$_4$/LiMn$_2$O$_4$

Manganese oxides have been deposited by ALD in many different manners, typically using [Mn(thd)$_3$ + O$_3$] or [Mn(CpEt)$_2$ + H$_2$O] as source of manganese [73, 74]. Both of these processes have been combined with several of the lithium-based processes for obtaining electroactive materials [38]. A highly electroactive spinel compound is obtained by combining [Mn(thd)$_3$ + O$_3$] and [Li(O^{t}Bu) + H$_2$O] showing capacities of near 200 mAh g^{-1}; see Figure 6.10. The same type of spinel compound, although with poorer electrochemical properties, was also obtained when [Mn(thd)$_3$ + O$_3$] was combined with the [Li(thd) + O$_3$] process. The [Li(O^{t}Bu) + H$_2$O] process was also tested in combination with the [Mn(EtCp)$_2$ + H$_2$O] process to obtain a Li–Mn–O material, but this rather resulted in the lack of incorporation of Li and an actual loss of Mn from the deposited film. A similar situation was also observed when the [Li(O^{t}Bu) + H$_2$O] process was replaced by [LiHMDS + H$_2$O]. Whether this was due to chemical etching of the MnO or by surface deactivation remains to be investigated. When the [LiHMDS + H$_2$O] process was combined with [Mn(thd)$_3$ + O$_3$], nonuniform and XRD amorphous films were obtained.

6.6.4 Subsequent Lithiation

It has also been proven possible to subsequently lithiate both MnO$_2$ and V$_2$O$_5$ by attempting to deposit an overlayer of Li$_2$CO$_3$ based on the [Li(thd) + O$_3$] and the Li(O^{t}Bu) + H$_2$O process [38]. Rather than formation of Li$_2$CO$_3$, the complete oxide films were converted to electroactive Li$_x$Mn$_2$O$_4$ and Li$_x$V$_2$O$_5$, respectively. This observation proves that the mobility of Li during deposition is rather high

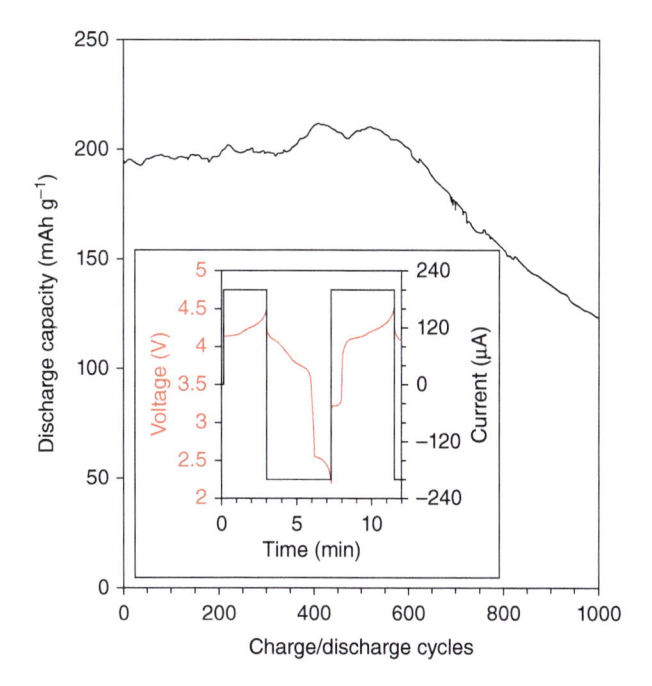

Figure 6.10 Discharge capacities and first-cycle potentiograms (insets) for an 86-nm MnO_2 treated with 200 cycles of $LiO^tBu + H_2O$, 1000 charge–discharge cycles with 200 µA. (Miikkulainen et al. 2013 [38]. Reproduced with permission of American Chemical Society.)

and that the traditional ALD view with self-saturating surface reactions is challenged. Similar attempts were also made with films of TiO_2, Al_2O_3, ZnO, Co_3O_4, Fe_2O_3, NiO, and MoO_3 deposited without observation of similar lithiation, pointing at the fact that the effect is material dependent. Lithiation of deposited films has also been achieved through solid-state reactions of films of Li_2CO_3 during subsequent annealing at higher temperatures.

6.6.5 LiFePO$_4$

Lithium iron phosphate was first reported to be grown by combining the processes $[Li(thd) + O_3]$, $[Fe(thd)_3 + O_3]$, and $[Me_3PO_4 + (H_2O + O_3)]$ [75]. The films are amorphous as deposited and crystallize to the $LiFePO_4$ phase upon annealing under 10% H_2 in Ar at 500 °C; however, electrochemical investigation of the deposited compound proved it to be rather inactive.

A more electroactive $LiFePO_4$ was reported using a combination of the $[FeCp_2 + O_3] + [TMPO + H_2O] + [LiO^tBu/H_2O]$ processes [76]. The film was deposited on CNT at 300 °C and was amorphous as deposited and crystallized to the orthorhombic structure when annealed at 700 °C in Ar. A discharge capacity of 150 mAh g^{-1} at 0.1 C (1 C = 170 mA g^{-1}) was achieved, and nearly 50% if its capacity was still obtainable at 60 C discharge rates.

Interestingly, it is possible to obtain an even more electroactive compound if the film is deposited as $FePO_4$ and rather lithiated in the assembled battery cell [48]. Electrodes of amorphous $FePO_4$ were deposited using the

[Fe(thd)$_3$ + O$_3$] + [TMPO + (H$_2$O+O$_3$)] combination at 246 °C, which showed excellent electrochemical properties with capacity reaching the theoretical limit of 178 mAh g^{-1} for 1 C charge–discharge rates and cycling properties proving stable capacity over at least 600 cycles.

Furthermore, this amorphous compound behaves as a pseudocapacitor below a critical electrode thickness, resulting in extremely facile kinetics, which enable the electrodes to deliver 50% of their theoretical capacity at 2560 C (1.4 s charge/discharge) displaying a specific power above 1 MW kg^{-1}. In addition, the amorphous electrodes could be cycled at 320 C (11 s charge/discharge) for 10 000 cycles, showing excellent cyclability, similarly to a supercapacitor.

FePO$_4$ has also been produced using the [FeCp$_2$ + O$_3$] + [TMP + H$_2$O] process at 200 to 350 °C on N-doped CNT. The structure functioned as a cathode material capable of delivering discharge capacity of 141 mAh g^{-1} at 1 C after 100 cycles [91]. The same coating has also been applied as protective coating to LiNi$_{0.5}$Mn$_{1.5}$O$_4$ as mentioned earlier in this chapter [28].

6.6.6 Sulfides

A rather alternative approach to the high potential oxides mentioned so far is to aim at lower potential materials with possibilities of utilizing several oxidation steps – frequently known as conversion materials. Thin films of copper and gallium sulfides have recently been explored as such conversion cathode materials [77–79]. The copper sulfides are deposited using bis(*N*,*N*'-di-*sec*-butylacetamidinato)dicopper(I) (CuAMD) and H$_2$S as precursors for deposition on SWCNTs. The structures show capacities in excess of 250 mAh g^{-1} for cycling between 0.01 and 3.00 V at a current density of 100 mA g^{-1} and no noticeable deterioration over 200 cycles. The GaS$_x$ was produced using hexakis-(dimethylamido)digallium and hydrogen sulfide in the temperature range of 125–225 °C. The deposition is uniform on high-aspect-ratio silicon structures and achieves a high specific capacity of 770 mAh g^{-1} at a current density of 320 mA g^{-1} in the voltage window of 0.01–2.00 V [78].

6.7 ALD for Anode Materials

The examples for deposition of anode materials by ALD are more limited and mainly relate to titanates and conversion materials. Lately, WN has also been suggested as a suitable anode material for lithium-ion batteries [80].

Pure TiO$_2$ is easily deposited by ALD and has been demonstrated as suitable anode material several times, as shown as follows. Titanium oxide exists in many crystalline phases where anatase is considered the most electroactive material and capable of inserting 0.5Li per TiO$_2$ formula unit with a potential toward Li/Li$^+$ of 1.55 V [81]. The major focus on the application of ALD-type TiO$_2$ for Li-ion batteries has been through coating on high-surface-area structures.

The areal capacity of TiO$_2$ has been increased by a factor of 10 by coating a structure of nanorods of aluminum [82]. An interpenetrating TiO$_2$ structure has also been formed using a 3D-porous anodic alumina structure [83]. TiO$_2$ has also been deposited as an anode material on Ni-plated Tobacco mosaic virus (TMV) [84], showing increased areal capacity and greatly improved cyclability. Hollow nanoribbons of TiO_2 with thickness of 15 nm have been produced using a sacrificial template of peptide assembly [85]. The hollow structure is particularly beneficial since it enables storage of electrolyte within the hollow structure as well. The size-dependent properties of TiO_2 have been studied by varying the thickness of TiO_2 layers coated inside anodic alumina templates [86]. The best overall properties were found for 5-nm-thick films, which achieved a capacity of 330 mAh g^{-1}, proving excellent performance.

Deposition of lithium-containing titanium oxide has been realized using [Li(O^{t}Bu + H$_2$O] and [Ti(O^{i}Pr)$_4$ + H$_2$O] [33, 34, 51] or [Li(O^{t}Bu + H$_2$O] and [TiCl$_4$ + H$_2$O] [51, 87]. When [Ti(O^{i}Pr)$_4$ + H$_2$O] was used, the composition of Li could be controlled well over a large range [51], and as-deposited crystalline films, which were seemingly air-stable, were obtained. However, by application of [TiCl$_4$ + H$_2$O], rather low concentration of Li was obtained, and the films were rather air sensitive [51].

In their oxidized state, cobalt oxides are typically regarded as cathode materials. However, they can also be used as conversion anodes in Li-ion batteries [88] with a potential capacity of 8Li per Co_3O_4 according to the following reaction (Equation 6.1):

$$Co_3O_4 + 8Li \rightleftharpoons 3Co + 4Li_2O, \quad \Delta E° = 1.87 V \text{ vs NHE} \qquad (6.1)$$

The films were deposited using [CoCp$_2$ + O$_2$ plasma], and 40-nm films of Co_3O_4 were capable of delivering a capacity of 1000 mAh g^{-1} for at least 70 cycles, proving the potential of ALD to stabilize conversion materials toward mechanical degradation.

Tungsten nitride was obtained by ALD using the [W(CO)$_6$ + NH$_3$] process with the aim of developing a new type of anode material [80]. The host material is highly conductive, and when deposited on stainless steel disks in thickness of about 53 nm, it provided stable capacity of about 5.5 µAh cm^{-2}. The capacity was increased to about 25 µAh cm^{-2} when deposited on CNT.

6.8 Outlook

The future of electrochemical energy storage materials looks bright from both fundamental research and development point of views. It is not a question whether new types of design and materials systems will be implemented, but rather in what direction the development will take place. This depends on both the possibilities available for energy and power densities but, perhaps, even more on outlooks for long-term electrochemical stability and operational safety. The consumer demands are so diverse that compromises will have to be made in all directions, leading to simultaneous development and implementation of several designs and material systems. A vital part in these designs will be the possibilities for depositing pinhole-free solid-state ionic electrolytes – a sweet spot for ALD.

Successful scale-up of ALD processes for passivation of powder-based materials [89] and porous surfaces will enable high-end bulk electrodes with prolonged lifetime. The field is still immature with respect to the choice of deposition technique and type of passivation layer, mainly due to the fact that it depends on the types of electrodes and electrolyte chemistry, among others.

The current choices in solid-state electrolytes are not sufficient for the production of efficient all-solid-state batteries. However, suggestions for new materials are constantly emerging [42]. These will undoubtedly be followed by suitable ALD deposition processes. Many of the processes presented in this chapter should be suitable to improving the sputtered 2D batteries currently available. When this is sufficiently demonstrated, 3D battery structures combining capacity with power will emerge. For this type of structure, ALD will begin to compete with the sputtering processes also for deposition of the electrode materials themselves, since much thinner layers are required. By enabling thin-film batteries originally intended for low-power applications to also deliver high power, one also opens the opportunity for rapid charging. One of the marked areas for such batteries is as energy storage inside smart cards. A high-power design will enable charging of the card itself during normal use without the awareness of the user. Through implementation of larger-scale ALD reactors, the aforementioned design can be developed even further to form pseudocapacitors [90], bridging the gap between batteries and capacitors.

One of the prerequisites for the current field has been the implementation of suitable lithium-based ALD processes (Table 6.1). These have proven to be more sensitive toward material combination than the regular ALD process, mainly due to their hygroscopic nature and the fact that lithium is, and should be, mobile during deposition. New lithium-based precursors and processes will undoubtedly emerge. Now we have also seen the very beginning of implementation of MLD materials as passivation layers, exploiting their rigid and flexible nature suitable for materials showing high volumetric changes. These materials will undoubtedly also be suggested for electrolytes and electroactive components in future designs.

The present focus has mainly been on lithium-based materials, driven by their high specific capacity. However, this can become secondary when considering thin-film batteries due to the required volume of the current collectors and packing material. ALD should also be suitable as passivation process for the emerging sodium batteries in addition to development of suitable solid electrolytes. The number of known good solid-state electrolytes for sodium is low when compared to the lithium-based analogies. This also applies for alternative technologies that we have presently not touched upon, such as the metal–air-based batteries.

There is no doubt that the future is interesting with respect to the combination of thin films and electrochemical storage. We are currently witnessing a race toward implementation of large-scale synthesis and choice of suitable materials for realization of an upcoming product range within the Internet of things, RFID, smart packaging, and more.

Table 6.1 Li compounds deposited by ALD.

Compound	Precursor combination	Result	References
Li_2CO_3	$[Li(thd) + O_3]$	Li_2CO_3 (185–300 °C), about 0.30 Å per cycle	[32]
	$[Li(O^tBu) + H_2O + CO_2]$	Li_2CO_3 (225 °C), QCM analysis	[33]
	$[LiHMDS + H_2O + CO_2]$	Li_2CO_3 (89–380 °C), about 0.35 Å per cycle	[36]
Li_2O $Li(OH)$	$[Li(O^tBu) + H_2O]$	Li_2CO_3 (225 °C), QCM analysis	[33]
Li–La–O	$[Li(thd) + O_3] + [La(thd)_3 + O_3]$	Li–La–O with carbonate formation (225 °C), good control of Li content	[32]
Li–Ti–O	$[Li(O^tBu) + H_2O] + [Ti(O^iPr)_4 + H_2O]$	$Li_4Ti_5O_{12}$ on N-CNT (250 °C), about 0.61 Å per cycle	[87]
	$[Li(O^tBu) + H_2O] + [Ti(O^iPr)_4 + H_2O]$	$Li_4Ti_5O_{12}$ (225 °C), 0.55 Å per cycle, good control of Li content. Crystalline as-deposited and air-stable films	[51]
	$[Li(O^tBu) + H_2O] + [TiCl_4 + H_2O]$	$Li_xTi_yO_z$ (225 °C), 1.5 Å per cycle, low concentration of Li and air-sensitive films	[51]
LLT	$[Li(O^tBu) + H_2O] + [TiCl_4 + H_2O] + [La(thd)_3 + O_3]$	$Li_{0.32}La_{0.30}TiO_z$ (225 °C), 0.45 Å per cycle. Nonlinear variation of Li content with Li pulsing	[50]
	$[Li(O^tBu) + H_2O] + [La(thd)_3 + O_3] + [TiCl_4 + H_2O]$	$Li_xLa_yTiO_z$ (225 °C), nonuniform films	[50]
Li–Al–O	$[LiO^tBu + H_2O] + [TMA + O_3]$	$LiAlO_2$ (225 °C), 2.8 Å per cycle, QCM analysis	[34]
	$[LiO^tBu + H_2O] + [TMA + H_2O]$	$LiAlO_2$ (225 °C), QCM analysis	[34]
	$[Li(O^tBu) + H_2O] + [TMA + H_2O]$	$LiAlO_2$ (225 °C), QCM + FTIR analysis	[35]
	$[Li(O^tBu) + H_2O] + [TMA + H_2O]$	$LiAlO_2$ (225 °C), coated on $LiNi_{0.5}Mn_{1.5}O_4$ powder, ionic conductivity of 5.6×10^{-8} S cm^{-1} at room temperature	[3]
	$[Li(O^tBu) + H_2O] + [TMA + O_3]$	$LiAlO_2$ (225 °C), showing large independence of composition and pulsing ratio	[37]

(continued)

Table 6.1 (Continued)

Compound	Precursor combination	Result	References
Li–Si–O	[LiHMDS + O_3]	Li_2SiO_3 (200–300 °C), QCM + MS analysis	[54]
	[LiHMDS + O_3]	Li_2SiO_3 (150–400 °C), 0.3–1.7 Å per cycle	[55]
Li–Al–Si–O	[Li(O^tBu) + H_2O] + [TMA + H_2O] + [TEOS + H_2O]	$LiAlSiO_4$ (290 °C), amorphous as-deposited, ionic conductivity in the range of 10^{-7}–10^{-9} S cm^{-1}	[41]
$LiNbO_3$	[LiHMDS + H_2O] + [Nb(OEt)$_5$ + H_2O]	$LiNbO_3$ (235 °C), 0.5 Å per cycle. Si-free films with good control of Li content	[60]
$LiTaO_3$	[Li(O^tBu) + H_2O] + [Ta(OEt)$_5$ + H_2O]	$LiTaO_3$ (225 °C), ionic conductivity was 2×10^{-8} S cm^{-1} at room temperature	[26, 37, 40]
Li_3PO_4	[Li(O^tBu) + TMPO]	Li_3PO_4 (225–300 °C), 0.7–1.0 Å per cycle	[61]
	[LiHMDS + TMPO]	Li_3PO_4 (275–350 °C), 0.4–1.3 Å per cycle.	[61]
Li_3N	[LiHMDS + NH_3]	Li_3N (167 °C), 0.95 Å per cycle	[36]
LiPON	[LiO^tBu + H_2O + TMP + $^PN_2$]	LiPON (250 °C), 1.05 Å s^{-1}, N content controlled by the $^PN_2$ pulse length. Ionic conductivity was 1.45×10^{-7} S cm^{-1} at room temperature	[65]
	[$H_2NP(O)(OC_2H_5)_2$] + LiHMDS]	LiPON (250–330 °C), 0.7–1.0 Å s^{-1}. Ionic conductivity was 6.6×10^{-7} S cm^{-1} at room temperature	[66]
LiF	[Li(thd) + TiF_4]	LiF (250–350 °C), 1.5–1.0 Å per cycle	[68]
	[Li(thd) + TiF_4] + [Mg(thd)$_2$ + TiF_4]	LiF (300–350 °C), 1.4 Å per cycle	[69]
$LiCoO_2$	[Li(O^tBu) + $^PO_2$] + [CoCp$_2$ + $^PO_2$]	$LiCoO_2$ (325 °C), 0.6 Å per cycle	[71, 72]

Material	Precursors	Description	Ref.
Li–Mn–O	[Li(thd) + O$_3$] + [Mn(thd)$_3$ + O$_3$]	LiMn$_2$O$_4$ (225 °C), about 0.5 Å per cycle	[38]
	[Li(thd)] + [Mn(thd)$_3$ + O$_3$]	LiMn$_2$O$_4$ (225 °C), nonuniform films	[38]
	[Li(O^{t}Bu) + H$_2$O] + [Mn(thd)$_3$ + O$_3$]	LiMn$_2$O$_4$ (250 °C), uniform films	[38]
	[Li(O^{t}Bu) + H$_2$O] + [Mn(EtCp)$_2$ + H$_2$O]	(250 °C), etching or poisoning of MnO growth	[38]
	[LiHMDS + H$_2$O] + [Mn(thd)$_3$ + O$_3$]	LiMn$_2$O$_4$ (200 °C), nonuniform and XRD amorphous films	[38]
	[LiHMDS + H$_2$O] + [Mn(EtCp)$_2$ + H$_2$O]	(250 °C), Etching or poisoning of MnO growth	[38]
LiFePO$_4$	[Li(thd) + O$_3$] + [Fe(thd)$_3$ + O$_3$] + [Me$_3$PO$_4$ + (H$_2$O+O$_3$)]	LiFePO$_4$ (250 °C), amorphous as-deposited, crystallized in 10% H$_2$ at 500 °C	[75]
	[Li(O^{t}Bu) + H$_2$O] + [Me$_3$PO$_4$ + H$_2$O] + [FeCp$_2$ + O$_3$]	LiFePO$_4$ (30 °C), amorphous as deposited, coated on CNT, 150 mAh g^{-1} at 0.1 C	[76]
FePO$_4$	[Fe(thd)$_3$ + O$_3$] + [Me$_3$PO$_4$ + (H$_2$O+O$_3$)]	FePO$_4$ (200–370 °C), 0.26 Å per cycle, amorphous as deposited, highly electroactive at high rates, 178 mAh g^{-1} for 1 C	[48]
	[FeCp$_2$ + O$_3$] + [TMP+H$_2$O]	FePO$_4$ (200–350 °C), 141 mAh g^{-1} at 1 C after 100 cycles	[28, 91]
V$_2$O$_5$	[VO(O^{i}Pr)$_3$ + H$_2$O]	V$_2$O$_5$ (65–125 °C), amorphous as deposited, crystallized to V$_2$O$_5$ after annealing in air at 400 °C, 455 mAh g^{-1} for a 200 nm film cycled at 4–1.5 V	[46, 70]
	[VO(O^{i}Pr)$_3$ + O$_3$]	V$_2$O$_5$ (170–190 °C), crystalline as deposited, higher capacity than the amorphous state	[46]
	[VO(thd)$_2$ + O$_3$]	V$_2$O$_5$ (235 °C), highly textured as deposited, capacity of 105 mAh g^{-1} when cycled at 2.75–3.80 V at 1 C, high rate capabilities	[47]

Acknowledgments

We would like to thank the Research Council of Norway for their funding (Project 220135 Nano-Materials for Improved Lithium Ion Batteries – Nanomilib and M-Era Net 233031 Laminated Lion ion batteries – LaminaLion).

References

1 Miikkulainen, V., Leskelä, M., Ritala, M., and Puurunen, R.L. (2013) *J. Appl. Phys.*, **113**, 021301.
2 Goodenough, J.B. and Kim, Y. (2009) *Chem. Mater.*, **22**, 587–603.
3 Park, J.S., Meng, X., Elam, J.W., Hao, S., Wolverton, C., Kim, C., and Cabana, J. (2014) *Chem. Mater.*, **26**, 3128–3134.
4 Yang, L., Takahashi, M., and Wang, B. (2006) *Electrochim. Acta*, **51**, 3228–3234.
5 Aaltonen, T., Miikkulainen, V., Gandrud, K.B., Pettersen, A., Nilsen, O., and Fjellvåg, H. (2011) *ECS Trans.*, **41**, 331–339.
6 Knoops, H., Donders, M., Van De Sanden, M., Notten, P., and Kessels, W. (2012) *J. Vac. Sci. Technol., A*, **30**, 010801.
7 Meng, X., Yang, X.Q., and Sun, X. (2012) *Adv. Mater.*, **24**, 3589–3615.
8 Riley, L.A., Cavanagh, A.S., George, S.M., Jung, Y.S., Yan, Y., Lee, S.H., and Dillon, A.C. (2010) *ChemPhysChem*, **11**, 2124–2130.
9 Wang, H.-Y. and Wang, F.-M. (2013) *J. Power Sources*, **233**, 1–5.
10 He, Y., Yu, X., Wang, Y., Li, H., and Huang, X. (2011) *Adv. Mater.*, **23**, 4938–4941.
11 Xiao, X., Lu, P., and Ahn, D. (2011) *Adv. Mater.*, **23**, 3911–3915.
12 Ahn, D. and Xiao, X. (2011) *Electrochem. Commun.*, **13**, 796–799.
13 Wang, D., Yang, J., Liu, J., Li, X., Li, R., Cai, M., Sham, T.-K., and Sun, X. (2014) *J. Mater. Chem. A*, **2**, 2306–2312.
14 Kang, E., Jung, Y.S., Cavanagh, A.S., Kim, G.H., George, S.M., Dillon, A.C., Kim, J.K., and Lee, J. (2011) *Adv. Funct. Mater.*, **21**, 2430–2438.
15 Lipson, A.L., Puntambekar, K., Comstock, D.J., Meng, X., Geier, M.L., Elam, J.W., and Hersam, M.C. (2014) *Chem. Mater.*, **26**, 935–940.
16 Jung, Y.S., Cavanagh, A.S., Riley, L.A., Kang, S.H., Dillon, A.C., Groner, M.D., George, S.M., and Lee, S.H. (2010) *Adv. Mater.*, **22**, 2172–2176.
17 Lotfabad, E.M., Kalisvaart, P., Kohandehghan, A., Cui, K., Kupsta, M., Farbod, B., and Mitlin, D. (2014) *J. Mater. Chem. A*, **2**, 2504–2516.
18 Lee, M.-L., Su, C.-Y., Lin, Y.-H., Liao, S.-C., Chen, J.-M., Perng, T.-P., Yeh, J.-W., and Shih, H.C. (2013) *J. Power Sources*, **244**, 410–416.
19 Lotfabad, E.M., Kalisvaart, P., Cui, K., Kohandehghan, A., Kupsta, M., Olsen, B., and Mitlin, D. (2013) *Phys. Chem. Chem. Phys.*, **15**, 13646–13657.
20 Lee, J.-H., Hon, M.-H., Chung, Y.-W., and Leu, C. (2011) *Appl. Phys. A*, **102**, 545–550.
21 Kohandehghan, A., Kalisvaart, P., Cui, K., Kupsta, M., Memarzadch, E., and Mitlin, D. (2013) *J. Mater. Chem. A*, **1**, 12850–12861.

22 Snyder, M.Q., Trebukhova, S.A., Ravdel, B., Wheeler, M.C., DiCarlo, J., Tripp, C.P., and DeSisto, W.J. (2007) *J. Power Sources*, **165**, 379–385.

23 Ahmed, B., Shahid, M., Nagaraju, D., Anjum, D.H., Hedhili, M.N., and Alshareef, H.N. (2015) *ACS Appl. Mater. Interfaces*, **7**, 13154–13163.

24 Yesibolati, N., Shahid, M., Chen, W., Hedhili, M., Reuter, M., Ross, F., and Alshareef, H. (2014) *Small*, **10**, 2849–2858.

25 Liu, J., Li, X., Cai, M., Li, R., and Sun, X. (2013) *Electrochim. Acta*, **93**, 195–201.

26 Li, X., Liu, J., Banis, M.N., Lushington, A., Li, R., Cai, M., and Sun, X. (2014) *Energy Environ. Sci.*, **7**, 768–778.

27 Park, J.S., Mane, A.U., Elam, J.W., and Croy, J.R. (2015) *Chem. Mater.*, **27**, 1917–1920.

28 Xiao, B., Liu, J., Sun, Q., Wang, B., Banis, M.N., Zhao, D., Wang, Z., Li, R., Cui, X., Sham, T.-K., and Sun, X. (2015) *Adv. Sci.*, **2**, 1500022.

29 Kannan, A., Rabenberg, L., and Manthiram, A. (2003) *Electrochem. Solid-State Lett.*, **6**, A16–A18.

30 Piper, D.M., Travis, J.J., Young, M., Son, S.B., Kim, S.C., Oh, K.H., George, S.M., Ban, C., and Lee, S.H. (2014) *Adv. Mater.*, **26**, 1596–1601.

31 Li, X., Lushington, A., Liu, J., Li, R., and Sun, X. (2014) *Chem. Commun.*, **50**, 9757–9760.

32 Putkonen, M., Aaltonen, T., Alnes, M., Sajavaara, T., Nilsen, O., and Fjellvåg, H. (2009) *J. Mater. Chem.*, **19**, 8767–8771.

33 Cavanagh, A.S., Lee, Y., Yoon, B., and George, S. (2010) *ECS Trans.*, **33**, 223–229.

34 Aaltonen, T., Nilsen, O., Magraso, A., and Fjellvag, H. (2011) *Chem. Mater.*, **23**, 4669–4675.

35 Comstock, D.J. and Elam, J.W. (2013) *J. Phys. Chem. C*, **117**, 1677–1683.

36 Østreng, E., Vajeeston, P., Nilsen, O., and Fjellvåg, H. (2012) *RSC Adv.*, **2**, 6315–6322.

37 Miikkulainen, V., Nilsen, O., Li, H., King, S.W., Laitinen, M., Sajavaara, T., and Fjellvåg, H. (2015) *J. Vac. Sci. Technol., A*, **33**, 01A101.

38 Miikkulainen, V., Ruud, A., Østreng, E., Nilsen, O., Laitinen, M., Sajavaara, T., and Fjellvåg, H. (2013) *J. Phys. Chem. C*, **118**, 1258–1268.

39 Nilsen, O., Miikkulainen, V., Gandrud, K.B., Østreng, E., Ruud, A., and Fjellvåg, H. (2014) *Phys. Status Solidi A*, **211**, 357–367.

40 Liu, J., Banis, M.N., Li, X., Lushington, A., Cai, M., Li, R., Sham, T.-K., and Sun, X. (2013) *J. Phys. Chem. C*, **117**, 20260–20267.

41 Perng, Y.-C., Cho, J., Sun, S.Y., Membreno, D., Cirigliano, N., Dunn, B., and Chang, J.P. (2014) *J. Mater. Chem. A*, **2**, 9566–9573.

42 Wang, Y., Richards, W.D., Ong, S.P., Miara, L.J., Kim, J.C., Mo, Y., and Ceder, G. (2015) *Nat. Mater.*, **14**, 1026–1031.

43 Nowak, S., Berkemeier, F., and Schmitz, G. (2015) *J. Power Sources*, **275**, 144–150.

44 Oudenhoven, J.F., Baggetto, L., and Notten, P.H. (2011) *Adv. Energy Mater.*, **1**, 10–33.

45 Le Van, K., Groult, H., Mantoux, A., Perrigaud, L., Lantelme, F., Lindström, R., Badour-Hadjean, R., Zanna, S., and Lincot, D. (2006) *J. Power Sources*, **160**, 592–601.

46 Chen, X., Pomerantseva, E., Gregorczyk, K., Ghodssi, R., and Rubloff, G. (2013) *RSC Adv.*, **3**, 4294–4302.

47 Østreng, E., Gandrud, K.B., Hu, Y., Nilsen, O., and Fjellvåg, H. (2014) *J. Mater. Chem. A*, **2**, 15044–15051.

48 Gandrud, K.B., Pettersen, A., Nilsen, O., and Fjellvåg, H. (2013) *J. Mater. Chem. A*, **1**, 9054–9059.

49 Shi, S., Qi, Y., Li, H., and Hector, L.G. Jr., (2013) *J. Phys. Chem. C*, **117**, 8579–8593.

50 Aaltonen, T., Alnes, M., Nilsen, O., Costelle, L., and Fjellvag, H. (2010) *J. Mater. Chem.*, **20**, 2877–2881.

51 Miikkulainen, V., Nilsen, O., Laitinen, M., Sajavaara, T., and Fjellvåg, H. (2013) *RSC Adv.*, **3**, 7537–7542.

52 Thangadurai, V. and Weppner, W. (2006) *Ionics*, **12**, 81–92.

53 min Lee, J., ho Kim, S., Tak, Y., and Yoon, Y.S. (2006) *J. Power Sources*, **163**, 173–179.

54 Tomczak, Y., Knapas, K., Sundberg, M., Leskelä, M., and Ritala, M. (2013) *J. Phys. Chem. C*, **117**, 14241–14246.

55 Hämäläinen, J., Munnik, F., Hatanpää, T., Holopainen, J., Ritala, M., and Leskelä, M. (2012) *J. Vac. Sci. Technol., A*, **30**, 01A106.

56 Nakagawa, A., Kuwata, N., Matsuda, Y., and Kawamura, J. (2010) *J. Phys. Soc. Jpn.*, **79**, 98–101.

57 Perentzis, G., Horopanitis, E., Pavlidou, E., and Papadimitriou, L. (2004) *Mater. Sci. Eng., B*, **108**, 174–178.

58 Özer, N. and Lampert, C.M. (1995) *Sol. Energy Mater. Sol. Cells*, **39**, 367–375.

59 Glass, A., Nassau, K., and Negran, T. (1978) *J. Appl. Phys.*, **49**, 4808–4811.

60 Østreng, E., Sønsteby, H.H., Sajavaara, T., Nilsen, O., and Fjellvåg, H. (2013) *J. Mater. Chem. C*, **1**, 4283–4290.

61 Hämäläinen, J., Holopainen, J., Munnik, F., Hatanpää, T., Heikkilä, M., Ritala, M., and Leskelä, M. (2012) *J. Electrochem. Soc.*, **159**, A259–A263.

62 Huggins, R.A. (1977) *Electrochim. Acta*, **22**, 773–781.

63 Culligan, S.D., Langmi, H.W., Reddy, V.B., and McGrady, G.S. (2010) *Inorg. Chem. Commun.*, **13**, 540–542.

64 Yu, X., Bates, J.B., Jellison, G.E., and Hart, F.X. (1997) *J. Electrochem. Soc.*, **144**, 524–532.

65 Kozen, A.C., Pearse, A.J., Lin, C.-F., Noked, M., and Rubloff, G.W. (2015) *Chem. Mater.*, **27**, 5324–5331.

66 Nisula, M., Shindo, Y., Koga, H., and Karppinen, M. (2015) *Chem. Mater.*, **27**, 6987–6993.

67 Oi, T. (1984) *Mater. Res. Bull.*, **19**, 451–457.

68 Mäntymäki, M., Hämäläinen, J., Puukilainen, E., Munnik, F., Ritala, M., and Leskelä, M. (2013) *Chem. Vap. Deposition*, **19**, 111–116.

69 Mäntymäki, M., Hämäläinen, J., Puukilainen, E., Sajavaara, T., Ritala, M., and Leskelä, M. (2013) *Chem. Mater.*, **25**, 1656–1663.

70 Lantelme, F., Mantoux, A., Groult, H., and Lincot, D. (2003) *J. Electrochem. Soc.*, **150**, A1202–A1208.
71 Donders, M., Arnoldbik, W., Knoops, H., Kessels, W., and Notten, P. (2013) *J. Electrochem. Soc.*, **160**, A3066–A3071.
72 Donders, M.E., Knoops, H.C., Kessels, W.M.M., and Notten, P.H. (2011) *ECS Trans.*, **41**, 321–330.
73 Burton, B., Fabreguette, F., and George, S. (2009) *Thin Solid Films*, **517**, 5658–5665.
74 Nilsen, O., Peussa, M., Fjellvåg, H., Niinistö, L., and Kjekshus, A. (1999) *J. Mater. Chem.*, **9**, 1781–1784.
75 Gandrud, K.B., Pettersen, A., Nilsen, O., and Fjellvåg, H. (2010) Baltic ALD 2010 & GerALD 2, Hamburg, Germany.
76 Liu, J., Banis, M.N., Sun, Q., Lushington, A., Li, R., Sham, T.K., and Sun, X. (2014) *Adv. Mater.*, **26**, 6472–6477.
77 Meng, X., Riha, S.C., Libera, J.A., Wu, Q., Wang, H.-H., Martinson, A.B., and Elam, J.W. (2015) *J. Power Sources*, **280**, 621–629.
78 Meng, X., Libera, J.A., Fister, T.T., Zhou, H., Hedlund, J.K., Fenter, P., and Elam, J.W. (2014) *Chem. Mater.*, **26**, 1029–1039.
79 Meng, X., He, K., Su, D., Zhang, X., Sun, C., Ren, Y., Wang, H.H., Weng, W., Trahey, L., and Canlas, C.P. (2014) *Adv. Funct. Mater.*, **24**, 5435–5442.
80 Nandi, D.K., Sen, U.K., Sinha, S., Dhara, A., Mitra, S., and Sarkar, S.K. (2015) *Phys. Chem. Chem. Phys.*, **17**, 17445–17453.
81 Hardwick, L.J., Holzapfel, M., Novák, P., Dupont, L., and Baudrin, E. (2007) *Electrochim. Acta*, **52**, 5357–5367.
82 Cheah, S.K., Perre, E., Rooth, M., Fondell, M., Hårsta, A., Nyholm, L., Boman, M., Gustafsson, T.R., Lu, J., and Simon, P. (2009) *Nano Lett.*, **9**, 3230–3233.
83 Wang, W., Tian, M., Abdulagatov, A., George, S.M., Lee, Y.-C., and Yang, R. (2012) *Nano Lett.*, **12**, 655–660.
84 Gerasopoulos, K., Chen, X., Culver, J., Wang, C., and Ghodssi, R. (2010) *Chem. Commun.*, **46**, 7349–7351.
85 Kim, S.-W., Han, T.H., Kim, J., Gwon, H., Moon, H.-S., Kang, S.-W., Kim, S.O., and Kang, K. (2009) *ACS Nano*, **3**, 1085–1090.
86 Panda, S.K., Yoon, Y., Jung, H.S., Yoon, W.-S., and Shin, H. (2012) *J. Power Sources*, **204**, 162–167.
87 Meng, X., Liu, J., Li, X., Banis, M.N., Yang, J., Li, R., and Sun, X. (2013) *RSC Adv.*, **3**, 7285–7288.
88 Donders, M., Knoops, H., Kessels, W., and Notten, P. (2012) *J. Power Sources*, **203**, 72–77.
89 Longrie, D., Deduytsche, D., and Detavernier, C. (2014) *J. Vac. Sci. Technol., A*, **32**, 010802.
90 Gandrud, K.B., Nilsen, O., and Fjellvåg, H. (2016) *J. Power Sources*, **306**, 454–458.
91 Liu, J., Xiao, B., Banis, M.N., Li, R., Sham, T.-K., and Sun, X. (2015) *Electrochim. Acta*, **162**, 275–281.

7

ALD-Processed Oxides for High-Temperature Fuel Cells

Michel Cassir, Arturo Meléndez-Ceballos, Marie-Hélène Chavanne, Dorra Dallel, and Armelle Ringuedé

PSL Research University, Chimie ParisTech – CNRS, Institut de Recherche de Chimie Paris, Paris Cedex 05, 75005 Paris, France

7.1 Brief Description of High-Temperature Fuel Cells

Fuel cells constitute electrochemical devices transforming the energy proceeding from electrochemical reactions into electrical and thermal energies, useful for transport, small portable applications, and stationary power. Although the concept of fuel cells (FC) was discovered in 1939 by Grove, it is only until the beginning of this century that commercialization has emerged in parallel with advances in material science and electrokinetics [1, 2].

Fuel cells can be classified by the nature of the electrolyte used or by their operating temperature (from 60 to 1000 °C). We can distinguish low-temperature fuel cells from high-temperature fuel cells (HTFC). At low temperatures, the fuel cells are alkaline (AFC), proton-exchange membrane fuel cells (PEMFCs), direct methanol fuel cells (DMFCs), and phosphoric acid fuel cells (PAFCs). Two types of high-temperature fuel cells, operating between 600 and 1000 °C, are developed: molten carbonate fuel cells (MCFCs) and solid oxide fuel cells (SOFCs). In this chapter, our interest is strictly focused on HTFC.

The main advantage of HTFCs is their ability is to reach high electrical yield (50–60%) to cogenerate heat and power at very high combined yields reaching 90% without the necessity of using expensive noble-metal electrocatalysts. Moreover, these devices can operate with a large variety of fuels (natural gas, hydrocarbons, alcohols, biomass, waste, and, of course, hydrogen and syngas). There are roughly two families of HTFCs: those with a molten carbonate (MC) electrolyte, called MCFCs, and those with a solid electrolyte (SO), called SOFCs [3]. Figure 7.1 shows the working principle of such fuel cells.

7.1.1 Solid Oxide Fuel Cells

SOFC is a fully solid-state electrochemical device constituted by two electronic or mixed ionic–electronic porous electrodes separated by a purely ionic dense electrolyte. The modularity of such systems and the elevated current densities (up to $1\,A\,cm^{-2}$), as well as their low sensitivity to impurities present in the

Atomic Layer Deposition in Energy Conversion Applications, First Edition. Edited by Julien Bachmann.
© 2017 Wiley-VCH Verlag GmbH & Co. KGaA. Published 2017 by Wiley-VCH Verlag GmbH & Co. KGaA.

fuel, are among their main features [3–5]. Electrical yields above 50% are easily reached. We may consider SOFCs as the best representative of next-generation fuel cells, and it is already available in some niche markets for residential and stationary applications, and for example, 50 MW was produced in 2013 [6]. Yttria-stabilized zirconia, YSZ, is the state-of-the-art ceramic electrolyte. The common cathode is strontium-doped lanthanum manganite, $La_xSr_{1-x}MnO_3$, which is a pure electronic conductor. The common anode is Ni-YSZ cermet, a mixed electronic–ionic conductor. The reactions occurring at the electrodes are as follows:

$$\text{Cathode}: O_2 + 4e^- \rightarrow 2O^{2-} \tag{7.1}$$

$$\text{Anode}: 2H_2 + 2O^{2-} \rightarrow 2H_2O + 4e^- \tag{7.2}$$

7.1.2 Molten Carbonate Fuel Cells

MCFC have reached maturity, with more than 250 MW produced in the last 3 years, and are initiating their market entry [6]. Although the current densities of such devices are lower than 200 mA cm^{-2}, they can reach electrical yields of 50% for several MW systems [3]. MCFC uses a molten alkali carbonate eutectic as electrolyte, commonly Li_2CO_3–K_2CO_3 or Li_2CO_3–Na_2CO_3, with or without additives in order to control its oxobasicity, defined by the partial pressure of CO_2 in the following self-ionization equilibrium relative to carbonate ions: $CO_3^{2-} \rightarrow CO_2 + O^{2-}$ [3, 7]. The electrolyte is supported by lithium aluminate. The anode is made of Ni reinforced mechanically by a few weight-percent of chromium or aluminum. The cathode, $Li_xNi_{1-x}O$, is constituted by porous nickel, oxidized, and lithiated *in situ* under the oxygen-rich atmosphere [8, 9]. Individual cells in series are separated by corrugated bipolar plates, ensuring the distribution of gases and the electrical continuity of the stack. The operation temperature of MCFC devices is in general 650 °C or slightly lower. The global reaction, involving the production of CO_2 at the anode and its consumption at the cathode, is the following:

$$H_2 + 1/2 O_2 + \underbrace{CO_2}_{\text{cathode}} \rightarrow H_2O + \underbrace{CO_2}_{\text{anode}} \tag{7.3}$$

7.2 Thin Layers in SOFC and MCFC Devices

7.2.1 General Features

In order to reach a sufficient conductivity for YSZ electrolyte and a good performance of the classical electrodes in SOFC systems, temperatures exceeding 850 °C are required. This is probably the major problem limiting the competitiveness of this technology, which led several research groups and developers to significantly lower this temperature to less than 700 °C, avoiding interfering chemical reactions at the electrolyte–electrode interfaces and allowing for the use of inexpensive stainless steel interconnects instead of the ceramic-based ones. However, lowering the temperature decreases electrolyte conductivity and electrode

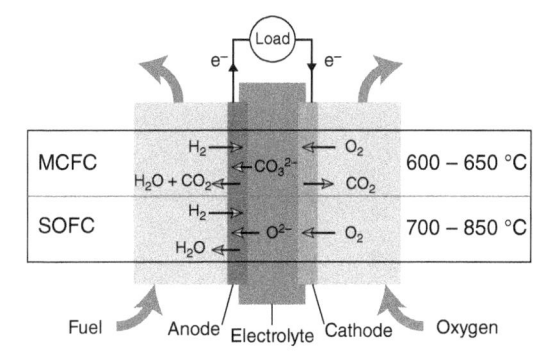

Figure 7.1 Working principle of high-temperature fuel cells SOFC and MCFC.
SOFC: Oxygen is dissociated at the cathode yielding O^{2-}, which migrates at high temperature through the electrolyte and is combined at the anode with hydrogen, to form water and release electrons. The reactions involved are Equations 7.1 and 7.2:

$$\text{Cathode} : O_2 + 4e^- \rightarrow 2O^{2-}$$
$$\text{Anode} : 2H_2 + 2O^{2-} \rightarrow 2H_2O + 4e^-$$

MCFC: Oxygen in the presence of CO_2 is consumed at the cathode producing carbonate ions (CO_3^{2-}), which migrate from the cathode to the anode. The reactions involved are the following:

$$\text{Cathode} : O_2 + 2CO_2 + 4e^- \rightarrow 2CO_3^{2-} \quad (7.4)$$
$$\text{Anode} : 2H_2 + 2CO_3^{2-} \rightarrow 2CO_2 + 4e^- \quad (7.5)$$

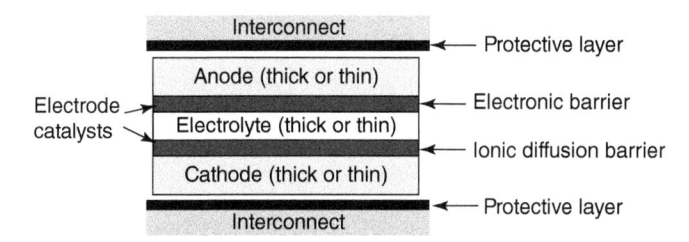

Figure 7.2 Role of thin layers in a high-temperature single cell, in particular SOFC [10].

kinetics, and it is, thus, compulsory to find either new materials or new architectures involving thin layers. Moreover, the use of thin layers not only lowers the electrolyte resistance but could also allow for better control of interfacial stress and mass transfer. High-quality thin layers from few micrometers to nanometers offer many possibilities, from low-resistance thin electrolyte layers (useful in micro-SOFCs) to catalysts (hydrogen and fuel oxidation), including interfacial films (diffusion or electronic barriers, bond, or protective layers) [10] (Figure 7.2).

Contrary to SOFC devices, the working temperature of MCFCs is not an important issue, knowing that it is the right temperature for reforming hydrocarbons into hydrogen and for good electrocatalytic properties at both electrodes. $Li_xNi_{1-x}O$ cathode formed through the contact of Ni with the oxidizing atmosphere and the lithium-containing molten carbonate eutectic presents a relatively high solubility in the electrolyte, which can provoke the formation of metallic nickel and short circuit between the anode and the cathode. Substitute cathodes should be more stable than $Li_xNi_{1-x}O$ in the carbonate medium with good electrical performance [11, 12]; however, the best solution is still to maintain the properties of the Ni cathode but to cover it with protective coatings, based on Co, Ce, or Ti oxides [13, 14]. The corrosion and dissolution of stainless steel bipolar plates in molten carbonates also require the use of protective coatings.

7.2.2 Interest of ALD

In order to process the coatings necessary for SOFC and MCFC systems, different deposition techniques have been used: sol–gel processes, sputtering, electrodeposition, as well as chemical gas-phase techniques such as electrochemical vapor deposition (EVD), chemical spray pyrolysis (CSP), and chemical vapor deposition (CVD). However, it is challenging to fabricate pinhole-free ultrathin materials with uniform thickness. Atomic layer deposition (ALD), which is a sequential CVD, is ideally suited to the mentioned application. It is a unique technique to deposit very thin (in order to reduce ohmic losses), conformal, and dense layers onto these materials in order to efficiently protect them from degradation or to create specific functionalities. Besides, the low deposition temperature (<250 °C) of ALD enables metastable phases and structures to be synthesized and fabricated.

ALD offers the opportunity to design and engineer surface structures at the atomic scale, targeting improved performance of not only fuel cells but also electrochemical sensors, electrolyzers, and pumps. Moreover, ALD permits production of high-quality, large-area flat-panel displays based on thin-film electroluminescence (TFEL) [15]. The important features of ALD in the desired applications are as follows: (i) the atomic-level control ensuring that ultrathin films and complex nanostructures can be processed [16] and (ii) the possibility of modifying the interfaces or doping a thin film at a level required, including delta-doping [17].

With respect to fuel cells, and SOFCs in particular, thin electrolyte membranes obtained by ALD exhibit enhanced surface exchange kinetics, reduced ohmic losses, and superior fuel cell performance as high as 1.34 W cm^{-2} at 500 °C.

One ALD cycle usually consists of four steps: (i) supplying the first precursor to introduce the first surface reactant into a reaction chamber, (ii) purging the unreacted precursor from the reaction chamber, (iii) supplying the second precursor (or oxidant) to introduce the second reactant into the reaction chamber, and (iv) purging the reaction chamber again.

The growth cycles should be repeated as many times as necessary to achieve the desired film thickness. At each half cycle, precursor molecules adsorb onto the underlayer and react to form a new layer. After saturation, no further adsorption or chemisorption takes place. This process may allow processing ternary compounds requiring a very strict control of stoichiometry, which addresses the problem of mixed oxides and complex structures very common for SOFC cathodes and electrolytes, as seen in Section 7.3.

In brief, because the deposition mechanism plays a predominant role with respect to the properties of the thin films, ALD is becoming an important tool in the new generation of fuel cells, including SOFCs and, probably in the near future, MCFCs, and other related techniques.

7.3 ALD for SOFC Materials

7.3.1 Electrolytes and Interfaces

Even though envisaging ALD-processed ultrathin layers (<500 nm) of pure electrolytes is mostly addressed to μ-SOFCs, the application is widespread if we consider their interface reactivity. The electrolytes considered are YSZ, doped ceria, and $LaGaO_3$.

7.3.1.1 Zirconia-Based Materials

Processing the individual oxides constituting YSZ by ALD has been known since the 1990s for ZrO_2 [18, 19] and Y_2O_3 [20, 21]. YSZ was processed with different sources, always $Y(thd)_3$ (thd = 2,2,6,6-tetramethyl-3,5-heptanedione), and $ZrCl_4$ (400 °C), $Zr(thd)_4$ (at 375–400 °C) or cyclopentadienyl Zr precursors: $Cp_2Zr(CH_3)_2$ or Cp_2ZrCl_2 [22, 23]. Growth rates of the films are highly related to the nature of the precursor: 0.9, 0.8, and 0.5 Å per cycle for Cp_2ZrCl_2, $Cp_2Zr(CH_3)_2$, and $Zr(thd)_4$, respectively [22–24]. Later on, electrochemical properties of YSZ overlayers, deposited on LSM at 300 °C using Cp_2ZrCl_2 and $Y(thd)_3$ precursors, were investigated by impedance spectroscopy, showing that, as expected, the resistance of the films decreased with the thickness (0.3–0.9 μm) and that the deposited films were crystalline without annealing treatments [25]. Single-cell tests on YSZ (<100 nm), produced from $Y[(MeCp)_3]$, $Zr[N(NMe_2)_4]$, and H_2O as oxidant, with two Pt current collectors on both sides, produced 270 mW cm^{-2} at 350 °C. This reduced temperature was obtained mainly because of a low electrolyte resistance [26]. The influence of small grain sizes (from 5 to 25 nm) was evidenced with ultrathin layers (around 30 nm) of YSZ, showing an electrical conductivity with an order of magnitude higher than that of the bulk material [27]. The geometry of the electrolyte also has an influence on the performance; for example, when depositing YSZ on prepatterned substrates, the power density can reach 861 mW cm^{-2} at only 450 °C (Figure 7.3) [28]. Nevertheless, the use of ultrathin electrolytes is limited, and the main interest of ALD for SOFC application is to process YSZ interfaces favoring mass and charge transfer. Chao *et al.*, by modifying bulk YSZ (8 mol%) with a 1-nm YSZ layer, with concentrations of yttria varying between 14 and 19 mol%, improved the power density of a single SOFC cell by 50% at 400 °C, which could be due to a higher incorporation rate of the oxide ion on the electrolyte surface [29]. Other authors demonstrated that the presence of an ALD-processed 80-nm YSZ interfacial layer between YSZ and LSM increased the adherence and reduced the interfacial resistance, more efficiently than with other deposition techniques (sputtering and dip coating) [30]. Nanostructuration of YSZ by

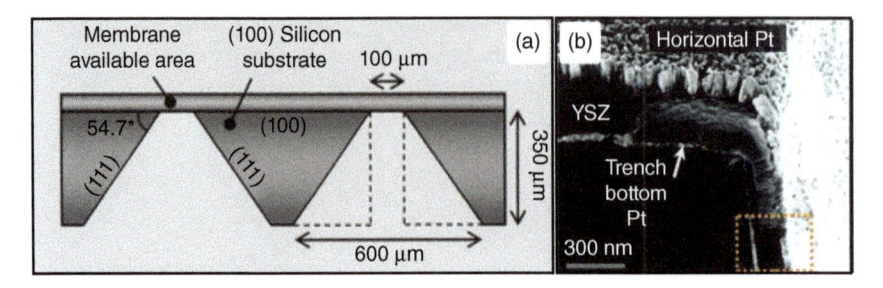

Figure 7.3 Influence of electrolyte geometry: (a) Scheme of the (100) silicon substrate, (b) SEM cross section of the ALD-processed YSZ electrolyte between two platinum layers. (Su *et al.* 2008 [28]. Reproduced with permission of American Chemical Society.)

nanosphere lithography and ALD produced corrugated thin films, reducing the polarization and ohmic losses, resulting in a μ-SOFC with 1.34 W cm^{-2} at 500 °C [31]. Recently, Ji *et al.* showed the benefit of using plasma-enhanced ALD, where the reactants are excited by a plasma source exhibiting a higher reactivity upon deposition; this technique allows for decreasing the minimum thickness of YSZ (maintaining gas-tightness and electronic insulation) required for a high open-circuit voltage (OCV) of 1.17 V at 500 °C, 70 nm instead of 180 nm for a non-plasma ALD process [32]. ALD-processed YSZ, with different concentrations of yttria, was found to be suitable for μ-SOFC applications, by analyzing its structure by HRTEM/XRD and its conductivity by cross-plane impedance spectroscopy; conductivity of several orders of magnitude higher than that with other deposition techniques and efficient fuel cell performance (OCV of 1 V at 100 °C) were achieved [33]. Other zirconia compounds were also considered, such as ZrO_2–In_2O_3 materials, which were processed by ALD at 300 °C, with gradual ionic to electronic composition, which improves the interfacial electrochemical properties; unfortunately, no single-cell tests were conducted to prove this concept; an illustration is given in Figure 7.4 [34].

7.3.1.2 Ceria-Based Materials

Lowering the temperature of SOFCs has opened an important field for ceria-based material electrolytes, because although they have a higher ionic conductivity with respect to zirconia compounds, they can also become electronic conductors in reducing atmospheres, which is not significant at temperatures lower than 650 °C. In any case, ceria compounds can be used as catalysts in the anode side for hydrogen oxidation and in double layer with ultrathin YSZ as electronic blocking film; an ALD-YSZ film between the GDC electrolyte and the anode enhances the OCV significantly [35]. Gadolinia-doped ceria (GDC), a good candidate, has been grown by ALD with Ce(thd)$_4$, Gd(thd)$_3$, and ozone precursors on planar and porous substrates [36, 37]. The films obtained were homogeneous and dense, but with low growth rates and a stoichiometry that is difficult to stabilize. Higher growth rate and a good stoichiometry were obtained with yttria-doped ceria (YDC) at 300 °C with thd precursors and ozone; YDC is a better conductor than YSZ at more than 400 °C [38]. Recently, some authors have shown the feasibility of

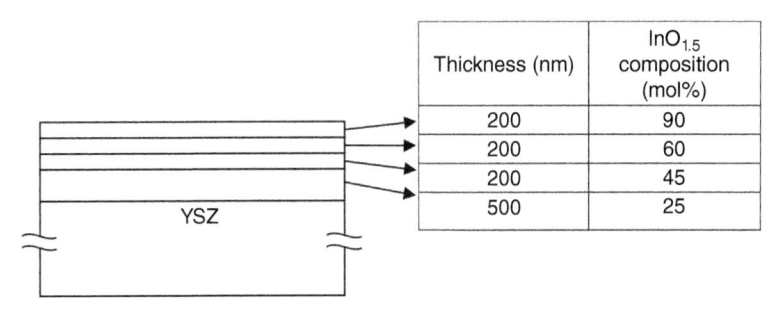

Figure 7.4 SEM cross section of the mixed conductivity system ZrO_2–In_2O_3 (1030 nm) deposited on an YSZ pellet by ALD at 300 °C. (Brahim et al. 2009 [34]. Reproduced with permission of Royal Society of Chemistry.)

producing epitaxial layers of ceria by ALD and showed, from both modeling and electrochemical measurements, that this structural feature favors the reduction power of CeO_2 and, as a consequence, the oxidation of hydrogen [39–42]. Such type of compounds could also be helpful for the direct oxidation of methane, which is an enormous challenge with fuel cells [43, 44].

7.3.1.3 Gallate Materials

Gallate-based compounds, because of their efficient O^{2-} conduction, stability over a large range of oxygen partial pressure, and the low volatility of Ga at lower temperature (<650 °C), are good candidates as SOFC electrolytes [45, 46]. Nevertheless, at the moment, it seems that only $LaGaO_3$ and Ga_2O_3 were processed by ALD [47, 48].

7.3.2 Electrodes and Current Collectors

7.3.2.1 Pt Deposits

Pt processed by ALD has been used by many authors to follow up the behavior of reference cells and for μ-SOFC systems where the cost of such material is not a dramatic issue. Pt can be anode, cathode, or current collector. Some authors found similar peak current densities between Pt anodes processed by ALD (using $MeCpPtMe_3$ precursor) and by DC sputtering with only the fifth of the amount of noble metal in the case of ALD [49]. Pt catalysts processed by both ALD and sputtering created nanostructures with high triple-phase-boundary densities and low-current-collecting resistance, resulting in a 90% increase in the SOFC peak power density with respect to catalyst structures deposited only by sputtering [50]. ALD films of Pt cathode of 25 nm allowed obtaining a peak power density of 110 mW cm^{-2} at only 450 °C [51].

7.3.2.2 Anode

In a strict sense, there is no development of ALD-processed anodes. Gallate-based materials could potentially be interesting as mixed ionic–electronic (MIEC) anodes under the form of La$_{0.9}$Sr$_{0.1}$Ga$_{0.8}$Mg$_{0.2}$O$_3$ (LSGM), which is still not easy to obtain by ALD [46]. However, we already mentioned the interest in CeO$_2$ anode catalysts, and in this direction, ALD-processed ruthenium deposited on sputtered Pt mesh has been tested in direct ethanol SOFCs [52]. In effect, Pt/Ru bimetallic catalyst allows oxidation of C≡O by Ru and deprotonation of ethanol and cleavage of C—C bonds by Pt. Even though the cost of such catalysts is elevated, the proof of concept of such systems is extremely promising.

7.3.2.3 Cathode

As for the anode, the structural complexity of SOFC cathodes has not allowed for an important research effort; however, it is a growing field, and recently, more papers have been published on this subject. Initially, because strontium and lanthanum may react with carbonate impurities, it was difficult to process the common La$_x$Sr$_{1-x}$MnO$_3$ cathode, but Holme et al., using cyclopentadienyl precursors with water, obtained crystalline as-deposited LSM [53]. However, the quality of this cathode was not sufficient for low-temperature operation. Another ALD-processed cathode, using (thd) precursor with ozone, is La$_{1-x}$Ca$_x$MnO$_3$ [54]. The role of interfaces has been outlined by Gong et al., who showed that an ALD conformal ZrO$_2$ film onto an MIEC cathode, La$_{0.6}$Sr$_{0.4}$CoO$_{3-\delta}$, decreased the degradation rate of the cathode material and the area-specific resistance by a factor close to 20 [55]. Contrarily, Al$_2$O$_3$, thin films processed by ALD, used to close defective pinholes in thin-film electrolytes at the interface with different cathode materials, had a detrimental effect on cell performance [56, 57]. Similar detrimental results were obtained with CeO$_2$ or SrO with MIEC cathodes [58].

7.4 Coatings for MCFC Cathodes and Bipolar Plates

As already mentioned, MCFC cathode suffers corrosion and dissolution in molten carbonates [7, 59, 60]. Corrosion of stainless steel bipolar plates is another important issue with regard to durability and performance of MCFCs [61, 62]. One of the most promising techniques to improve the dissolution and corrosion resistance of cathode and bipolar plates is through protective coatings of metal oxides. Coating techniques such as electrodeposition, sol–gel and plasma, and laser sputtering have proven to be of interest, but they have some drawbacks such as material adhesion, layer homogeneity, crystallinity of the

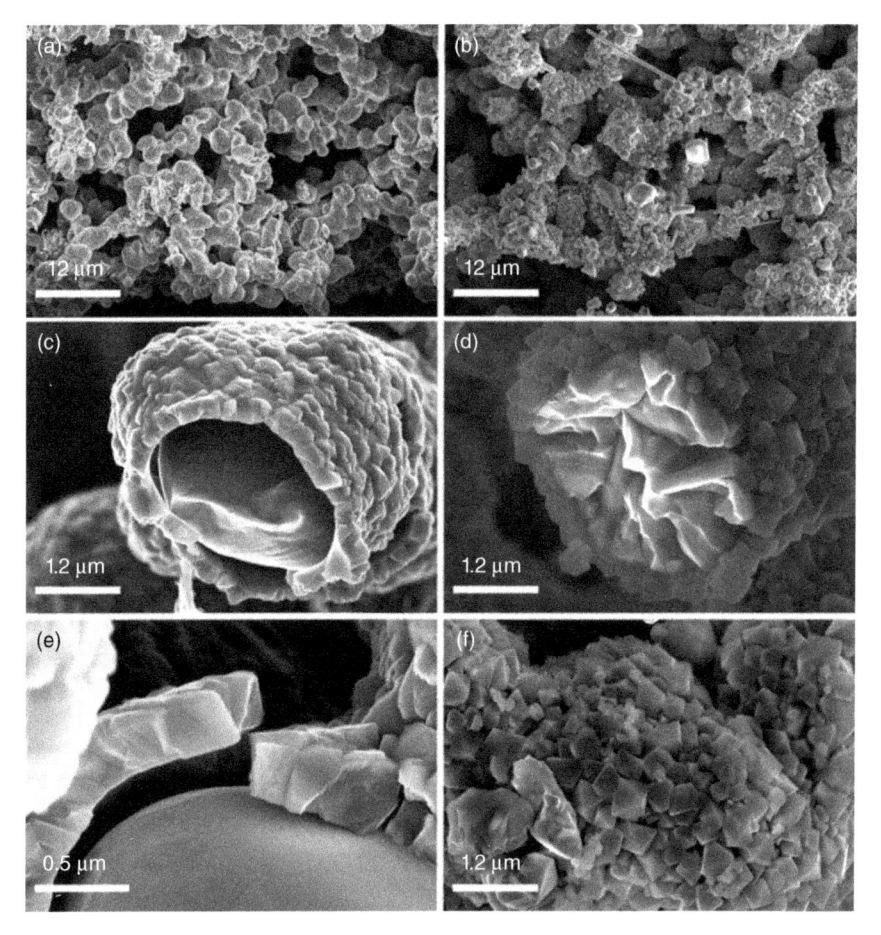

Figure 7.5 ALD-processed TiO$_2$ (300 nm) on porous Ni cathode as-deposited (a, c, e) and after immersion (b, d, f) in molten Li–K carbonates during 230 h at 650 °C. (Meléndez-Ceballos *et al.* 2013 [65]. Reproduced with permission of Elsevier.)

oxides, layer thickness control, and porosity [63–65]. ALD seems particularly adapted to process ultrathin and fine layers with protective features but without affecting the basic properties of the substrates with either cathode or bipolar plate. Until now, only Meléndez-Ceballos *et al.* used ALD-processed CeO$_2$, TiO$_2$, Co$_2$O$_3$, and Nb$_2$O$_5$ films on porous Ni substrates for MCFC applications [14, 65, 66]. Most of the deposits had a significant effect on cathode protection against dissolution, maintaining the electrochemical performance. Figure 7.5 shows deposits of TiO$_2$ on porous Ni cathode before and after immersion in molten carbonates. TiO$_2$ is crystalline as-deposited, dense, and conformal (a, c, e) and after 230 h of immersion in molten Li–K eutectic (b, d, f), the coated surface evolved into a Li–Ti–Ni–O mixed phase mainly composed of Li$_2$TiO$_3$. Ni solubility decreased by a factor 2 with the protective coating (8 against 15 wt.-ppm), which is a better result than that obtained with TiO$_2$ coated by other techniques [67].

7.5 Conclusion and Emerging Topics

ALD is becoming an extraordinary and scalable tool for functionalizing materials and interfaces in many fields and, more recently, in energy devices where numerous extensions become possible. MIEC cathodes and anodes for SOFC devices are facing important challenges due to the complexity of ALD processing with three or more precursors; nevertheless, some studies are promising, and new prospects will appear very soon. Epitaxial catalysts for direct oxidation of methane in both SOFC and MCFC could become a strategic topic in the near future. In a similar field, new cerate-based electrolyte materials for proton electrolyte fuel cell (PEFC)) could be processed by ALD, thereby increasing the reactivity of this device, which operates at temperatures lower than 500 °C. In the case of MCFC, protection of cathode and anode bipolar plate could be enhanced against corrosion by ALD homogeneous and dense films; this field is quite large, and research efforts are predictable. Another topic is processing functional catalytic or sulfur sorbent layers at MCFC anode [64]. Thin layers can also play similar roles in hybrid direct carbon fuel cells (HDCFC), which are systems combining the features of an SOFC system with a reservoir filled with molten carbonates, enhancing transport of the carbon fuel at the anode side [68].

Finally, high-temperature water electrolysis in reverse SOFC (or MCFCs) is gaining an increasing interest since high-temperature water vapor pressure is produced in industrial processes (nuclear) [69].

References

1 Grove, W.R. and Phil, M. (1842) *London Edinburgh Philos. Mag. J. Sci.*, **21**, 417–420.
2 Kordesch, K.V. and Simader, G.R. (1996) *Fuel Cells & Their Applications*, VCH, Weinheim.
3 Cassir, M., Jones, D., Lair, V., and Ringuedé, A. (2013) in *Handbook of Membrane Reactors*, Part III, Chapter 20, (ed. A. Basile), Woodhead Publishing Limited, pp. 553–605.
4 Williams, M.C., Starkey, J.P., Surdoval, W.A., and Wilson, L.C. (2006) *Solid State Ionics*, **177**, 2039–2044.
5 Steele, B.C.H. and Heinzel, A. (2015) *Nature*, **414**, 345–352.
6 Carter, D. and Wing, J. (2013) The Fuel Cell Industry Review 2013. Fuel Cell Today, pp. 36–37.
7 Scaccia, S. (2005) *J. Mol. Liq.*, **116**, 67–71.
8 Janowitz, K., Kah, M., and Wendt, H. (1999) *Electrochim. Acta*, **45**, 1025–1037.
9 Fukui, T., Okawa, H., and Tsunooka, T. (1998) *J. Power Sources*, **71**, 239–243.
10 Cassir, M., Ringuedé, A., and Niinistö, L. (2010) *J. Mater. Chem.*, **20**, 8987–8993.
11 Mohamedi, M., Hisamitsu, Y., Kihara, K., Kudo, T., Itoh, T., and Ushida, I. (2001) *J. Alloys Compd.*, **315**, 224–233.

12 Belhomme, C., Gourba, E., Cassir, M., and Tessier, C. (2001) *J. Electroanal. Chem.*, **503**, 69–77.
13 Kulkarni, A. and Giddey, S. (2012) *J. Solid State Electrochem.*, **16**, 3123–3146.
14 Meléndez-Ceballos, A., Albin, V., Ringuedé, A., Fernandez-Valverde, S.M., and Cassir, M. (2014) *Int. J. Hydrogen Energy*, **39**, 12233–12241.
15 Suntola, T. and Antson, J. (1977) US Patent 4 058 430.
16 Knez, M., Nielsch, K., and Niinisto, L. (2007) *Adv. Mater.*, **19**, 3425–3438.
17 Lehto, S., Lappalainen, R., Viirola, H., and Niinisto, L. (1996) *Fresenius J. Anal. Chem.*, **355**, 129–134.
18 Ritala, M. and Leskela, M. (1994) *Appl. Surf. Sci.*, **75**, 333–340.
19 Cassir, M., Goubin, F., Bernay, C., Vernoux, P., and Lincot, D. (2002) *Appl. Surf. Sci.*, **193**, 120–128.
20 Molsa, H., Niinisto, L., and Utriainen, M. (1994) *Adv. Mater. Opt. Electron.*, **4**, 389–400.
21 Putkonen, M., Sajavaara, T., Johansson, L.S., and Niinistö, L. (2001) *Chem. Vap. Deposition*, **7**, 44–50.
22 Cassir, M., Lincot, D., Goubin, F., and Bernay, C. (2002) Patent WO020537981.
23 Bernay, C., Ringuede, A., Colomban, P., Lincot, D., and Cassir, M. (2003) *J. Phys. Chem. Solids*, **64**, 1761–1770.
24 Putkonen, M., Sajavaara, T., Niinisto, J., Johansson, L.-S., and Niinisto, L. (2002) *J. Mater. Chem.*, **12**, 442–448.
25 Brahim, C., Ringuedé, A., Cassir, M., Putkonen, M., and Niinistö, L. (2007) *Appl. Surf. Sci.*, **253**, 3962–3968.
26 Shim, J.-H., Chao, C.-C., Huang, H., and Prinz, F.B. (2007) *Chem. Mater.*, **19**, 3850–3854.
27 Ginestra, C.N., Sreenivasan, R., Karthikeyan, A., Ramanathan, S., and McIntyre, P.C. (2007) *Electrochem. Solid-State Lett.*, **10**, B161–B165.
28 Su, P.C., Chao, C.-C., Shim, J.H., Fashings, R., and Prinz, F.B. (2008) *Nano Lett.*, **8**, 2289–2292.
29 Chao, C.-C., Kim, Y.B., and Prinz, F.B. (2009) *Nano Lett.*, **9**, 3626–3628.
30 Benamira, M., Ringuedé, A., Cassir, M., Horwat, D., Pierson, J.F., Lenormand, P., Ansart, F., Bassat, J.M., and Fullenwarth, J. (2009) *Open Fuels Energy Sci. J.*, **2**, 32–44.
31 Chao, C.-C., Hsu, C.-M., Cui, Y., and Prinz, F.B. (2011) *ACS Nano*, **5**, 5692–5696.
32 Ji, S., Cho, G.Y., Yu, W., Su, P.-C., Lee, M.H., and Cha, S.W. (2015) *ACS Appl. Mater. Interfaces*, **7**, 2998–3002.
33 Jang, D.Y., Kim, H.K., Kim, J.W., Bae, K., Schlupp, M.V.F., Park, S.W., Prestat, M., and Shim, J.H. (2015) *J. Power Sources*, **274**, 611–618.
34 Brahim, C., Chauveau, F., Ringuedé, A., Cassir, M., Putkonen, M., and Niinistö, L. (2009) *J. Mater. Chem.*, **760**, 760–766.
35 Ji, S., Chang, I., Lee, Y.H., Park, J., Paek, J.Y., Lee, M.H., and Cha, S.W. (2013) *Nanoscale Res. Lett.*, **8**, 48.
36 Gourba, E., Ringuedé, A., Cassir, M., Billard, A., Paivasaari, J., Niinisto, J., Putkonen, M., and Niinisto, L. (2003) *Ionics*, **9**, 15–20.

37 (a) Gourba, E., Ringuede, A., Cassir, M., Paivasaari, J., Niinisto, J., Putkonen, M., and Niinisto, L. (2003) Proceedings of the 8th International Symposium on Solid Oxide Fuel Cells, Paris, April 27–May 2; (b) Singhal, S.C. and Dokiya, M. (eds) (2003), vol. **7**, The Electrochemical Society, Pennington, NJ, pp. 267–274.

38 Ballée, E., Ringuedé, A., Cassir, M., Putkonen, M., and Niinistö, L. (2009) *Chem. Mater.*, **21**, 4614–4619.

39 Coll, M., Gazquez, J., Palau, A., Varela, M., Obradors, X., and Puig, T. (2012) *Chem. Mater.*, **24**, 3732–3737.

40 Marizy, A., Roussel, P., and Ringuedé, A. (2015) *J. Mater. Chem. A*, **19**, 10498–10503.

41 Désaunay, T., Ringuedé, A., Cassir, M., Labat, F., and Adamo, C. (2012) *Surf. Sci.*, **606**, 305–311.

42 Désaunay, T., Bonura, G., Chiodo, V., Freni, S., Couzine, J.-P., Bourgon, J., Ringuedé, A., Labat, F., Adamo, C., and Cassir, M. (2013) *J. Catal.*, **297**, 193–201.

43 Steele, B.C. and Heinzel, A. (2001) *Nature*, **414**, 345–352.

44 Gorte, R.J. and Vohs, J.M. (2003) *J. Catal.*, **216**, 477–486.

45 Huang, P. and Petric, A. (1996) *J. Electrochem. Soc.*, **143**, 1644–1648.

46 Chen, F. and Meilin, L. (1998) *J. Solid State Electrochem.*, **13**, 7–14.

47 Nieminen, M., Lehto, S., and Niinistö, L. (2001) *J. Mater. Chem.*, **11**, 3148–3153.

48 Dezelah, C.L., Niinistö, J., Arstila, K., Niinistö, L., and Winter, C.H. (2006) *Chem. Mater.*, **18**, 471–475.

49 Jiang, X., Huang, H., Prinz, F.B., and Bent, S.F. (2008) *Chem. Mater.*, **20**, 3897–3905.

50 Chao, C.-C., Motoyama, M., and Prinz, F.B. (2012) *Adv. Energy Mater.*, **2**, 651–654.

51 Ji, S., Chang, I., Cho, G.Y., Lee, Y.H., Shim, J.H., and Cha, S.W. (2014) *Int. J. Hydrogen Energy*, **39**, 12402–12408.

52 Jeong, H.J., Kim, J.W., Jang, D.Y., and Shim, J.H. (2015) *J. Power Sources*, **291**, 239–245.

53 Holme, T.M., Lee, C., and Prinz, F.B. (2008) *Solid State Ionics*, **179**, 1540–1546.

54 Nilsen, O., Rauwal, E., Fjallvag, F., and Kjekshus, A. (2007) *J. Mater. Chem.*, **17**, 1466–1475.

55 Küngas, R., Yu, A.S., Levine, J., Vohs, J.M., and Gorte, R. (2013) *J. Electrochem. Soc.*, **160**, F205–F211.

56 Kim, E.-H., Jung, H.-J., An, K.-S., Park, J.-Y., Lee, J., Hwang, I.-D., Kim, J.-Y., Lee, M.-J., Kwon, Y., and Hwang, J.-H. (2014) *Ceram. Int.*, **40**, 7817–7822.

57 Yu, A.S., Küngas, R., Vohs, J.M., and Gorte, R.J. (2014) *J. Electrochem. Soc.*, **160**, F1225–F1231.

58 Gong, Y., Palacio, D., Song, X., Patel, R.L., Liang, X., Zhao, X., Goodenough, J.B., and Huang, K. (2013) *Nano Lett.*, **13**, 4340–4365.

59 Ota, K., Mitsushima, S., Kato, S., Asano, S., Yoshitake, H., and Kamiya, N. (1992) *J. Electrochem. Soc.*, **139**, 667–671.

60 Brenscheidt, T., Nitschké, F., Söllner, O., and Wendt, H. (2001) *Electrochim. Acta*, **46**, 783–797. doi: 10.1016/S0013-4686(00)00665-4
61 Cassir, M. and Belhomme, C. (1999) *Plasmas Ions*, **2**, 3–15.
62 Yuh, C., Johnsen, R., Farooque, M., and Maru, H. (1995) *J. Power Sources*, **56**, 1–10.
63 Agll, A.A.A., Hamad, Y.M., Hamad, T.A., Thomas, M., Bapat, S., Martin, K.B., and Sheffield, J.W. (2013) *Appl. Therm. Eng.*, **59**, 634–638.
64 Albin, V., Goux, A., Belair, S., Lair, V., Ringuedé, A., and Cassir, M. (2007) *ECS Trans.*, **3**, 205–213.
65 Meléndez-Ceballos, A., Fernández-Valverde, S.M., Barrera-Díaz, C., Albin, V., Lair, V., Ringuedé, A., and Cassir, M. (2013) *Int. J. Hydrogen Energy*, **38**, 13443–13452.
66 Meléndez-Ceballos, A., Albin, V., Fernández-Valverde, S.M., Ringuedé, A., and Cassir, M. (2014) *Electrochim. Acta*, **140**, 174–181.
67 Hong, M.Z., Lee, H.S., Kim, M.H., Park, E.J., Ha, H.W., and Kim, K. (2006) *J. Power Sources*, **156**, 158–165.
68 Nabae, Y., Pointon, K.D., and Irvine, J.T.S. (2008) *Energy Environ. Sci.*, **1**, 148–155.
69 Yildiz, B. and Kazimi, M.S. (2006) *Int. J. Hydrogen Energy*, **31**, 77–92.
70 Putkonen, M. and Niinisto, L. (2001) *J. Mater. Chem.*, **11**, 3141.
71 Putkonen, M., Niinistö, J., Kukli, K., Sajavaara, T., Karppinen, M., Yamauchi, H., and Niinistö, L. (2003) *Chem. Vap. Deposition*, **9**, 207–212.

Part IV

ALD in Photoelectrochemical and Thermoelectric Energy Conversion

8

ALD for Photoelectrochemical Water Splitting

Lionel Santinacci

Aix Marseille Univ, CNRS, CINAM, Marseille, France

8.1 Introduction

More energy from sunlight strikes Earth in 1 h than all of the energy consumed by humans in an entire year [1]. However, the sunlight's intermittent nature is one of the issues limiting widespread harvesting of solar energy for societal power infrastructure [2]. A leading approach is to store the energy produced by this discontinuous renewable energy source as chemical fuels. Since the chemical fuels are produced from solar energy, they are called solar fuels [3]. Photoelectrochemical water splitting, evidenced first in 1972 by Fujishima and Honda [4], has enabled promising perspectives to produce solar hydrogen. This method could lead to a large-scale electrochemical energy production with minimal global-warming gas emission since no CO_2 is emitted during the process. Since the photoelectrochemical process exhibits strong similarities to photosynthesis, the concept of artificial photosynthesis (APS) has emerged (several reviews are dedicated to this field, see, e.g., [3, 5, 6]). APS consists of producing solar fuels from water or CO_2 using photoelectrochemical processes. Nocera has even proposed the concept of an "artificial leaf" that is directly inspired by the natural processes occurring in chlorophyll plants [7]. The principle of such an artificial leaf is depicted in Figure 8.1. The photosynthetic membrane is replaced by a Si junction, which performs the light capture and conversion to a wireless current. The oxygen-evolving complex (OEC) and ferredoxin reductase of the photosynthetic membrane are replaced by Co-OEC and oxygen evolution reaction (OER) and hydrogen evolution reaction (HER) catalysts, respectively, to perform water splitting.

Among the various solar fuels such as methanol, formic acid, or formaldehyde, molecular hydrogen is a very good candidate because it has a very high energy density and its combustion or oxidation is carbon free. In addition to be a renewable H_2 production process, the photoelectrochemical dissociation of water generates high-purity hydrogen that can be directly used in fuel cells. H_2 is, however, not widely used as fuel because it is mainly produced from fossil resources [8]. Although solar-driven water splitting is a sustainable and abundant

Atomic Layer Deposition in Energy Conversion Applications, First Edition. Edited by Julien Bachmann.
© 2017 Wiley-VCH Verlag GmbH & Co. KGaA. Published 2017 by Wiley-VCH Verlag GmbH & Co. KGaA.

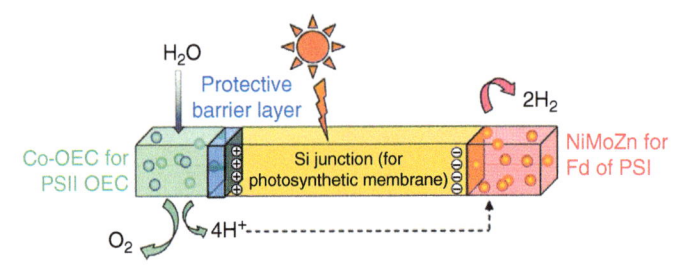

Figure 8.1 Schematic description of the artificial leaf concept. (Nocera 2012 [7]. Reproduced with permission of American Chemical Society.)

energy storage, the development of efficient, stable, and cost-effective photoelectrochemical cells (PECs) is required for fabrication of commercial devices [9].

Nowadays, the efficiency of the PECs ranges from 12% to 18%, depending on the materials and the type of cells (single- or multiple junction) while the theoretical limit is 24.4% and 30% for tandem and multijunction cells, respectively [10]. High efficiency at a high cost was shown from multijunction cells in 2001 [11], but no sufficient improvements leading to a market compliant PEC have been reported yet. A comprehensive techno-economic evaluation [9, 12] and other investigations [13] have confirmed this issue. The photoelectrochemical technology remains at a low technology readiness level (TRL 1 to 2), and the main issue is the actual production cost of H_2. To solve this issue, it is mandatory to significantly reduce the costs and to increase the photoconversion efficiency. Many research groups are currently investigating different routes to fabricate PECs that could respond to the market demand. To improve efficiency, stability, and price, one has, of course, to select a cost-effective photosensitive material and the appropriate cell design. It has been recently shown that micro- or nanostructuring and/or surface functionalization of the photoelectrodes can lead to higher performances. Among the numerous approaches and techniques that have been used since nanosciences and nanotechnologies have emerged, atomic layer deposition (ALD) has recently demonstrated its high effectiveness in the fabrication of both two- and three-dimensional (2D, 3D) nano-objects. Energy storage and production are part of the fields of applications in which ALD has shown highly promising perspectives [14]. It is shown in this chapter that, during the last 5 years, ALD has been effectively integrated in various fabrication strategies of photoelectrodes.

The aim of this chapter is to report the different types of uses of ALD to manufacture photoelectrodes with improved performance. The description of the principle of the technique itself is therefore out of scope. The reader can refer to Chapter 1 of this book and to the numerous review articles and books that have been recently published in the field [15–19]. Section 8.2 is dedicated to the description of the PEC. First, the principle of the PEC is reported briefly. The different types of materials employed to build up the photoelectrodes are then listed. The section ends with the current trends, including ALD, to improve the PECs. The last and most detailed section (Section 8.3) reviews the different uses of ALD in the field of solar fuel production: active materials, surface state passivation, and corrosion protection.

8.2 Photoelectrochemical Cell: Principle, Materials, and Improvements

The first part of this section aims at a brief description of the field of PECs. The reader is strongly recommended to refer to comprehensive reviews in the field to obtain an in-depth knowledge (see, e.g., Refs [10, 20–25]). To improve the PEC performances, three main research topics have been identified in the literature [9]. (i) The material composition of the photoelectrodes determines the energetics of the semiconductor/electrolyte junction while (ii) its geometry plays a crucial role in the light absorption, charge carrier separation and transport, and, finally, (iii) the coating (or functionalization) is often necessary to protect and stabilize the electrode or to passivate the surface. These three fields of investigation are then related in the following parts of this section.

8.2.1 Principle of the PEC

The photocatalytic water splitting consists of separating water according to two half-reactions. Molecular H_2 is produced by reduction (HER) at the photocathode (Equation 8.1), and O_2 is generated at the photoanode (Equation 8.2) by oxidation (OER).

$$2H^+ + 2e^- \rightarrow H_2 \quad E° = 0\,V \text{ vs NHE} \tag{8.1}$$

$$2H_2O \rightarrow O_2 + 4H^+ + 4e^- \quad E° = 0\,V \text{ vs NHE} \tag{8.2}$$

Although these reactions look very simple, they require a semiconductive material that converts the absorbed light into electron/hole (e^-/h^+) pairs with an appropriate thermodynamics. As described in Figure 8.2, the wavelength of the incident photon should correspond to energy greater than 1.23 eV ($\lambda < 1000$ nm) in order to oxidize H_2O in O_2 using the photogenerated h^+ in the valence band (VB) and to reduce the H^+ to H_2 using the e^- excited in the conduction band (CB). The reverse reaction, that is, the water formation from H_2 and O_2, as well as the e^-/h^+ recombination, should be avoided. In some cases, a co-catalyst is required to promote the H_2 evolution at the photocathode. Different energy schemes can be used. One-step water photosplitting is achieved when the photogenerated charges used for both oxidation and reduction are created in the same semiconductor (Figure 8.2a). It is, however, possible to use a counter electrode that will collect one of the photocarriers in the counter reaction (Figure 8.2b,c). The second approach consists of a double excitation process also known as Z-scheme (Figure 8.2d). It requires a tandem cell design in which the photons are absorbed in two different semiconductors. HER and OER proceed exclusively at one of the two photoelectrodes, and a redox couple mediates the charge between the semiconductors.

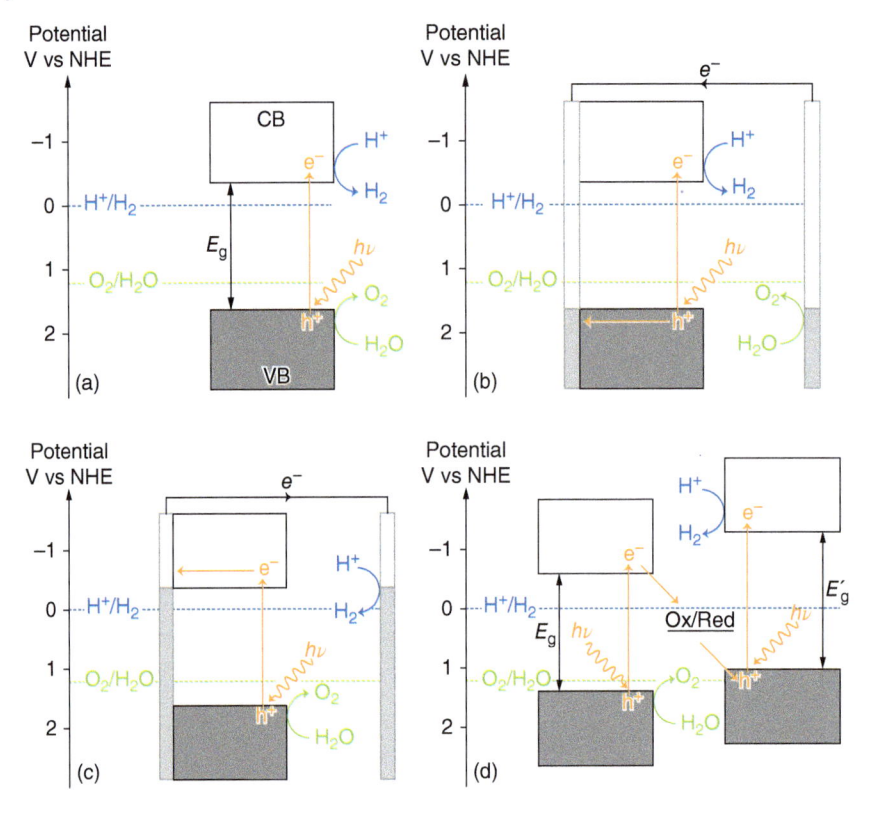

Figure 8.2 Energy diagrams of photocatalytic water splitting. (a) Single-excitation process with a single semiconductive photoelectrode, (b) single-excitation process with a combination of a semiconductive photocathode and a metallic anode, (c) single-excitation process with a combination of a semiconductive photoanode and a metallic cathode, and (d) double-excitation process or Z-scheme.

8.2.2 Photoelectrode Materials

TiO_2 has been initially used to demonstrate the process [4]. Other semiconductive metal oxides, silicon, III–V and II–VI semiconductors such as InP, GaAs, GaP, CdTe, CdS have also been utilized to split water [20]. In order to achieve the highest photoelectrochemical performances, the semiconductive electrode should fulfill the following requirements:

- The band-gap energy should allow the maximum light absorption in the range of the solar radiation.
- It should have a CB- and VB-edge energy that straddles the electrochemical potentials $E°(H^+/H_2)$ and $E°(O_2/H_2O)$ and that can drive the HER and OER using e^-/h^+ generated under illumination.
- The penetration depth of the light into the materials should be as deep as possible in order to use the largest part of the photoelectrode. It should be mentioned that for thin electrodes, the penetration depth of light should be in the range of the material thickness.

- The diffusion length of the minority charge carriers (or their carrier lifetime) should be the longest as possible to avoid the electron–hole recombination from which no current can be collected.

8.2.2.1 Metal Oxides

After the discovery of water photosplitting using TiO_2 in 1972 [4], transition metal oxides have been largely investigated because they are earth-abundant and thus affordable. Since they are mainly n-type semiconductors, they serve essentially as photoanodes. They are generally chemically stable in the harsh conditions of water oxidation [26]. Due to their large band gap, metal oxides absorb light mostly in the UV range, and the theoretical efficiency is then limited. In addition, they often exhibit a slow kinetics of water oxidation. TiO_2 has been intensively studied, but the lack of success in increasing the energy conversion efficiency of this material using sunlight to above ~1% over the last 30 years has resulted in skepticism concerning the potential of this approach [27]. Shortly after TiO_2, the ability of Fe_2O_3 [28] and WO_3 [29] for water photo-oxidation has been revealed. Although these three materials remain the most studied nowadays, other n-type metal oxides have shown promising photoelectrochemical properties, such as ZnO, $BiVO_4$, $SrTiO_3$, SnO_2, and MnO_2 [20, 23].

In comparison to the large amount of reports on the use of n-type metal oxides for water oxidation, very little work has been published on photocathodes for water reduction [26]. This is due to low number of p-type metal oxides such as CuO [30], Cu_2O [31], Bi_2O_3 [30], p-Fe_2O_3 [32], or $CaFe_2O_4$ [33]. These materials have also been studied during the 1970s, but most of them were unstable under the photoreduction conditions. Hardee and Bard reported the photocathodic response of both CuO and Bi_2O_3 in 1977 [30], but it is only in 1982 that p-type Fe_2O_3 was proposed as a stable photocathode [32]. Later, the photoelectrochemical properties of $CaFe_2O_4$ and Cu_2O have also been reported, but, again, the materials were not stable [31, 33, 34]. Recently, it has been shown that CoO nanoparticles can be used to efficiently photoreduce water [35]. Attempts to use NiO have also been reported [36], but now it is mainly used as coating on other photoactive materials [37]. Several reviews on the use of metal-oxide-based photoelectrodes for solar fuel production have been recently published (see, e.g., Refs [10, 38, 39]), but, until now, the best solar fuel production efficiency has been achieved with elemental or compound semiconductors.

8.2.2.2 Elemental and Compound Semiconductors

Si is a very good candidate for water splitting because it complies with all the cited requirements. In contrast to the metal oxides that absorb mainly in the UV range, silicon has a band-gap energy of 1.1 eV, allowing a high light absorption in the sun radiation wavelengths. The positions of the band edges fit with both $E°(O_2/H_2O)$ and $E°(H^+/H_2)$. Thus, n-type and p-type Si can be used for OER and HER, respectively [40, 41]. Silicon can thus be used for single or tandem PECs. It shows a very deep penetration of the light (11 µm for $\lambda = 800$ nm [42]) and a very long diffusion length of the minority charge carriers (up to several 100 µm for holes in p-Si [43]). Last but not least, Si is the most abundant element after oxygen in the earth's crust, and it is environmentally benign. In order to be widely spread, several drawbacks must be addressed before p- and n-type Si can be used in PEC devices. The surface of Si must be protected under ambient conditions to avoid the rapid formation of silicon oxide at the surface, which may lead to the complete quenching of its photoactivity. Furthermore, an electrocatalyst promoting the HER must be introduced at the Si surface to overcome the kinetic barriers associated with proton reduction. Although the use of noble metal catalysts has been efficient, it is undesirable due to the scarce reserves and high costs. Similarly, when it is used as photocathode, n-Si suffers from two other main drawbacks: surface recombination that hinders the OER kinetics and photocorrosion [44]. Simple, inexpensive, and scalable methods for preparing protected Si are therefore required to pave the way for Si in PEC devices.

Compound semiconductors have shown the ability to photodissociate water. III–V semiconductors such as GaP [45], InP [46], GaAs [47], GaN [48], and GaInP$_2$ [49] have been mainly used as photocathodes, but some of them have also been employed as photoanodes or both. They have demonstrated the best efficiency. Using a GaInP/GaAs-based multijunction device, Khaselev et al. have reported, for instance, a yield of 16.4% [50]. II–VI materials are preferentially used in photovoltaics devices, but they can be implemented in PECs as well. CdS [51], CdSe [52], CdTe [53], Cd$_{1-x}$Zn$_x$S [54], and CdInGaSe (GIGS) can efficiently photodissociate water. These materials have demonstrated very good properties, but similarly to Si, they are prone to corrode under illumination, and their use is limited by their cost or toxicity in some cases. Recently, extremely thin absorbers (ETAs) such as Sb$_2$S$_3$ are integrated in all solid-state dye-sensitized solar cells (sDSSC) as replacement of the dye [55, 56], and they can also be used for solar fuel generation [57].

8.2.2.3 Nitrides

Compared to transition metal oxides and elemental or compound semiconductors, nitrides are much less used as photoelectrodes. Because the electronegativity of nitrogen is lower than that of oxygen, metal nitrides are prone to anodic photocorrosion. However, they have considerable potential as photocathodes [10, 23]. They are usually synthesized by nitridation of their corresponding oxides. Ta$_3$N$_5$ and W$_2$N can, for instance, be obtained from Ta$_2$O$_5$ and WO$_3$ using C$_3$N$_4$ and NH$_3$, respectively [58, 59]. A nonmetallic nitride, C$_3$N$_4$, has also recently been successfully used to produce solar hydrogen [60].

8.2.3 Geometry of the Photoelectrodes: Micro- and Nanostructuring

The previous section has shown the large diversity of materials that can be used as photoelectrodes. To improve the cost-effectiveness of PEC devices, it is necessary to also consider the cell design and the photoelectrode geometry. It has indeed been demonstrated during the past years that these two parameters could drastically increase the performances of the PECs. There is a competition between various types of PEC designs: single- or dual-bed suspension, fixed-panel arrays, and concentrator arrays. Although the last geometry is not in the core of the photoelectrochemical issues, the choice between suspension and panel technologies could be questionable [9]. Panel arrays have efficiency in the same range as suspensions, but they are more expensive. Nevertheless, colloidal solutions suffer

Figure 8.3 Schematic of a water-splitting device concept utilizing structured solar absorbers and a proton-permeable membrane for ion transport. The high aspect ratio structures can improve light absorption for semiconductor materials with short minority carrier diffusion lengths, and the high surface area can enhance catalyst loading. (Warren et al. 2014 [62]. Reproduced with permission of American Chemical Society.)

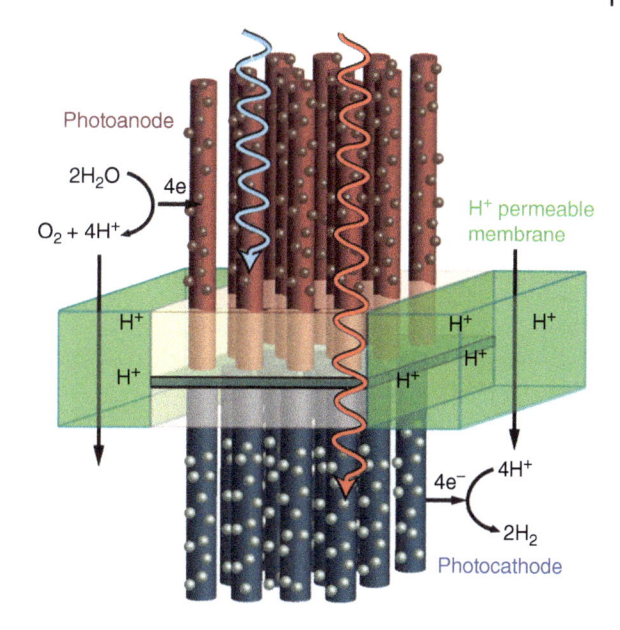

from critical safety issues. It explains the huge number of research works carried out on panel-designed PECs.

In opposition to the high-efficiency multijunction PEC proposed by Turner et al. [47, 50], there are more and more emerging concepts in which a single piece is used to perform both HER and OER. This PEC geometry, sometimes called "wireless cell" [61] because there is no external bias, is thought to be a way to decrease the complexity of the device and thus the costs. Unfortunately, the efficiency is still low (2.5%) [61]. Other compact devices in which the photocathode and photoanode are connected by a proton-conducting membrane are also intensively investigated. Figure 8.3 schematically shows such type of devices made of p-type and n-type Si [62]. Further improvements have been proposed by depositing bioinspired molecular co-catalysts (cubane-like clusters, Mo_3S_4) on the wire surface [63] or by using metal oxides instead of Si [3]. Based on a similar approach, Kibria et al. have reported the visible-light-driven efficient wireless oxygen- and hydrogen-evolving reactions at neutral pH using p-type metal-nitride nanowire arrays [64]. CoO nanoparticles synthesized from micropowders by laser ablation and ball milling have also been employed to split water (H_2 and O_2 generation) without external bias or sacrificial reagents [35]. Due to its simplicity and potential low price, this approach appears to be highly valuable. Liu et al. have also proposed a cheap and simple concept in which freestanding nanowire mesh networks composed of $BiVO_4$ and $Rh-SrTiO_3$ can achieve overall water splitting [65]. In all these recent works, the common evolution is the micro- or nanostructuring of the photoelectrodes. Indeed, it seems that the list of available photoactive materials is now well established, and it is possible to select the appropriate material for the targeted application. Unfortunately, the photoconversion efficiency is not yet satisfactory, and further developments are required. It appears that micro- and nanostructuring are interesting ways to improve the cost-efficiency balance since it is a means to drastically enhance the photoconversion yield of the common active materials.

Silicon is a good material to illustrate the impact of materials micro- and nanostructuring. The planar Si electrodes can themselves be substantially improved through optimization of doping, antireflective surface texture, and catalyst placement, but several groups have proposed the surface micro- or nanostructuring in order to further increase the photoelectrochemical properties of the Si-based PECs. Among the numerous types of Si surface micro- and nanostructures, micro- and nanowires (SiNWs and SiMWs) have been the most used. There are, indeed, only few reports on the use of other nanostructures such as porous Si that leads to high-yield devices (see, e.g., Refs [66–68]). The wire arrays have, obviously, a high light absorption, and it should be possible to double the energy-conversion efficiencies to values that approach those achieved for optimized single-crystal photocathodes (i.e., >10%) [69]. SiMW or SiNW arrays have also shown their ability to improve the PEC performances by a better charge carrier collection, a better light scattering, a quantum confinement inducing a better charge transfer, an appropriate band-edge position, a surface-area-enhanced charge transfer, a multiple exciton generation increasing the photovoltage, a better fill factor, and a better internal quantum efficiency [42, 69]. In brief, the wired structure ensures effective harvesting of the photons and effective photocurrent collection (by virtue of the length available for absorption combined with short, radial, minority-carrier transport distances) while also exposing a large surface area for the catalytic reactions [63]. According to Kayes *et al.* [70], micro- or nanostructured Si allows the use of lower-purity Si with shorter minority-carrier diffusion lengths. Recently, it has been demonstrated that SiNWs can be grown, using inexpensive metallurgical-grade or upgraded silicon (MG-Si or UMG-Si), by metal-assisted chemical etching (MaCE) [71]. These self-purified Si nanostructures named nanoMG-Si exhibited highly promising performances for water reduction [71]. It is further confirmed by another group that has assembled a solar cell with UMG-SiNWs showing an efficiency about 12% [72].

The photoelectrode micro- and nanostructuring can obviously be extended to other materials. For that matter, there are numerous examples of the use of micro- or nanostructured photoelectrodes with other active materials such as Fe_2O_3, WO_3, $BiVO_4$, GaP [65, 73–78]. It has been mentioned earlier that hematite is one of the most used materials for water photo-oxidation but, similarly to silicon, it exhibits some limitations such as large overpotential, slow water oxidation kinetics, relatively short light penetration depth, poor charge-carrier mobility, and short diffusion length of minority carriers [26]. It has been reported recently that nanostructuring could increase the photocurrent because it is a means to enlarge the active area, to improve the charge separation and transport as well as the light absorption, and to quicken the kinetics [10].

Kayes *et al.* have developed a radial p–n junction model to assess the interest of fabricating nanorod-based solar devices. It shows that such photoelectrode design can drastically enhance the cell performances of materials exhibiting a deep light absorption but a short minority-carrier diffusion length [70]. Several photoelectrode architectures have thus been proposed. In all cases, the main idea is to orthogonalize the paths of the light absorption and the transport of the charge carriers. Photoelectrodes exhibiting cylindrical geometries such as wires, rods, or tubes are thus suitable candidates to achieve enhanced photoconversion. Among them, core–shell structures are interesting because it is possible to select the appropriate materials for light absorption and for charge transport. In such systems, an ETA is sandwiched between two transparent wide-band-gap semiconductors. The light absorption occurs within the ETA layer while the p-type and n-type semiconductors transport the charges to the current collectors [79].

8.2.4 Coating and Functionalization of the Photoelectrodes

It has been indicated earlier that depending on their chemical nature, the photoelectrodes exhibit limitations such as slow kinetics, surface recombination, or photocorrosion. In many cases, those restrictions can be overcome by coating the photoelectrode with an appropriate material.

A fundamental problem with using low-band-gap semiconductors such as Si as photocathode for H_2 generation is that the difference between the Si valence band edge and the H^+/H_2 redox level is small, which limits the photovoltage. To address this problem and thus to increase the HER kinetics, expensive Pt particles are usually deposited onto the p-Si surface [80]. Strategies to replace the noble metal are thus required. N^+/p-Si junctions have been initially suggested to increase the photovoltage by enhancing band bending [81], but it is more reasonable to use metal oxide heterojunctions [82].

Surface recombinations in semiconductors hinder the photoelectrochemical activity. It is even drastically enhanced when the electrode is nanostructured because, in this case, the effective area is greatly expanded. The use of noble metal clusters can suppress this effect [83], but it is now believed that a passivation of the whole surface with a catalytically acting transition metal oxide is more suitable to decrease the surface recombinations [84–86].

Many photoelectrode materials corrode during the water photosplitting. It happens for photocathodes such as CdS, InP, GaP, Cu_2O as well as for photoanodes such as Si. Organic functional groups have thus been proposed to protect the Si [42]; however, water-splitting devices should last for decades, and it is not known whether these organics can remain protective for that long [83]. The deposition of catalytically acting transition metal oxides such as Ni/NiO_x has also been reported and appears to be more stable [82, 87].

8.3 Interest of ALD for PEC

In the previous sections, the main challenges in PEC technology have been listed and the leading strategies to improve the photoelectrodes have been drawn. It appears that the synthesis of new electrode materials can be a solution, but the more promising way to enhance the performance is the fabrication of nanostructured photoelectrodes as well as their functionalization. Due to its numerous key advantages over other deposition methods (control of the thickness and composition, ability to coat complex 3D nanostructures, fabrication of core–shell structures, etc.), ALD appears to be a powerful tool to build the next-generation PECs. Recent reviews have shown how this deposition technique can be beneficial in the field of photocatalysis and dye-sensitized solar cells (DSSCs) [88, 89]. The target of this chapter is therefore to be complementary to these previous reports.

ALD has been successfully used to grow 2D and mainly 3D complex layers that are either active or passive films. Active layer means that the film takes part in the core photoconversion process (i.e., absorbers, charge separators: gap engineering, doping, optimized band structure, etc.) while passive layer accounts for layers that are not directly involved in the charge creation, separation, and transport but plays a key role in the device stability and performances (photocorrosion inhibitor, passivation of the surface states, blocking and catalytic layers, transparent conductive oxides, etc.). It is therefore intended to depict various uses of ALD in both active and passive layers. In this section, it is shown how ALD has been used to synthesize active materials and to fabricate nanostructured photoelectrodes. The following parts depict the use of ALD to deposit catalysts, to passivate and modify the junctions, and, finally, to protect the electrodes against corrosion.

8.3.1 Synthesis of Electrode Materials

ALD can be envisioned to synthesize electrode materials and to tailor the composition with a high accuracy. Among the various types of materials that are used as active layers, ALD has been mainly utilized to grow metal oxides. Due to its layer-by-layer growth process, ALD is not a suitable method to synthesize bulk materials. Other solution-based processes such as sol–gel, spray pyrolysis, or hydrothermal routes are more suitable for this purpose. It is though interesting to mention that it is possible to use ALD to grow polymorphic phases with unusual properties. As presented in Figure 3.7, Emery *et al.* have reported the synthesis of metastable β-Fe_2O_3 via isomorphic epitaxy for photoassisted water oxidation [90]. The β-phase grown by ALD has indeed demonstrated an unexpected stability under strong alkaline conditions and by comparison with the α-Fe_2O_3, the band gap is narrowed ($\Delta E_g = 0.1$ eV), the photocurrent onset potential is improved (~0.1 V), and the photoconversion efficiency is even enhanced in the red part of the solar spectrum ($\lambda > 600$ nm). This work demonstrates that ALD can lead to the synthesis of active materials with promising properties.

Similarly, it has been shown that α-Fe_2O_3 grown by ALD could exhibit specific photoelectrochemical properties depending on the substrate on which the active material was deposited [91, 92]. The photo-oxidation of water performed on the ultrathin hematite is drastically enhanced when underlayers such as Ga_2O_3, Nb_2O_5, WO_3, or indium tin oxide (ITO) are previously deposited onto the fluorine-doped tin oxide (FTO) because the crystallinity of the Fe_2O_3 is increased.

These examples show two interesting uses of ALD for PEC applications. Additional examples can be found in Section 3.3, where ALD of various light absorbers is reported for photovoltaics application. However, this technique is highly efficient to conformally deposit pinhole-free thin films. It seems therefore more relevant to implement ALD processes to synthesize existing active materials onto nanostructured scaffolds or to enhance, passivate, and protect operating layers. These aspects are related, in more details, in the following sections.

8.3.2 Nanostructured Photoelectrodes

The performances of the planar photoelectrodes can substantially be improved through selecting the appropriate material, optimization of doping, antireflective texturing, and catalyst deposition. However, it has been shown in Section 8.2.3 that nanostructured electrodes can potentially exhibit much better photodissociation abilities. Although ALD is used to synthesize materials and to tailor the composition with a high accuracy, it is the fabrication of 3D nano-architectured materials via thin-film deposition for which it is of most interest.

ALD can be used to grow thin Fe_2O_3 layer in nanoporous Al_2O_3 membranes that are used as templates [93]. The resulting ordered parallel Fe_2O_3 nanotubes are then annealed and electrochemically roughened. Such simple nanostructures are very promising because the current densities measured for water oxidation are three orders of magnitude higher than that of the equivalent planar electrodes. Although the current density does not depend linearly on the pore length, this enhancement is directly ascribed to the nanotubular geometry. A comparison with other oxide-based materials indicates that such type of nanostructured photoanode is highly valuable because high performances are reached with an extremely abundant element.

Similarly to the nanoporous Al_2O_3 membranes, opals can be used as ordered templates to form complex and efficient nanostructures. Close-packed polystyrene (PS) spheres are uniformly coated with TiO_2 by ALD. The PS is then removed by calcination, leading to the inverse opals. These hollow nanostructures are finally sensitized with CdS quantum dots (QDs) to serve as photoanodes (Figure 8.4) [94]. The same research group has proposed a derived method where a thin ZnO layer is deposited by ALD on the TiO_2 inverse opals and converted to ZnSe by an anion-exchange reaction. Again, the structure is finally sensitized by CdSe QDs [95]. In both cases, the highly ordered and percolated 3D structures act as effective and rapid charge-transport pathways and lead to a substantial enhancement of the photoelectrochemical activity. Further improvements are envisioned by photonic-band-gap engineering to achieve back reflection, surface resonant modes, slow photons through modulation of the cavity diameter, and, of course, the use of other QD sensitizers.

In opposition to the nanoporous Al_2O_3 membranes that are inert templates, core–shell structures are very interesting because they offer a large variety of designs in which both materials play a specific role. ALD is naturally envisioned to fabricate core–shell nanostructures because it allows to conformally coat complex three-dimensional nano-objects while physical and chemical deposition methods are neither conformal nor self-limited (i.e., no perfect control of the coating thickness on the whole nanostructure). Many research groups have thus reported ALD-based core–shell design fabrication processes with various materials and numerous geometries such as nanotubes (NTs), nanonets (NNs), nanowires (NWs), nanoparticles (NPs), and nanopores. The resulting

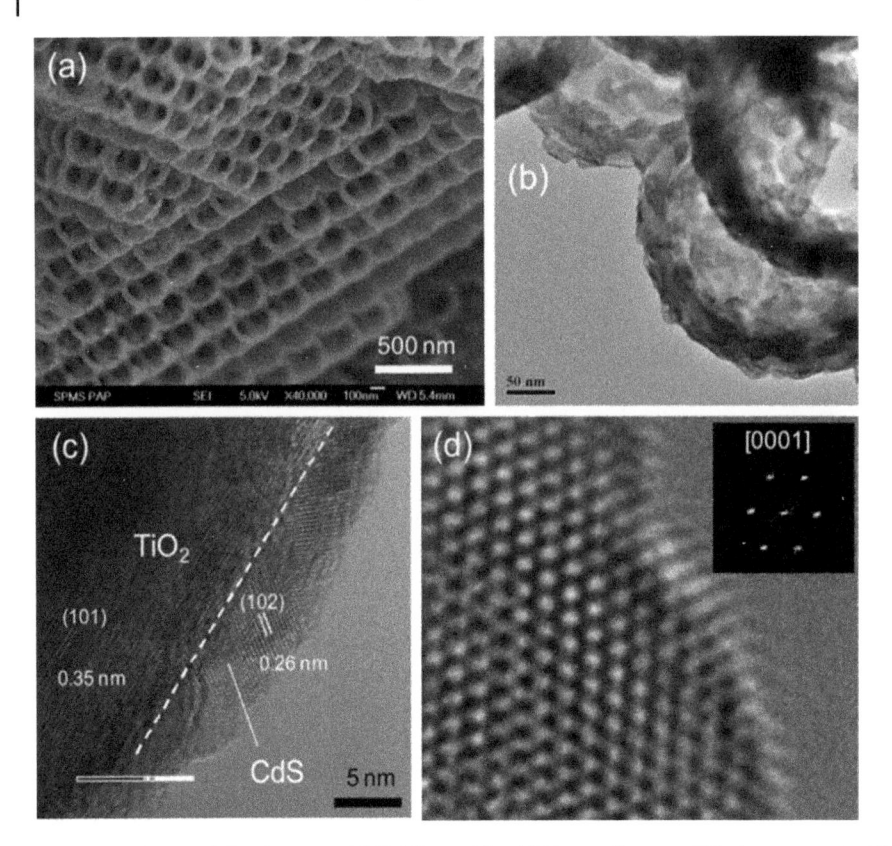

Figure 8.4 SEM and TEM images of CdS QD-sensitized 288-nm-diameter TiO_2 inverse opals. (a) A 20° tilted view at the crack area. (b) TEM image of several QD-coated inverse TiO_2 spheres. (c) HRTEM image of the CdS/TiO_2 interface. (d) Magnified HRTEM image of the CdS QD in (c). The inset is the corresponding Fourier-transform electron diffraction (ED) pattern of the QDs observed in zone axis [0001]. (Cheng *et al*. 2011 [94]. Reproduced with permission of John Wiley and Sons.)

improvements arise from diverse aspects of the nanostructure construction and demonstrate the variety of the palette offered to the researchers to enhance the efficiency of the photoelectrodes.

A first approach consists of coating a nanostructured $TiSi_2$ scaffold with the photoactive material. The $TiSi_2$ core structure acts, here, as a conductive nanonet that collects the charge generated in the conformal photoactive layer that can be TiO_2, WO_3, or Fe_2O_3 [96–98]. Only ALD can be used to fully cover such tangled design without pinhole. A schematic description of the system is presented in Figure 8.5 for $TiSi_2/Fe_2O_3$, and the principle is similar in the cases of $TiSi_2/WO_3$ and $TiSi_2/TiO_2$. In this core–shell system, the highly conductive $TiSi_2$ nanonet is combined to an n-type metal oxide semiconductor. When the electrolyte–semiconductor junction is established, the whole oxide layer is under depletion condition (Figure 8.5a). This leads to the efficient separation of the photogenerated electron–hole pairs. The e^- are gathered by the $TiSi_2$ core and quickly transported away while the h^+ are injected in the solution for the water oxidation

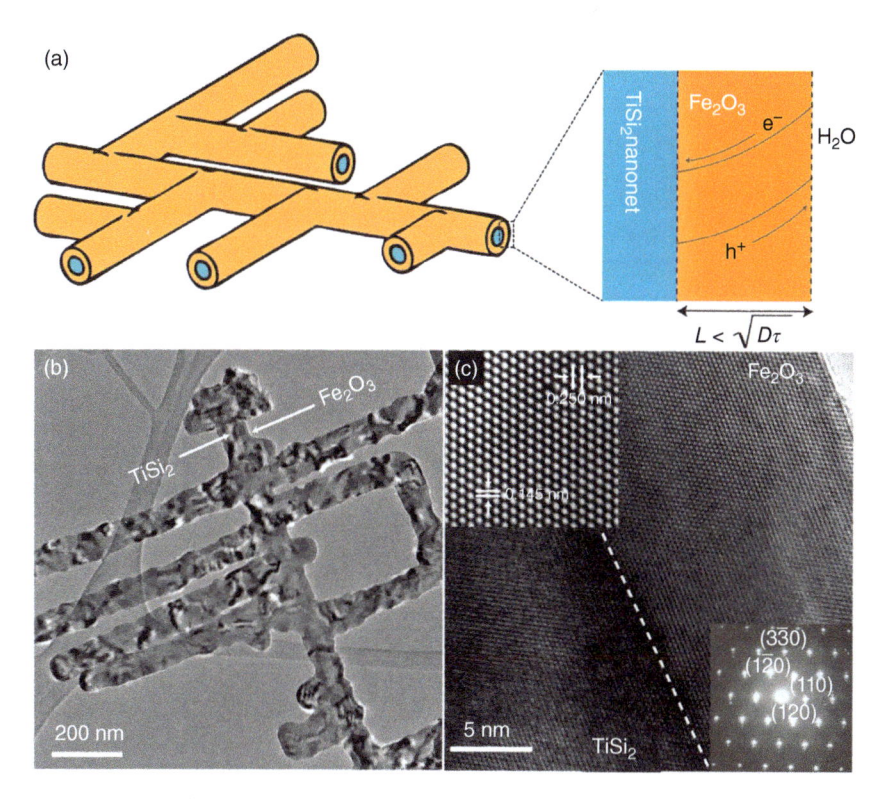

Figure 8.5 (a) Schematic illustration of the design principle, which involves the use of a highly conductive TiSi$_2$ nanonet as an effective charge collector. The electronic band structure is shown in the enlarged cross-sectional view. Efficient charge collection is achieved when the hematite thickness is smaller than the charge-diffusion distance. (b) Low-magnification TEM image showing the structural complexity of a typical hetero-nanostructure and its TiSi$_2$ core/hematite shell nature. (c) HRTEM data. A dashed line has been added at the interface as a guide to the eyes. Insets: (left) lattice-fringe-resolved HRTEM image showing the hematite lattice spacings for (110) (0.250 nm) and ($\bar{3}$30) (0.145 nm); (right) ED pattern of hematite. (Lin et al. 2011 [98]. Reproduced with permission of American Chemical Society.)

reaction (it is the reverse situation for H$_2$ production over TiSi$_2$/TiO$_2$). This photoelectrode architecture is very interesting for several reasons: (i) the oxide/liquid junction area is significantly enlarged, (ii) the charge transport is facilitated by the high conductivity of the TiSi$_2$, and (iii) the e$^-$/h$^+$ separation is facilitated in the thin oxide since it is fully depleted. Although the results are highly satisfying for hematite [98], further improvements are also possible. The deposition of Mn co-catalyst onto WO$_3$ expands the stability of the oxide and allows the full water splitting into O$_2$ and H$_2$ [97] while the codeposition of W together with TiO$_2$ enhances the photon absorption [96]. These heterostructures are therefore very promising since they demonstrate that the use of abundant materials within a nanoscale design pushes back the existing limits of planar electrodes.

Other groups have proposed equivalent designs. Noh et al. have fabricated ITO core–TiO$_2$ shell nanowires on stainless steel mesh for flexible PECs [99]. The highly conductive ITO nanowires are grown by vapor transport method

(VTM) onto the micrometric metallic mesh while the TiO_2 conformal thin film is deposited by ALD. This work demonstrates again that the geometry matters since the optimized wire length and shell thickness have been experimentally assessed. Due to its high effective working area, better light harvesting, and efficient charge separation, this nanostructured electrode exhibits a photocurrent four times higher than that of the reference system.

The function of the core materials can be different. In the case of α-Fe_2O_3-coated SiNWs, the silicon scaffold is, of course, used as a charge collector, but it also acts as a complementary absorber material [100]. Hematite absorbs light mainly in the UV-vis range while photons with a higher wavelength (λ from 600 to 1100 nm) can interact with silicon. In addition to the expected properties related to the wired geometry (high surface area, charge separation enhancement, etc.), it is the dual-absorber design that leads to one of the lowest potential onsets ($U = 0.6$ V vs RHE) for water oxidation onto hematite.

Although titanium dioxide is one of the most studied materials for water photosplitting, it still suffers from a short minority-carrier diffusion length, a long optical penetration depth, and an inefficient electron collection. It is possible to improve the performance of TiO_2 using nanoporous geometry. The TiO_2 thin film is grown by ALD onto a transparent Sb-doped tin oxide colloid film supported by a transparent conductive glass (FTO) [101]. The resulting photocurrent is three times higher than that of the equivalent TiO_2 layer deposited on FTO because the interconnected $Sb:SnO_2/TiO_2$ network geometrically separates the light absorption and charge collection while maintaining efficient electron collection. The electrochemical investigations indicate that the photocurrent is related to the TiO_2 shell thickness and the nanoporous core structure lowers the charge recombination by accelerating the transport of photoelectrons. A very similar approach is proposed by Stefik *et al.* [102]. The nanoporous scaffold is obtained from commercial TiO_2 nanoparticles. It is covered by a host–guest system composed of a transparent conductive oxide, $Nb:SnO_2$, and a photoactive layer, α-Fe_2O_3. These layers are successively deposited by ALD and atmospheric-pressure chemical vapor deposition (APCVD) [103], respectively. The TiO_2 NPs act as a high-surface-area substrate but also as a light scattering structure in order to increase the optical path length. The $Nb:SnO_2$ host collects and transports the charge carriers that are photogenerated in the α-Fe_2O_3 guest. More recently, ultrathin hematite layers (about 10 nm) have been deposited by ALD onto TiO_2 nanowires grown by hydrothermal synthesis and combined with an affordable $Ni(OH)_2$ catalyst [104]. In this core–shell nanostructure, the thin α-Fe_2O_3 film acts as the photoactive material while the TiO_2 serves as a dopant source and as a hole-blocking material. This low-temperature process leads to photoelectrochemical performances never measured for such thin active films.

Finally, it is possible to combine both approaches, that is, using an inert template to build ordered complex core–shell structures such as coaxial nanocylindrical arrays [56]. ALD is used to successively grow an n-type semiconductor, an ETA, and a p-type semiconductor within nanoporous Al_2O_3 membranes. As mentioned in Section 8.2.3, this concept ascribes the light absorption, charge transport, and electrocatalytic activity to separate materials, each of which can be tailored distinctly (physical–chemical properties and geometry). Wu *et al.* have

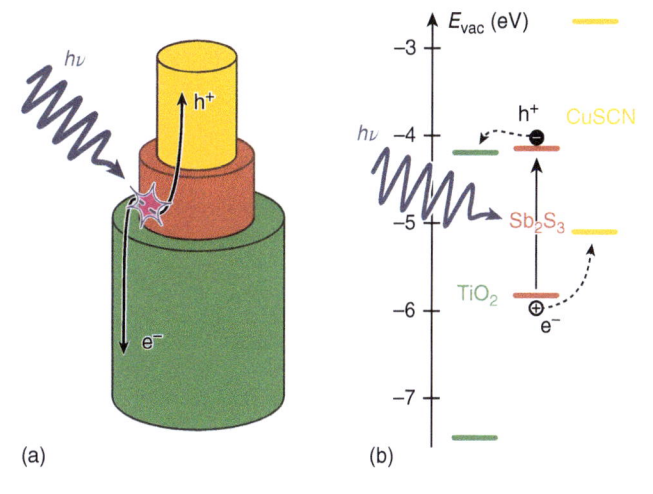

Figure 8.6 Functional principle of the coaxial nanocylindrical solar cell. (a) Schematic view of the geometry of one individual coaxial p–i–n junction among the large numbers of parallel cylinders constituting the solar cell device. (b) Band diagram of the semiconductors involved. (Wu et al. 2015 [56]. Reproduced with permission of The Royal Society of Chemistry.)

proposed such photoelectrode design for solar cell application, but it can easily be transposed to solar fuel production. The two wide-band-gap semiconductors, TiO_2 (n-type) and CuSCN (p-type), are separated by Sb_2S_3, the light absorber (Figure 8.6). The photons are absorbed in the Sb_2S_3 layer along the "vertical" light absorption path (i.e., the axis of the initial Al_2O_3 nanopores) while the charge carrier separation takes place in the TiO_2 and CuSCN layers along the "lateral" width.

8.3.3 Catalyst Deposition

ALD is used to deposit noble metals because it offers a precise control of the catalyst loading and size to minimize the overall raw material cost. Weber et al. have reported the synthesis of bimetallic nanoparticles by selective ALD of Pd and Pt [105]. This demonstrates the potential of this approach to tailor the morphology and composition of catalysts. As mentioned in Section 8.2.4, noble metals can be deposited onto photoactive materials such as Si to facilitate the photoelectrochemical reactions. Dasgupta et al. proposed the use of ALD to functionalize high-aspect-ratio SiNWs with Pt nanoparticles that exhibit a narrow size distribution [106]. Owing to the precise control of the metal amount it can potentially reduce the costs. Other groups have followed this concept. Figure 8.7a,b illustrate the added value of ALD to homogeneously coat SiNWs with noble metal NPs while other methods would cover only the apex of the nanowires [107]. The photoreduction of water occurs at the Pt nanoparticle locations while the charges are trapped at the rest of the NW surface. When the metal particles are uniformly distributed over the Si surface, the reaction is of course unleashed and the measured photocurrent is drastically increased. In the case of decorated SiNWs, the charge collection is also facilitated in comparison with a metal deposited in

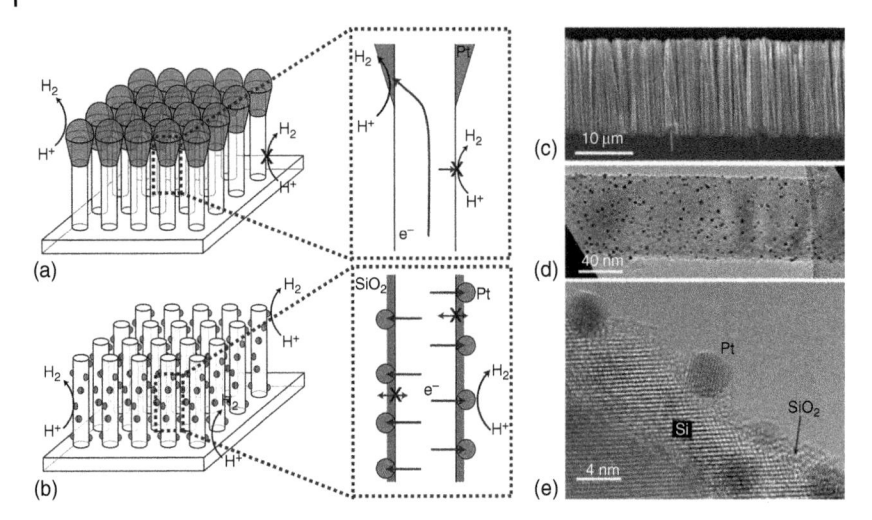

Figure 8.7 The differences between Pt nanoparticle catalysts produced by (a) other techniques and by (b) ALD. Electron micrographs of SiNWs decorated with ALD Pt. (c) Cross-sectional SEM showing no obvious catalyst aggregation. (d) Low-magnification TEM showing the distribution and uniformity of ALD Pt. (d) HRTEM showing the crystalline quality of Pt particles grown on the SiNW surface by ALD. The existence of amorphous SiO$_2$ is indicated by an arrow. (Dai et al. 2013 [107]. Reproduced with permission of John Wiley and Sons.)

the top region of the nanowires for which the diffusion path is much longer. As expected, the HER is significantly enhanced in comparison to bare SiNWs and also with regard to electrochemically deposited Pt NPs [107]. This approach should be applicable to other catalytic materials such as MoS$_x$ and Ni-Mo.

ALD is also used to enhance the activity of catalysts that are grown by another deposition method. A thin conformal TiO$_2$ layer is indeed grown onto p-type InP nanopillar arrays prior to Ru sputtering [108]. ALD is required to achieve a conformal and pinhole-free coating of TiO$_2$ to passivate the surface while heterogeneous Ru film is obtained by physical vapor deposition (PVD). The combination of the surface nanotexturing with the InP protection and the presence of co-catalyst lead to a higher current densities ($j = 30$ mA cm^{-2}) and a potential onset of 0.23 V toward positive direction.

In the field of catalysis, there are more and more examples of ALD of ultrathin oxide layers to encapsulate catalytic metallic NPs. Those oxide films are thin enough (1–3 nm) to prevent the activity of catalysts, but they protect them and consequently extend their lifetime [109]. This strategy has not been implemented yet in the field of water photosplitting, but it could potentially be beneficial for future developments.

8.3.4 Passivation and Modification of the Junction

Micro- and nanostructuring are critical to enhance the performances of PECs, but further functionalization processes are required to obtain the best photoelectrochemical properties over long operation durations. Thermal treatments can improve the performances of hematite by suppressing the detrimental surface states [92]. Several reports have proposed different methods to deposit various materials on semiconductor surfaces. PVD methods have naturally been used to deposit metallic and metal oxide thin films such as Ni/NiO$_x$ [82, 87] on planar electrodes. ALD has also been used to grow Al$_2$O$_3$ and TiO$_2$ on flat p-type semiconductors to prevent the photocathode deterioration [85, 86]. Others have used the same approach on Si photoanodes to optimize their photocatalytic activity and to circumvent their photocorrosion [84, 110, 111]. However, only few techniques are appropriated to fully cover three-dimensional micro- or nanostructured electrodes. As stated earlier in this chapter, ALD is a powerful technique to grow highly conformal and pinhole-free layers with a perfect control of the thickness and composition. Other deposition methods have shown interesting perspectives such as sol–gel for NiO$_x$ [37] and drop casting for TiNi alloy [112], but the results achieved with ALD look very promising. In photovoltaic technology, ALD is thus largely employed at different locations in the solar cells. Among those numerous uses, it has been mentioned in Section 2.2 that ALD thin films can successfully passivate surfaces and interfaces in solar cells. Surface passivation by ALD can therefore be transposed to PEC technology. This section is devoted to the use of ALD to passivate or to modify the electronic distribution of the photoelectrode/electrolyte junction. The thin-film deposition to prevent the photocorrosion is described thereinafter (Section 8.3.5).

Today, the passivation of α-Fe$_2$O$_3$ by ALD of Al$_2$O$_3$ is largely investigated because hematite is an abundant and efficient electrode material that exhibits some drawbacks that can potentially be circumvented by the deposition of a thin alumina layer. One of the earliest uses of ALD to passivate nanostructured α-Fe$_2$O$_3$ is recent (dating from 2011) [113]. The authors have grown Al$_2$O$_3$ onto hematite formed by APCVD. The photoelectrochemical measurements have revealed an enhancement of the photoelectrochemical properties while TiO$_2$-covered samples were not improved. The characterizations carried out by photoluminescence and electrochemical impedance spectroscopies demonstrate that it is not a catalytic effect but clearly a passivation of the surface states. Investigations performed by synchrotron radiation photoemission (SR-PES) and X-ray absorption spectroscopy (XAS) have shown that the exposure to trimethyl aluminum (TMA), the main precursor for Al$_2$O$_3$ deposition, induces a positive modification of the electronic properties of the hematite [114]. The interaction of TMA with the α-Fe$_2$O$_3$ leads to an electron transfer to the hematite that corresponds to a polaron and changes the Fe–O bonds. After TMA treatment, a higher photocurrent density is measured. It is attributed to a better charge transport. It is also reported that thermal treatments of α-Fe$_2$O$_3$ lead to better photoelectrode properties [115]. Although no consensus has been yet established, this effect could be ascribed to the oxygen vacancies (V$_O$) when annealing is performed in O$_2$-free atmospheres. A recent study of the absorption properties of Al$_2$O$_3$ grown by ALD onto α-Fe$_2$O$_3$ and α-Fe$_2$O$_{3-x}$ has given new insights into the understanding of these phenomena [115]. It was already shown that ALD of Al$_2$O$_3$ on hematite negatively shifts the potential onset for water oxidation [103] as well as the decrease of the capacitance [116], but in this last work [115], it is reported the ALD coating contributes to the inhibition of the trap-mediated surface electron–hole recombination. The ultrathin Al_2O_3 layer passivates only the surface trap states and leaves the bulk interband trap states of α-Fe_2O_{3-x} available for the photogenerated electrons. Since the electron and hole populations are spatially separated, no trap-mediated recombination can proceed. The electrons are thus collected by the external circuit. This leads to a negative shift of 0.2 V of the onset potential for water oxidation.

Similarly to hematite, tungsten trioxide is a very promising electrode material, but it also suffers from a fast charge recombination due to the surface and the formation of peroxo-species at the surface. Optical absorption characterization has again revealed that an ALD of a thin Al_2O_3 layer drastically reduces the electron trapping at the WO_3 surface and could also promote the electron transfer to the collecting circuit [117]. Al_2O_3 shell can also be successfully used on TiO_2 NTs. In the case of nanotubes, the surface area is drastically expanded. As mentioned in Section 8.2.3, such geometries are usually beneficial, but surface detrimental processes such as electron–hole recombination at defect locations are enhanced as well. The deposition of Al_2O_3 onto TiO_2 NTs decreases the surface defect concentration and leads to better photoelectrochemical properties [118]. As schematically drawn in Figure 8.8, this phenomenon is coupled to a field-effect passivation due to the electron accumulation in the alumina layer.

Surprisingly, it is also interesting to coat hydrothermal rutile TiO_2 NWs grown on FTO with an ALD of amorphous, anatase, or rutile TiO_2 thin film [119]. This

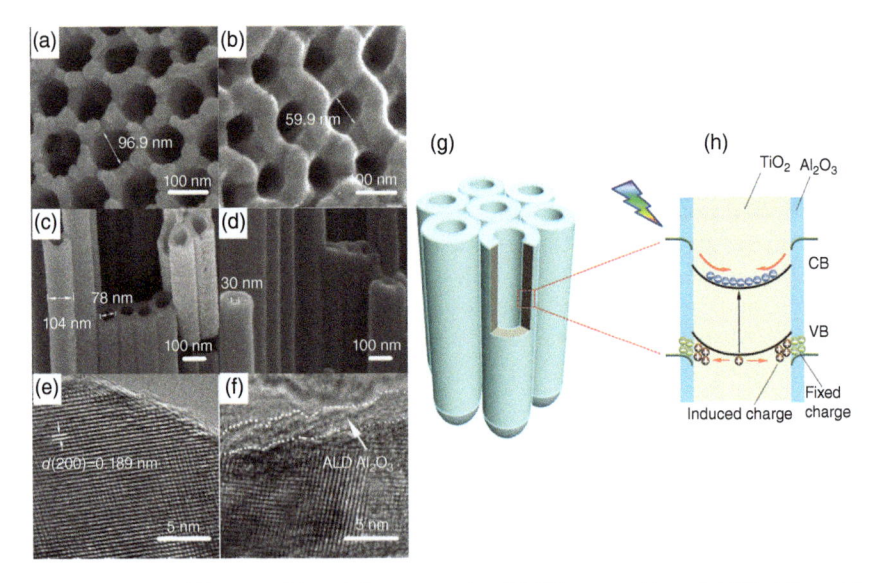

Figure 8.8 (a and b) Top and (c and d) cross-sectional SEM images of (a and c) TiO_2 NTs and (b and d) Al_2O_3-covered TiO_2 NTs (180 cycles). TEM images of (e) TiO_2 NTs and (f) Al_2O_3-covered TiO_2 NTs (25 cycles). (g and h) Schematic diagram of the field-effect passivation by Al_2O_3 coating. (g) Illustration of the structure of the TiO_2 NTs deposited with (powder blue) Al_2O_3 shell. (h) Energy band diagram of TiO_2 NTs coated with Al_2O_3 shell. Under UV irradiation, photogenerated holes are trapped at the surface due to the presence of the negative charges located in the Al_2O_3 film, leaving behind unpaired electrons in the center of tube wall. (Gui et al. 2014 [118]. Reproduced with permission of American Chemical Society.)

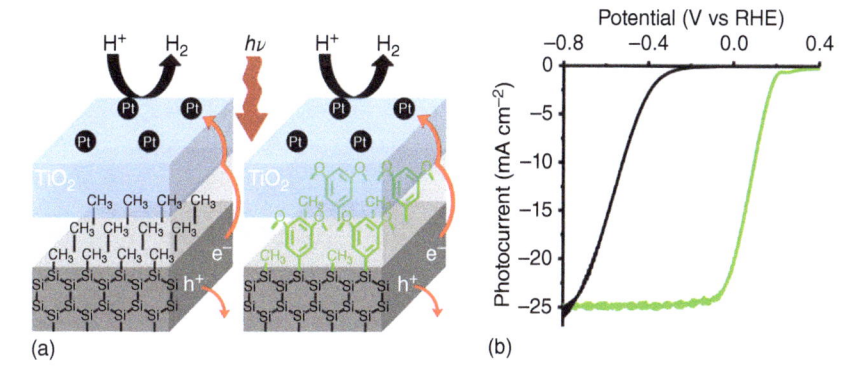

Figure 8.9 (a) Schematic view of the functionalized of Cl-terminated p-Si (111) substrates with CH_3 and 3,5-dimethoxyphenyl, followed by successive ALD of TiO_2 and Pt. (b) Corresponding current–voltage curves for CH_3 (black) and 3,5-dimethoxyphenyl (green) in 0.5 M H_2SO_4 under AM 1.5G 100 mW cm^{-2} at 0.1 V s^{-1}. (Seo et al. 2015 [120]. Reproduced with permission of American Chemical Society.)

approach reveals the crucial role of the interface since no photocurrent enhancement is observed when the NWs are covered by amorphous or anatase TiO_2 while rutile shells lead to a photoelectrochemical activity 1.5 times higher. The performance improvement is also ascribed to the surface area expansion (because the TiO_2 deposit is slightly rough) and, again, the passivation of the surface defects.

Owing to the low barrier height of molecular grafted p-type Si, the H_2 production is still challenging on such surfaces. To address this issue, it has been proposed to functionalize the photoelectrodes with a hybrid organic/inorganic coating [120]. Al_2O_3 or TiO_2 thin films are thus grown by ALD onto p-Si (111) previously modified by organic groups (e.g., methyl, phenyl, naphthalene, etc.). The last step is the ALD of Pt particles on the top of the hybrid multilayer (Figure 8.9a). Depending on the nature of the grafted molecules, the band edge is modulated, and its position can be tuned to optimize the onset of water photoreduction (Figure 8.9b). The effect of the metal oxide composition (Al_2O_3 or TiO_2) is limited because this ultrathin film acts as a tunneling barrier. However, TiO_2 in combination with a methoxy group exhibits the best performance because the series resistance is minimized. To further improve this photoelectrode design, the authors propose the use of earth-abundant metals as replacement of Pt. A similar study with very close conclusions has been performed with additional characterization methods [121]. In this later work, TiO_2 and Pt catalysts have been deposited onto CH_3-terminated p-Si (111).

It is also possible to deposit a material other than alumina. Thin Co_3O_4 layers grown onto anodic TiO_2 NTs have shown improved photoelectrochemical performances that have been attributed to the enhanced light absorption when the NTs are covered by the cobalt oxide and a better photocarrier separation [122]. Owing to their geometry, the TiO_2 NTs can partially absorb light in the visible range [123]. However, the lower band gap of Co_3O_4 ($E_g \sim 2.07$ eV) drastically enlarges this absorption between $\lambda = 400$ and 650 nm. The photocurrent density is directly related to the thickness of the ALD coating. It increases up to

4 nm of Co_3O_4 and decreases after this value. The photogenerated charges are easily separated because the electrons are transferred to the TiO_2 and Ti contact while the holes are injected in the electrolyte through the Co_3O_4. Similarly, a TiO_2 film can act as complementary photoactive material to Si. This approach has been proposed for SiNWs by Yang *et al.* [124] and [125] for black silicon (macropore array) by Ao *et al.* later. In such systems, the TiO_2 layer optimizes the charge distribution at the interface and absorbs photons in the UV range while Si cannot.

Two studies performed on SnO_2- and TiO_2-based photoelectrodes used in DSSCs report valuable information on the field of water photosplitting. Prasittichai and Hupp have grown an ultrathin Al_2O_3 layer on electrodes in contact with an I_3^-/I^--containing electrolyte. The layer acts as a tunnel barrier for thickness greater than 2 Å because the current is exponentially dependent on the depth of the film [126]. A single ALD cycle for Al_2O_3, ZrO_2, or TiO_2 depositions on SnO_2 or TiO_2 causes a significant increase in the injected electron lifetime, improvement of the fill factor and the open-circuit photovoltage as well as a moderate rise of the short-circuit photocurrent. Those enhancements are ascribed to the passivation of the reactive and low-energy surface states in combination with a shift of the band edges. The authors have extended their study to another substrate (TiO_2) and have compared the effect of different functional layers (ZrO_2, TiO_2, and Al_2O_3) [127].

8.3.5 Photocorrosion Protection

Owing to its ability to perfectly cover three-dimensional nanostructures, ALD can also be used to prevent the corrosion of the photoelectrodes. Similarly to passivation, the deposition of the protecting films has initially been carried out on planar substrates by PVD and by then ALD, and, finally, solely ALD has been used for three-dimensional electrodes. Although it is not a unique technique to enhance the stability of the PECs, ALD is one of the best tools to preserve these devices because it has the capability to synthesize pinhole-free and highly conformal thin film on any type of photoelectrode. It is applied on 2D as well as 3D photoanodes and photocathodes. In addition to inhibiting the deterioration of the core materials, the ALD shell induces an improvement of the photoelectrode performances.

8.3.5.1 Protection of Planar Photoanodes

As mentioned in Section 8.2.4, many semiconductors corrode during the photoreduction or photooxidation. Kenney *et al.* have proposed to protect n-type silicon by a thin Ni layer grown by PVD [82]. In this work, it is shown that the nickel layer prevents the Si photocorrosion and acts as an earth-abundant metal co-catalyst for the oxidation. The thinnest layers (2 nm) show the best activity with a current density stable over 80 h. This method has been slightly modified and applied to other systems. Sputtered NiO_x transparent layers are formed onto amorphous or crystalline Si (a-Si, c-Si) and n-CdTe [87]. In all cases, the stability and activity of the photoelectrodes are improved. It demonstrates therefore that thin-film deposition onto such electrodes is potentially valuable for PEC applications. In this context, ALD has naturally been envisioned to grow protective thin films.

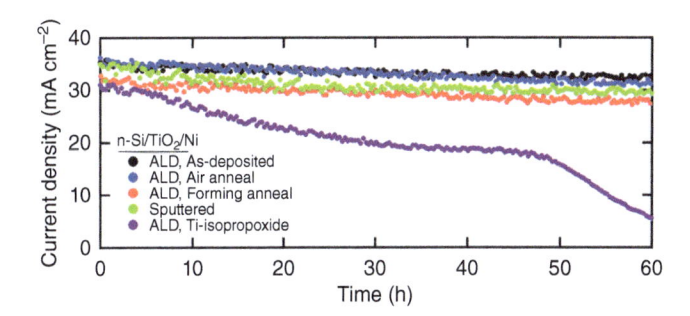

Figure 8.10 Potentiostatic stability tests of various n-Si/TiO$_2$/Ni samples during water oxidation in 1.0 M KOH(aq.) under simulated 1-sun illumination. The electrodes were held at 1.85 V versus RHE. All films were ~100 nm thick except for the TTIP-ALD sample, for which the thickness varied within the sample from ~50 to ~150 nm. (McDowell et al. 2015 [129]. Reproduced with permission of American Chemical Society.)

One of the first attempts to use ALD to protect photoelectrode consists of growing a thin layer of TiO$_2$ onto n-type silicon. A subsequent electron-beam deposition of the co-catalyst, iridium, is carried out on this planar electrode [110]. In opposition to ineffective previous works performed in 1977 by CVD [128], the thin ALD TiO$_2$ layer (2 nm) stabilizes the Si-based electrode in both acidic and alkaline media for several hours of operations. Due to its large band gap, the TiO$_2$ is almost transparent to the visible photons and allows the charge transport via tunneling processes. More recently, the Lewis' group has demonstrated that nickel, an affordable metal, could replace iridium [129]. This study also indicates that the Ti-containing precursor plays a crucial role in the stability of the photoelectrode. As shown in Figure 8.10, the TiO$_2$ layers grown using tetrakis-dimethylamidotitanium (TDMAT) are stable over 60 h of O$_2$ production while titanium tetraisopropoxide (TTIP) leads to unstable photoelectrodes. Thermal treatments of the protecting layers (annealing in air or forming gas) have no significant effect on the stability, but they have a positive effect on the open-circuit voltage. This strategy has been applied to silicon and other compound semiconductors such as GaAs and GaP [84]. The thickness of the TiO$_2$ film was varied from 4 to more than 140 nm and the Ni overlayer was either continuous or composed of nanoparticles. A constant water oxidation photocurrent is measured for more than 100 h, indicating, again, a perfect photocorrosion inhibition. The same group has also implemented this approach on an n-type CdTe [130]. This II–VI semiconductor is now widely used in large photovoltaic power stations, although it is not used in PEC devices because it is chemically instable. This work demonstrates that, in combination with the appropriate tunneling protecting layer, CdTe could be an interesting candidate for photoelectrochemical O$_2$ generation. Or, recently, ALD of amorphous TiO$_2$ has been implemented in the fabrication process of a monolithically tandem cell to protect the III–V semiconductor-based multilayered photoanode [131]. This wireless PEC is very promising since it exhibits 10.5% efficiency for more than 40 h.

Other protective materials have been grown by ALD on different photoanodes. MnO$_2$ has been efficiently used on Si and glassy carbon [111, 132] while CoO$_x$

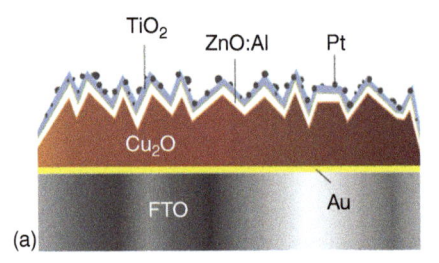

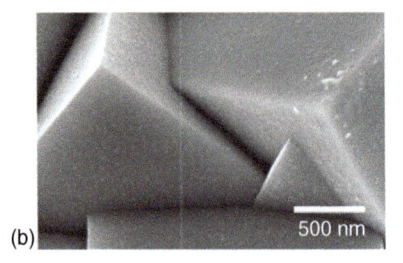

Figure 8.11 (a) Schematic representation of the electrode structure. (b) Scanning electron micrograph showing a top view of the electrode after ALD of 5 × (4 nm ZnO/0.17 nm Al_2O_3)/11 nm TiO_2 followed by electrodeposition of Pt nanoparticles. (Paracchino et al. 2011 [135]. Reproduced with permission of American Chemical Society.)

and Ta_2O_5 have been deposited on $BiVO_4$ and ZnO, respectively [133, 134]. In both cases, the coated photoelectrodes exhibit a long stability during operations and the photocurrent is also drastically increased. In addition to preventing corrosion, the CoO_x layer contributes, indeed, to the incident light absorption and acts as a co-catalyst while the Ta_2O_5 ultrathin layer passivates the surface defects of the ZnO substrate.

8.3.5.2 Protection of Planar Photocathodes

Similarly to photoanodes, ALD can, obviously, be applied to the protection of the photocathodes. Numerous semiconductors such as Si, InP, WSe_2, Cu_2O, and GaP suffer, indeed, from cathodic decomposition during water photoreduction. The Grätzel's group first reported, in 2011, the sequential depositions of transparent conductive Al:doped ZnO and protective TiO_2 on p-type Cu_2O [135] (Figure 8.11). To enhance the water reduction kinetics, Pt nanoparticles are electrodeposited on the TiO_2. This strategy leads to a stable photocathode exhibiting a high photocurrent on metal-oxide-based electrode with a Faradaic efficiency of 100%.

TiO_2 thin layers can also be used to prevent the corrosion of p-type InP for solar H_2 generation [86, 136]. As expected, the ALD coating inhibits the InP degradation, but the charge carrier distribution at the p-InP/n-TiO_2 junction is also strongly beneficial to the photoreduction of water. The conduction bands of the two materials are aligned while the valence band of titania is largely lower than the VB edge of indium phosphide. This situation leads to a facile photoelectron transfer from InP toward TiO_2 where the electrons are further injected into the electrolyte for the water reduction, whereas the built-in potential induced by the VB offset drives the holes back from the surface. The surface recombination is therefore strongly limited and the photocurrent efficiency is largely enhanced.

8.3.5.3 Protection of Nanostructured Photoelectrodes

The natural evolution of this approach is to protect nanostructured photoelectrodes by ALD of thin conformal layers. As mentioned in Section 8.3.3, ALD of TiO_2 on textured p-InP passivates the surface defects and promotes, therefore, the photoelectrochemical activity [108]. This titania also acts as conformal and pinhole-free anticorrosion layer. Choi *et al.* have proposed a similar evolution for their strategy. They have first used ALD to grow a protective Al_2O_3 film onto planar p-Si [85], and then, they have successfully applied this method on nanoporous silicon. Again, in addition to hindering corrosion, the coating improves the photocathode efficiency. The physical model developed by Kayes *et al.* [70] that compares the performances of planar and radial photoelectrode designs clearly indicates that micro- and nanostructured electrodes should deliver more energy. Following this concept, Das *et al.* have covered unstable SiMWs with TiO_2 [137]. Since ALD is a conformal deposition process, it yields the same improvements observed in planar photocathodes. As expected, it is also possible to protect nanostructured photoanodes such as ZnO nanorods with an ALD coating. It has been demonstrated using a thin TiO_2 shell [138].

8.4 Conclusion and Outlook

In this chapter, it clearly appears that ALD can be highly beneficial to the solar fuel production. Although it has been used, at the laboratory scale, for less than 5 years, it seems that this technique is now a major part of the PEC development roadmap. The strong research effort should be intensified to optimize the electrode architectures, to better understand how the ALD thin films modify the properties of the junctions and to widen the panel of electrode designs. The slow deposition rate of ALD could be considered as a hindrance, but the recent technological progresses have demonstrated that ALD could be used to rapidly grow 2D layers on large-scale substrates (larger than $1\,m^2$). High-speed deposition on 3D nanostructured electrodes is still challenging and therefore requires an intense research work.

References

1. Lewis, N.S. (2007) Toward cost-effective solar energy use. *Science*, **315**, 798–801.
2. Lewis, N.S. and Nocera, D.G. (2006) Powering the planet: chemical challenges in solar energy utilization. *Proc. Natl. Acad. Sci. U.S.A.*, **103**, 15729–15735.
3. Gray, H.B. (2009) Powering the planet with solar fuel. *Nat. Chem.*, **1**, 7–12.
4. Fujishima, A. and Honda, K. (1972) Electrochemical photolysis of water at a semiconductor electrode. *Nature*, **238**, 37–38.
5. Tachibana, Y., Vayssieres, L., and Durrant, J.R. (2012) Artificial photosynthesis for solar water-splitting. *Nat. Photon.*, **6**, 511–518.
6. Ronge, J., Bosserez, T., Martel, D., Nervi, C., Boarino, L., Taulelle, F., Decher, G., Bordiga, S., and Martens, J.A. (2014) Monolithic cells for solar fuels. *Chem. Soc. Rev.*, **43**, 7963–7981.
7. Nocera, D.G. (2012) The artificial leaf. *Acc. Chem. Res.*, **45**, 767–776.
8. Armaroli, N. and Balzani, V. (2011) The hydrogen issue. *ChemSusChem*, **4**, 21–36.

9 Pinaud, B.A., Benck, J.D., Seitz, L.C., Forman, A.J., Chen, Z., Deutsch, T.G., James, B.D., Baum, K.N., Baum, G.N., Ardo, S., Wang, H., Miller, E., and Jaramillo, T.F. (2013) Technical and economic feasibility of centralized facilities for solar hydrogen production via photocatalysis and photoelectrochemistry. *Energy Environ. Sci.*, **6**, 1983–2002.

10 Osterloh, F.E. (2013) Inorganic nanostructures for photoelectrochemical and photocatalytic water splitting. *Chem. Soc. Rev.*, **42**, 2294–2320.

11 Licht, S., Wang, B., Mukerji, S., Soga, T., Umeno, M., and Tributsch, H. (2001) Over 18% solar energy conversion to generation of hydrogen fuel; theory and experiment for efficient solar water splitting. *Int. J. Hydrogen Energy*, **26**, 653–659.

12 Miller, E. (2011) Advanced Materials for Water Photolysis. *Task 26 Annual Report*, US DOE, Washington, DC.

13 McKone, J.R., Lewis, N.S., and Gray, H.B. (2014) Will solar-driven water-splitting devices see the light of day? *Chem. Mater.*, **26**, 407–414.

14 Marichy, C., Bechelany, M., and Pinna, N. (2012) Atomic layer deposition of nanostructured materials for energy and environmental applications. *Adv. Mater.*, **24**, 1017–1032.

15 Leskela, M. and Ritala, M. (2003) Atomic layer deposition chemistry: recent developments and future challenges. *Angew. Chem. Int. Ed.*, **42**, 5548–5554.

16 Knez, M., Nielsch, K., and Niinistö, L. (2007) Synthesis and surface engineering of complex nanostructures by atomic layer deposition. *Adv. Mater.*, **19**, 3425–3438.

17 George, S.M. (2010) Atomic layer deposition: an overview. *Chem. Rev.*, **110**, 111–131.

18 Detavernier, C., Dendooven, J., Sree, S.P., Ludwig, K.F., and Martens, J.A. (2011) Tailoring nanoporous materials by atomic layer deposition. *Chem. Soc. Rev.*, **40**, 5242–5253.

19 Miikkulainen, V., Leskela, M., Ritala, M., and Puurunen, R.L. (2013) Crystallinity of inorganic films grown by atomic layer deposition: overview and general trends. *J. Appl. Phys.*, **113**, 021301.

20 Walter, M.G., Warren, E.L., McKone, J.R., Boettcher, S.W., Mi, Q., Santori, E.A., and Lewis, N.S. (2010) Solar water splitting cells. *Chem. Rev.*, **110**, 6446–6473.

21 Cook, T.R., Dogutan, D.K., Reece, S.Y., Surendranath, Y., Teets, T.S., and Nocera, D.G. (2010) Solar energy supply and storage for the legacy and nonlegacy worlds. *Chem. Rev.*, **110**, 6474–6502.

22 Maeda, K. (2011) Photocatalytic water splitting using semiconductor particles: history and recent developments. *J. Photochem. Photobiol., C*, **12**, 237–268.

23 Li, Z., Luo, W., Zhang, M., Feng, J., and Zou, Z. (2013) Photoelectrochemical cells for solar hydrogen production: current state of promising photoelectrodes, methods to improve their properties, and outlook. *Energy Environ. Sci.*, **6**, 347–370.

24 Hisatomi, T., Kubota, J., and Domen, K. (2014) Recent advances in semiconductors for photocatalytic and photoelectrochemical water splitting. *Chem. Rev. Soc.*, **43**, 7520–7535.

25 Cho, S., Jang, J.-W., Lee, K.-H., and Lee, J.S. (2014) Research update: strategies for efficient photoelectrochemical water splitting using metal oxide photoanodes. *APL Mater.*, **2**, 010703.

26 Prévot, M.S. and Sivula, K. (2013) Photoelectrochemical tandem cells for solar water splitting. *J. Phys. Chem. C*, **117**, 17879–17893.

27 Nowotny, J., Bak, T., Nowotny, M.K., and Sheppard, L.R. (2007) Titanium dioxide for solar-hydrogen I. Functional properties. *Int. J. Hydrogen Energy*, **32**, 2609–2629.

28 Hardee, K.L. and Bard, A.J. (1976) Semiconductor electrodes. V. The application of chemically vapor deposited iron oxide films to photosensitized electrolysis. *J. Electrochem. Soc.*, **123**, 1024–1026.

29 Hodes, G., Cahen, D., and Manassen, J. (1976) Tungsten trioxide as a photoanode for a photoelectrochemical cell (PEC). *Nature*, **260**, 312–313.

30 Hardee, K.L. and Bard, A.J. (1977) Semiconductor electrodes. X. Photoelectrochemical behavior of several polycrystalline metal oxide electrodes in aqueous solutions. *J. Electrochem. Soc.*, **124**, 215–224.

31 Hara, M., Kondo, T., Komoda, M., Ikeda, S., Kondo, J.N., Domen, K., Shinohara, K., and Tanaka, A. (1998) Cu_2O as a photocatalyst for overall water splitting under visible light irradiation. *Chem. Commun.*, 357–358.

32 Leygraf, C., Hendewerk, M., and Somorjai, G.A. (1982) Photocatalytic production of hydrogen from water by a p- and n-type polycrystalline iron oxide assembly. *J. Phys. Chem.*, **86**, 4484–4485.

33 Matsumoto, Y., Sugiyama, K., and Sato, E.I. (1988) Photocathodic hydrogen evolution reactions at p-type $CaFe_2O_4$ electrodes with Fermi level pinning. *J. Electrochem. Soc.*, **135**, 98–104.

34 de Jongh, P.E., Vanmaekelbergh, D., and Kelly, J.J. (1999) Cu_2O: a catalyst for the photochemical decomposition of water? *Chem. Commun.*, 1069–1070.

35 Liao, L., Zhang, Q., Su, Z., Zhao, Z., Wang, Y., Li, Y., Lu, X., Wei, D., Feng, G., Yu, Q., Cai, X., Zhao, J., Ren, Z., Fang, H., Robles-Hernandez, F., Baldelli, S., and Bao, J. (2014) Efficient solar water-splitting using a nanocrystalline CoO photocatalyst. *Nat. Nanotechnol.*, **9**, 69–73.

36 Barr, M.K.S., Assaud, L., Wu, Y., Laffon, C., Parent, P., Bachmann, J., and Santinacci, L. (2015) Engineering a three-dimensional, photoelectrochemically active p-NiO/i-Sb_2S_3 junction by atomic layer deposition. *Electrochim. Acta*, **179**, 504–511.

37 Sun, K., Park, N., Sun, Z., Zhou, J., Wang, J., Pang, X., Shen, S., Noh, S.Y., Jing, Y., Jin, S., Yu, P.K.L., and Wang, D. (2012) Nickel oxide functionalized silicon for efficient photo-oxidation of water. *Energy Environ. Sci.*, **5**, 7872–7877.

38 Sivula, K. (2013) Metal oxide photoelectrodes for solar fuel production, surface traps, and catalysis. *J. Phys. Chem. Lett.*, **4**, 1624–1633.

39 Awad, N.K., Ashour, E.A., and Allam, N.K. (2014) Recent advances in the use of metal oxide-based photocathodes for solar fuel production. *J. Renew. Sustain. Energy*, **6**, 022702.

40 Nakato, Y., Tsumura, A., and Tsubomura, H. (1982) Efficient photoelectrochemical conversion of solar energy with n-type silicon semiconductor

electrodes surface-doped with IIIA-group elements. *Chem. Lett.*, **11**, 1071–1074.

41 Ueda, K., Nakato, Y., Sakamoto, H., Sakai, Y., Matsumura, M., and Tsubomura, H. (1987) Efficient solar to chemical conversion with an n-type amorphous silicon/p-type crystalline silicon heterojunction electrode. *Chem. Lett.*, **16**, 747–750.

42 Boettcher, S.W., Spurgeon, J.M., Putnam, M.C., Warren, E.L., Turner-Evans, D.B., Kelzenberg, M.D., Maiolo, J.R., Atwater, H.A., and Lewis, N.S. (2010) Energy-conversion properties of vapor-liquid-solid–grown silicon wire-array. *Science*, **327**, 185–187.

43 Tyagi, M.S. and Van Overstraeten, R. (1983) Minority carrier recombination in heavily-doped silicon. *Solid State Electron.*, **26**, 577–597.

44 Matsumura, M. and Morrison, S.R. (1983) Anodic properties of n-Si and n-Ge electrodes in HF solution under illumination and in the dark. *J. Electroanal. Chem. Interfacial Electrochem.*, **147**, 157–166.

45 Tomkiewicz, M. and Woodall, J.M. (1977) Photoassisted electrolysis of water by visible irradiation of a p-type gallium phosphide electrode. *Science*, **196**, 990–991.

46 Heller, A. and Vadimsky, R.G. (1981) Efficient solar to chemical conversion: 12% efficient photoassisted electrolysis in the [P-type InP(Ru)]/HCL-KCL/Pt(Rh) cell. *Phys. Rev. Lett.*, **46**, 1153–1156.

47 Khaselev, O. and Turner, J.A. (1998) A monolithic photovoltaic-photoelectrochemical device for hydrogen production via water splitting. *Science*, **280**, 425–427.

48 Waki, I., Cohen, D., Lal, R., Mishra, U., DenBaars, S.P., and Nakamura, S. (2007) Direct water photoelectrolysis with patterned n-GaN. *Appl. Phys. Lett.*, **91**, 093519.

49 Kocha, S.S., Turner, J.A., and Nozik, A.J. (1994) Study of the schottky barrier and determination of the energetic positions of band edges at the n- and p-type gallium indium phosphide electrode/electrolyte interface. *J. Electroanal. Chem.*, **367**, 27–30.

50 Khaselev, O., Bansal, A., and Turner, J.A. (2001) High-efficiency integrated multijunction photovoltaic/electrolysis systems for hydrogen production. *Int. J. Hydrogen Energy*, **26**, 127–132.

51 Kaneko, M., Yao, G.-J., and Kira, A. (1989) Efficient water cleavage with visible light by a system mimicking photosystem II. *J. Chem. Soc., Chem. Commun.*, (18), 1338–1339. doi: 10.1039/C39890001338

52 Gerrard, W.A. and Owen, J.R. (1977) Stable photo-electrochemical solar-cell employing a CdSe photoanode. *Mater. Res. Bull.*, **12**, 677–684.

53 Mathew, X., Bansal, A., Turner, J.A., Dhere, R., Mathews, N.R., and Sebastian, P.J. (2002) Photoelectrochemical characterization of surface modified CdTe for hydrogen production. *J. New Mater. Electrochem. Syst.*, **5**, 149–154.

54 Xing, C., Zhang, Y., Yan, W., and Guo, L. (2006) Band structure-controlled solid solution of $Cd_{1-x}Zn_xS$ photocatalyst for hydrogen production by water splitting. *Int. J. Hydrogen Energy*, **31**, 2018–2024.

55 Peng, G., Wu, J., Zhao, Y., Xu, X., Xu, G., and Star, A. (2014) Ultra-small TiO_2 nanowire forests on transparent conducting oxide for solid-state semiconductor-sensitized solar cells. *RSC Adv.*, **4**, 46987–46991.

56 Wu, Y., Assaud, L., Kryschi, C., Capon, B., Detavernier, C., Santinacci, L., and Bachmann, J. (2015) Antimony sulfide as a light absorber in highly ordered, coaxial nanocylindrical arrays: preparation and integration into a photovoltaic device. *J. Mater. Chem. A*, **3**, 5971–5981.

57 Kim, J., Sohn, Y., and Kang, M. (2013) New fan blade-like core-shell Sb_2Ti_xSy photocatalytic nanorod for hydrogen production from methanol/water photolysis. *Int. J. Hydrogen Energy*, **38**, 2136–2143.

58 Yuliati, L., Yang, J.-H., Wang, X., Maeda, K., Takata, T., Antonietti, M., and Domen, K. (2010) Highly active tantalum(V) nitride nanoparticles prepared from a mesoporous carbon nitride template for photocatalytic hydrogen evolution under visible light irradiation. *J. Mater. Chem.*, **20**, 4295–4298.

59 Chakrapani, V., Thangala, J., and Sunkara, M.K. (2009) WO_3 and W_2N nanowire arrays for photoelectrochemical hydrogen production. *Int. J. Hydrogen Energy*, **34**, 9050–9059.

60 Wang, X., Maeda, K., Thomas, A., Takanabe, K., Xin, G., Carlsson, J.M., Domen, K., and Antonietti, M. (2009) A metal-free polymeric photocatalyst for hydrogen production from water under visible light. *Nat. Mater.*, **8**, 76–80.

61 Reece, S.Y., Hamel, J.A., Sung, K., Jarvi, T.D., Esswein, A.J., Pijpers, J.J.H., and Nocera, D.G. (2011) Wireless solar water splitting using silicon-based semiconductors and earth-abundant catalysts. *Science*, **334**, 645–648.

62 Warren, E.L., Atwater, H.A., and Lewis, N.S. (2014) Silicon microwire arrays for solar energy-conversion applications. *J. Phys. Chem. C*, **118**, 747–759.

63 Hou, Y., Abrams, B.L., Vesborg, P.C.K., Björketun, M.E., Herbst, K., Bech, L., Setti, A.M., Damsgaard, C.D., Pedersen, T., Hansen, O., Rossmeisl, J., Dahl, S., Nørskov, J.K., and Chorkendorff, I. (2011) Bioinspired molecular co-catalysts bonded to a silicon photocathode for solar hydrogen evolution. *Nat. Mater.*, **10**, 434–438.

64 Kibria, M.G., Chowdhury, F.A., Zhao, S., AlOtaibi, B., Trudeau, M.L., Guo, H., and Mi, Z. (2015) Visible light-driven efficient overall water splitting using p-type metal-nitride nanowire arrays. *Nat. Commun.*, **6**, 6797.

65 Liu, B., Wu, C.-H., Miao, J., and Yang, P. (2014) All inorganic semiconductor nanowire mesh for direct solar water splitting. *ACS Nano*, **8**, 11739–11744.

66 Koshida, N. and Echizenya, K. (1991) Characterization studies of p-type porous Si and its photoelectrochemical activation. *J. Electrochem. Soc.*, **138**, 837–841.

67 Jung, J.-Y., Choi, M.J., Zhou, K., Li, X., Jee, S.-W., Um, H.-D., Park, M.-J., Park, K.-T., Bang, J.H., and Lee, J.-H. (2014) Photoelectrochemical water splitting employing a tapered silicon nanohole array. *J. Mater. Chem. A*, **2**, 833–842.

68 Chandrasekaran, S., Macdonald, T.J., Mange, Y.J., Voelcker, N.H., and Nann, T. (2014) A quantum dot sensitized catalytic porous silicon photocathode. *J. Mater. Chem. A*, **2**, 9478–9481.

69 Boettcher, S.W., Warren, E.L., Putnam, M.C., Santori, E.A., Turner-Evans, D., Kelzenberg, M.D., Walter, M.G., McKone, J.R., Brunschwig, B.S., Atwater, H.A., and Lewis, N.S. (2011) Photoelectrochemical hydrogen evolution using Si microwire arrays. *J. Am. Chem. Soc.*, **133**, 1216–1219.

70 Kayes, B.M., Atwater, H.A., and Lewis, N.S. (2005) Comparison of the device physics principles of planar and radial p-n junction nanorod solar cells. *J. Appl. Phys.*, **97**, 114302.

71 Li, X., Xiao, Y., Bang, J.H., Lausch, D., Meyer, S., Miclea, P.-T., Jung, J.-Y., Schweizer, S.L., Lee, J.-H., and Wehrspohn, R.B. (2013) Upgraded silicon nanowires by metal-assisted etching of metallurgical silicon: a new route to nanostructured solar-grade silicon. *Adv. Mater.*, **25**, 1521–4095.

72 Zhang, J., Song, T., Shen, X., Yu, X., Lee, S.-T., and Sun, B. (2014) A 12%-efficient upgraded metallurgical grade silicon–organic heterojunction solar cell achieved by a self-purifying process. *ACS Nano*, **8**, 11369–11376.

73 van de Krol, R., Liang, Y., and Schoonman, J. (2008) Solar hydrogen production with nanostructured metal oxides. *J. Mater. Chem.*, **18**, 2311–2320.

74 Mor, G.K., Shankar, K., Paulose, M., Varghese, O.K., and Grimes, C.A. (2005) Enhanced photocleavage of water using titania nanotube arrays. *Nano Lett.*, **5**, 191–195.

75 Yang, X.Y., Wolcott, A., Wang, G.M., Sobo, A., Fitzmorris, R.C., Qian, F., Zhang, J.Z., and Li, Y. (2009) Nitrogen-doped ZnO nanowire arrays for photoelectrochemical water splitting. *Nano Lett.*, **9**, 2331–2336.

76 Wang, D.F., Pierre, A., Kibria, M.G., Cui, K., Han, X.G., Bevan, K.H., Guo, H., Paradis, S., Hakima, A.R., and Mi, Z.T. (2011) Wafer-level photocatalytic water splitting on GaN nanowire arrays grown by molecular beam epitaxy. *Nano Lett.*, **11**, 2353–2357.

77 Wang, H.L., Deutsch, T., and Turner, J.A. (2008) Direct water splitting under visible light with nanostructured hematite and WO_3 photoanodes and a $GaInP_2$ photocathode. *J. Electrochem. Soc.*, **155**, F91–F96.

78 Mishra, P.R., Shukla, P.K., and Srivastava, O.N. (2007) Study of modular PEC solar cells for photoelectrochemical splitting of water employing nanostructured TiO_2 photoelectrodes. *Int. J. Hydrogen Energy*, **32**, 1680–1685.

79 Tena-Zaera, R., Ryan, M.A., Katty, A., Hodes, G., Bastide, S.P., and Lévy-Clément, C. (2006) Fabrication and characterization of ZnO nanowires/CdSe/CuSCN eta-solar cell. *C.R. Chim.*, **9**, 717–729.

80 Oh, I., Kye, J., and Hwang, S. (2012) Enhanced photoelectrochemical hydrogen production from silicon nanowire array photocathode. *Nano Lett.*, **12**, 298–302.

81 Nakato, Y., Egi, Y., Hiramoto, M., and Tsubomura, H. (1984) Hydrogen evolution and iodine reduction on an illuminated n-p junction silicon electrode and its application to efficient solar photoelectrolysis of hydrogen iodide. *J. Phys. Chem. C*, **88**, 4218–4222.

82 Kenney, M.J., Gong, M., Li, Y., Wu, J.Z., Feng, J., Lanza, M., and Dai, H. (2013) High-performance silicon photoanodes passivated with ultrathin nickel films for water oxidation. *Science*, **342**, 836–840.

83 Seger, B., Pedersen, T., Laursen, A.B., Vesborg, P.C.K., Hansen, O., and Chorkendorff, I. (2013) Using TiO_2 as a conductive protective layer for photocathodic H2 evolution. *J. Am. Chem. Soc.*, **135**, 1057–1064.

84 Hu, S., Shaner, M.R., Beardslee, J.A., Lichterman, M., Brunschwig, B.S., and Lewis, N.S. (2014) Amorphous TiO_2 coatings stabilize Si, GaAs, and GaP photoanodes for efficient water oxidation. *Science*, **344**, 1005–1009.

85 Choi, M.J., Jung, J.-Y., Park, M.-J., Song, J.-W., Lee, J.-H., and Bang, J.H. (2014) Long-term durable silicon photocathode protected by a thin Al_2O_3/SiO_x layer for photoelectrochemical hydrogen evolution. *J. Mater. Chem. A*, **2**, 2928–2933.

86 Lin, Y., Kapadia, R., Yang, J., Zheng, M., Chen, K., Hettick, M., Yin, X., Battaglia, C., Sharp, I.D., Ager, J.W., and Javey, A. (2015) Role of TiO_2 surface passivation on improving the performance of p-InP photocathodes. *J. Phys. Chem. C*, **119**, 2308–2313.

87 Sun, K., Saadi, F.H., Lichterman, M.F., Hale, W.G., Wang, H.-P., Zhou, X., Plymale, N.T., Omelchenko, S.T., He, J.-H., Papadantonakis, K.M., Brunschwig, B.S., and Lewis, N.S. (2015) Stable solar-driven oxidation of water by semiconducting photoanodes protected by transparent catalytic nickel oxide films. *Proc. Natl. Acad. Sci. U.S.A.*, **112**, 3612–3617.

88 Wang, T., Luo, Z., Li, C., and Gong, J. (2014) Controllable fabrication of nanostructured materials for photoelectrochemical water splitting via atomic layer deposition. *Chem. Rev. Soc.*, **43**, 7469–7484.

89 Bakke, J.R., Pickrahn, K.L., Brennan, T.P., and Bent, S.F. (2011) Nanoengineering and interfacial engineering of photovoltaics by atomic layer deposition. *Nanoscale*, **3**, 3482–3508.

90 Emery, J.D., Schlepütz, C.M., Guo, P., Riha, S.C., Chang, R.P.H., and Martinson, A.B.F. (2014) Atomic layer deposition of metastable β-Fe_2O_3 via isomorphic epitaxy for photoassisted water oxidation. *ACS Appl. Mater. Interfaces*, **6**, 21894–21900.

91 Zandi, O., Beardslee, J.A., and Hamann, T. (2014) Substrate dependent water splitting with ultrathin alpha-Fe_2O_3 electrodes. *J. Phys. Chem. C*, **118**, 16494–16503.

92 Zandi, O. and Hamann, T.W. (2014) Enhanced water splitting efficiency through selective surface state removal. *J. Phys. Chem. Lett.*, **5**, 1522–1526.

93 Haschke, S., Wu, Y., Bashouti, M., Christiansen, S., and Bachmann, J. (2015) Engineering nanoporous iron(III) oxide into an effective water oxidation electrode. *ChemCatChem*, **7**, 2455–2459.

94 Cheng, C., Karuturi, S.K., Liu, L., Liu, J., Li, H., Su, L.T., Tok, A.I.Y., and Fan, H.J. (2012) Quantum-dot-sensitized TiO_2 inverse opals for photoelectrochemical hydrogen generation. *Small*, **8**, 37–42.

95 Luo, J., Karuturi, S.K., Liu, L., Su, L.T., Tok, A.I.Y., and Fan, H.J. (2012) Homogeneous photosensitization of complex TiO_2 nanostructures for efficient solar energy conversion. *Sci. Rep.*, **2**, 451–456.

96 Lin, Y., Zhou, S., Liu, X., Sheehan, S., and Wang, D. (2009) $TiO_2/TiSi_2$ heterostructures for high-efficiency photoelectrochemical H_2O splitting. *J. Am. Chem. Soc.*, **131**, 2772–2773.

97 Liu, R., Lin, Y., Chou, L.-Y., Sheehan, S.W., He, W., Zhang, F., Hou, H.J.M., and Wang, D. (2011) Water splitting by tungsten oxide prepared by atomic layer deposition and decorated with an oxygen-evolving catalyst. *Angew. Chem. Int. Ed.*, **50**, 499–502.

98 Lin, Y., Zhou, S., Sheehan, S.W., and Wang, D. (2011) Nanonet-based hematite heteronanostructures for efficient solar water splitting. *J. Am. Chem. Soc.*, **133**, 2398–2401.

99 Noh, J.H., Ding, B., Han, H.S., Kim, J.S., Park, J.H., Park, S.B., Jung, H.S., Lee, J.-K., and Hong, K.S. (2012) Tin doped indium oxide core—TiO_2 shell nanowires on stainless steel mesh for flexible photoelectrochemical cells. *Appl. Phys. Lett.*, **100**, 084104.

100 Mayer, M.T., Du, C., and Wang, D. (2012) Hematite/Si nanowire dual-absorber system for photoelectrochemical water splitting at Low applied potentials. *J. Am. Chem. Soc.*, **134**, 12406–12409.

101 Peng, Q., Kalanyan, B., Hoertz, P.G., Miller, A., Kim, D.H., Hanson, K., Alibabaei, L., Liu, J., Meyer, T.J., Parsons, G.N., and Glass, J.T. (2013) Solution-processed, antimony-doped tin oxide colloid films enable high-performance TiO_2 photoanodes for water splitting. *Nano Lett.*, **13**, 1481–1488.

102 Stefik, M., Cornuz, M., Mathews, N., Hisatomi, T., Mhaisalkar, S., and Grätzel, M. (2012) Transparent, conducting $Nb:SnO_2$ for host–guest photoelectrochemistry. *Nano Lett.*, **12**, 5431–5435.

103 Cesar, I., Kay, A., Gonzalez Martinez, J.A., and Grätzel, M. (2006) Translucent thin film Fe_2O_3 photoanodes for efficient water splitting by sunlight: nanostructure-directing effect of Si-doping. *J. Am. Chem. Soc.*, **128**, 4582–4583.

104 Steier, L., Luo, J., Schreier, M., Mayer, M.T., Sajavaara, T., and Grätzel, M. (2015) Low-temperature atomic layer deposition of crystalline and photoactive ultrathin hematite films for solar water splitting. *ACS Nano*, **9**, 11775–11783.

105 Weber, M.J., Mackus, A.J.M., Verheijen, M.A., van der Marel, C., and Kessels, W.M.M. (2012) Supported core/shell bimetallic nanoparticles synthesis by atomic layer deposition. *Chem. Mater.*, **24**, 2973–2977.

106 Dasgupta, N.P., Liu, C., Andrews, S., Prinz, F.B., and Yang, P. (2013) Atomic layer deposition of platinum catalysts on nanowire surfaces for photoelectrochemical water reduction. *J. Am. Chem. Soc.*, **135**, 12932–12935.

107 Dai, P., Xie, J., Mayer, M.T., Yang, X., Zhan, J., and Wang, D. (2013) Solar hydrogen generation by silicon nanowires modified with platinum nanoparticle catalysts by atomic layer deposition. *Angew. Chem. Int. Ed.*, **52**, 11119–111223.

108 Lee, M.H., Takei, K., Zhang, J., Kapadia, R., Zheng, M., Chen, Y.-Z., Nah, J., Matthews, T.S., Chueh, Y.-L., Ager, J.W., and Javey, A. (2012) P-type InP nanopillar photocathodes for efficient solar-driven hydrogen production. *Angew. Chem. Int. Ed.*, **51**, 10760–10764.

109 Lu, J., Elam, J.W., and Stair, P.C. (2013) Synthesis and stabilization of supported metal catalysts by atomic layer deposition. *Acc. Chem. Res.*, **46**, 1806–1815.

110 Chen, Y.W., Prange, J.D., Duehnen, S., Park, Y., Gunji, M., Chidsey, C.E.D., and McIntyre, P.C. (2011) Atomic layer-deposited tunnel oxide stabilizes silicon photoanodes for water oxidation. *Nat. Mater.*, **10**, 539–544.

111 Strandwitz, N.C., Comstock, D.J., Grimm, R.L., Nichols-Nielander, A.C., Elam, J., and Lewis, N.S. (2013) Photoelectrochemical behavior of n-type Si(100) electrodes coated with thin films of manganese oxide grown by atomic layer deposition. *J. Phys. Chem. C*, **117**, 4931–4936.

112 Lai, Y.-H., Park, H.S., Zhang, J.Z., Matthews, P.D., Wright, D.S., and Reisner, E. (2015) A Si photocathode protected and activated with a Ti and Ni composite film for solar hydrogen production. *Chem. Eur. J.*, **21**, 3919–3923.

113 Le Formal, F., Tetreault, N., Cornuz, M., Moehl, T., Gratzel, M., and Sivula, K. (2011) Passivating surface states on water splitting hematite photoanodes with alumina overlayers. *Chem. Sci.*, **2**, 737–743.

114 Tallarida, M., Das, C., Cibrev, D., Kukli, K., Tamm, A., Ritala, M., Lana-Villarreal, T., Gomez, R., Leskela, M., and Schmeisser, D. (2014) Modification of hematite electronic properties with trimethyl aluminum to enhance the efficiency of photoelectrodes. *J. Phys. Chem. Lett.*, **5**, 3582–3587.

115 Forster, M., Potter, R.J., Ling, Y., Yang, Y., Klug, D.R., Li, Y., and Cowan, A.J. (2015) Oxygen deficient alpha-Fe_2O_3 photoelectrodes: a balance between enhanced electrical properties and trap-mediated losses. *Chem. Sci.*, **6**, 4009–4016.

116 Klahr, B. and Hamann, T. (2014) Water oxidation on hematite photoelectrodes: insight into the nature of surface states through in situ spectroelectrochemistry. *J. Phys. Chem. C*, **118**, 10393–10399.

117 Kim, W., Tachikawa, T., Monllor-Satoca, D., Kim, H.-I., Majima, T., and Choi, W. (2013) Promoting water photooxidation on transparent WO3 thin films using an alumina overlayer. *Energy Environ. Sci.*, **6**, 3732–3739.

118 Gui, Q., Xu, Z., Zhang, H., Cheng, C., Zhu, X., Yin, M., Song, Y., Lu, L., Chen, X., and Li, D. (2014) Enhanced photoelectrochemical water splitting performance of anodic TiO_2 nanotube arrays by surface passivation. *ACS Appl. Mater. Interfaces*, **6**, 17053–17058.

119 Hwang, Y.J., Hahn, C., Liu, B., and Yang, P. (2012) Photoelectrochemical properties of TiO_2 nanowire arrays: a study of the dependence on length and atomic layer deposition coating. *ACS Nano*, **6**, 5060–5069.

120 Seo, J., Kim, H.J., Pekarek, R.T., and Rose, M.J. (2015) Hybrid organic/inorganic band-edge modulation of p-Si(111) photoelectrodes: effects of R, metal oxide, and Pt on H-2 generation. *J. Am. Chem. Soc.*, **137**, 3173–3176.

121 Kim, H.J., Kearney, K.L., Le, L.H., Pekarek, R.T., and Roses, M.J. (2015) Platinum-enhanced electron transfer and surface passivation through ultra-thin film aluminum oxide (Al_2O_3) on Si(111)-CH_3 photoelectrodes. *ACS Appl. Mater. Interfaces*, **7**, 8572–8584.

122 Huang, B., Yang, W., Wen, Y., Shan, B., and Chen, R. (2015) Co_3O_4-modified TiO_2 nanotube arrays via atomic layer deposition for improved visible-light photoelectrochemical performance. *ACS Appl. Mater. Interfaces*, **7**, 422–431.

123 Dai, G., Yu, J., and Liu, G. (2011) Synthesis and enhanced visible-light photoelectrocatalytic activity of p–n junction BiOI/TiO$_2$ nanotube arrays. *J. Phys. Chem. C*, **115**, 7339–7346.

124 Hwang, Y.J., Boukai, A., and Yang, P. (2009) High density n-Si/N-TiO$_2$ core/shell nanowire arrays with enhanced photoactivity. *Nano Lett.*, **9**, 410–415.

125 Ao, X., Tong, X., Kim, D.S., Zhang, L., Knez, M., Müller, F., He, S., and Schmidt, V. (2012) Black silicon with controllable macropore array for enhanced photoelectrochemical performance. *Appl. Phys. Lett.*, **101**, 111901.

126 Prasittichai, C. and Hupp, J.T. (2010) Surface modification of SnO$_2$ photoelectrodes in dye-sensitized solar cells: significant improvements in photovoltage via Al$_2$O$_3$ atomic layer deposition. *J. Phys. Chem. Lett.*, **1**, 1611–1615.

127 Prasittichai, C., Avila, J.R., Farha, O.K., and Hupp, J.T. (2013) Systematic modulation of quantum (electron) tunneling behavior by atomic layer deposition on nanoparticulate SnO$_2$ and TiO$_2$ photoanodes. *J. Am. Chem. Soc.*, **135**, 16328–16331.

128 Kohl, P.A., Frank, S.N., and Bard, A.J. (1977) Semiconductor electrodes. XI. Behavior of n- and p-type single crystal semiconductors covered with thin films. *J. Electrochem. Soc.*, **124**, 225–229.

129 McDowell, M.T., Lichterman, M.F., Carim, A.I., Liu, R., Hu, S., Brunschwig, B.S., and Lewis, N.S. (2015) The influence of structure and processing on the behavior of TiO$_2$ protective layers for stabilization of n-Si/TiO$_2$/Ni photoanodes for water oxidation. *ACS Appl. Mater. Interfaces*, **7**, 15189–15199.

130 Lichterman, M.F., Carim, A.I., McDowell, M.T., Hu, S., Gray, H.B., Brunschwig, B.S., and Lewis, N.S. (2014) Stabilization of n-cadmium telluride photoanodes for water oxidation to O-2(G) in aqueous alkaline electrolytes using amorphous TiO$_2$ films formed by atomic-layer deposition. *Energy Environ. Sci.*, **7**, 3334–3337.

131 Verlage, E., Hu, S., Liu, R., Jones, R.J.R., Sun, K., Xiang, C., Lewis, N.S., and Atwater, H.A. (2015) A monolithically integrated, intrinsically safe, 10% efficient, solar-driven water-splitting system based on active, stable earth-abundant electrocatalysts in conjunction with tandem III–V light absorbers protected by amorphous TiO$_2$ films. *Energy Environ. Sci.*, **8**, 3166–3172.

132 Pickrahn, K.L., Gorlin, Y., Seitz, L.C., Garg, A., Nordlund, D., Jaramillo, T.F., and Bent, S.F. (2015) Applications of ALD MnO to electrochemical water splitting. *Phys. Chem. Chem. Phys.*, **17**, 14003–14011.

133 Lichterman, M.F., Shaner, M.R., Handler, S.G., Brunschwig, B.S., Gray, H.B., Lewis, N.S., and Spurgeon, J.M. (2013) Enhanced stability and activity for water oxidation in alkaline media with bismuth vanadate photoelectrodes modified with a cobalt oxide catalytic layer produced by atomic layer deposition. *J. Phys. Chem. Lett.*, **4**, 4188–4191.

134 Li, C., Wang, T., Luo, Z., Zhang, D., and Gong, J. (2015) Transparent ALD-grown Ta$_2$O$_5$ protective layer for highly stable ZnO photoelectrode in solar water splitting. *Chem. Commun.*, **51**, 7290–7293.

135 Paracchino, A., Laporte, V., Sivula, K., Grätzel, M., and Thimsen, E. (2011) Highly active oxide photocathode for photoelectrochemical water reduction. *Nat. Mater.*, **10**, 456–461.

136 Qiu, J., Zeng, G., Ha, M.-A., Ge, M., Lin, Y., Hettick, M., Hou, B., Alexandrova, A.N., Javey, A., and Cronin, S.B. (2015) Artificial photosynthesis on TiO_2-passivated InP nanopillars. *Nano Lett.*, **15**, 6177–6181.

137 Das, C., Tallarida, M., and Schmeisser, D. (2015) Si microstructures laminated with a nanolayer of TiO_2 as long-term stable and effective photocathodes in PEC devices. *Nanoscale*, **7**, 7726–7733.

138 Liu, M., Nam, C.-Y., Black, C.T., Kamcev, J., and Zhang, L. (2013) Enhancing water splitting activity and chemical stability of zinc oxide nanowire photoanodes with ultrathin titania shells. *J. Phys. Chem. C*, **117**, 13396–13402.

9

Atomic Layer Deposition of Thermoelectric Materials

Maarit Karppinen and Antti J. Karttunen

Aalto University, Department of Chemistry and Materials Science, Kemistintie 1, 02150 Espoo, Finland

9.1 Introduction

9.1.1 Thermoelectric Energy Conversion and Cooling

Thermoelectric (TE) materials can be exploited as both power generators and heat pumps. The continuously increasing worldwide demand for energy and the resultant depletion of fossil fuels are the strong driving forces to develop efficient TE power generation technologies for capturing unused or released and otherwise wasted heat from various naturally occurring (solar radiation, geothermal processes, etc.) and human-activity based (industrial processes, transportation, home heating, etc.) heat sources and converting it directly into electricity [1]. On the other hand, the continuous downscaling of microelectronic devices calls for more efficient cooling systems for portable devices – an issue that could be solved by utilizing localized embeddable thin-film TE coolers [2].

A thermoelectric generator or heat pump is an all-solid-state device that can directly convert heat into electricity or *vice versa*. Such devices can in principle be used to improve the energy efficiency of any application or process where heat is involved, thus providing us with excellent means for next-generation sustainable energy exploitation. Thermoelectric devices do not contain any mechanically moving parts and are extremely reliable, durable, silent, and spatially compact/scalable and thus most suitable for, for example, ubiquitous uses. In terms of energy volumes, huge gains could be expected for TE generators attached to, for example, heat exchangers of high-energy industrial processes such as glass or steel production where energy accounts for a major part of the total production cost. Another application of high future potential is in the automotive field, where considerable enhancements in fuel efficiency are anticipated by using TE modules to harvest energy from engine exhaust gases. An interesting option would also be to combine TE power generation with photovoltaic systems. A yet totally different example of an area in which thermoelectrics would be of considerable benefit is small wireless and/or wearable device where the problem of energy storage for the device could be solved by powering the device with local TE modules. Here, in particular, the demand is for novel flexible thin-film TE materials that could

Atomic Layer Deposition in Energy Conversion Applications, First Edition. Edited by Julien Bachmann.
© 2017 Wiley-VCH Verlag GmbH & Co. KGaA. Published 2017 by Wiley-VCH Verlag GmbH & Co. KGaA.

harvest, for example, human body heat to power wearable sensors and other electronic devices [3]. It is important to emphasize that even though such applications (e.g., medical sensing) are currently strongly emerging, the devices are still mostly powered by batteries that require frequent recharging. This is also an area where ALD-fabricated TE modules could most readily show superiority.

9.1.2 Designing and Optimizing Thermoelectric Materials

The thermoelectric (Seebeck–Peltier) phenomena behind the TE devices were already discovered in the early 1800s. The major obstacle in implementing the thermoelectrics into large-scale practical and economically feasible uses is the energy-conversion efficiency of the TE material [1, 4]. This efficiency is evaluated with the scale of figure of merit, $ZT \equiv S^2 T/\rho\kappa$, at the given operation temperature T, and the difficulty lies in the fact that the individual factors for Z, that is, electrical conductivity ($\sigma = 1/\rho$), Seebeck coefficient (S), and thermal conductivity (κ), are not independently tunable, but an enhancement in one parameter often occurs at the expense of the others. Practical thermoelectric devices consist of modules made of solid n- and p-type TE materials in a serial connection. This poses another challenge for the material design as the n- and p-type material components should also be mutually compatible.

An efficient TE material, n- or p-type, should play the dual role of being concomitantly both a good electrical conductor and a poor thermal conductor [5]. In practice, S, σ, and the electronic part of thermal conductivity κ_{el} ($\equiv \kappa_{tot} - \kappa_{lat}$) depend, according to the Wiedemann–Franz law, on the carrier concentration (n) in a contradictory manner (see Figure 9.1). Another contradictory aspect of the thermoelectrics to be optimized concerns the effective mass (m^*) and the mobility (μ) of carriers [6]. Namely, in ionic compounds, the effective mass may be high enough but the carrier mobility is low, whereas in the more covalent compounds, a high carrier mobility is realized but the effective mass is often too low. Through crystal and band-structure engineering, TE materials have been tailored to exhibit a large enough m^* value without drastically reducing the μ value.

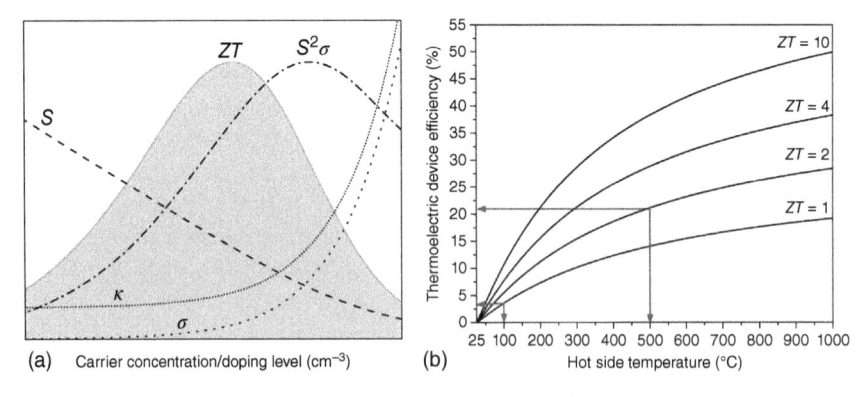

Figure 9.1 (a) Dependencies of the individual factors of the thermoelectric figure of merit Z, that is, S, σ, and κ, on the carrier concentration, and (b) the dependence of the TE conversion efficiency on the heat-source/hot-side temperature (cold-side temperature: 25 °C).

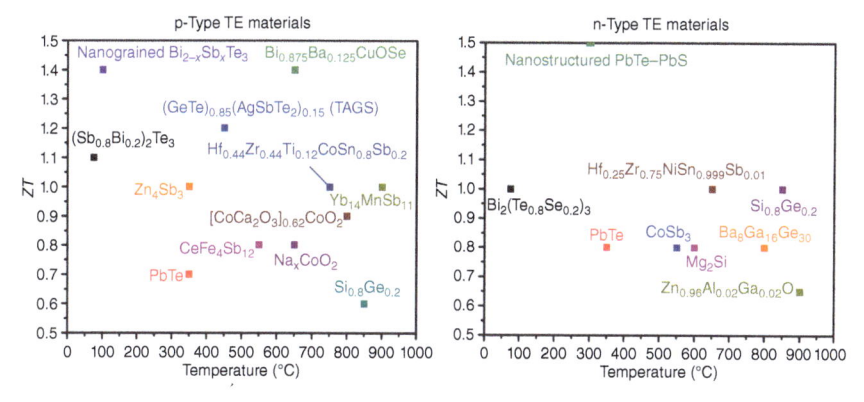

Figure 9.2 Thermoelectric figure-of-merit values, ZT, at different temperatures for representative thermoelectric material families.

The TE materials currently employed, such as Bi_2Te_3 and its alloys, are heavily doped semiconductors with an optimized ZT value at the carrier concentration in the range of 10^{19}–10^{21} carriers per cubic centimeter; they then show ZT values slightly above unity near room temperature. An obvious disadvantage of these materials is that they are composed of rare, expensive, and environmentally not-so-friendly constituents. Another drawback is that they are stable only at relatively low temperatures, typically below 200 °C. This latter aspect is problematic also for the sake of energy conversion efficiency as the TE materials fundamentally work more efficiently at high temperatures (see Figure 9.1).

In the past two decades or so, a number of material families have been investigated as new material candidates for thermoelectrics (see Figure 9.2). A large/complex unit cell typically efficiently scatters phonons, thus decreasing the lattice part of thermal conductivity (κ_{lat}). For example, the so-called filled skutterudites derived from $CoSb_3$ and the so-called Zintl phases such as $Yb_{14}MnSb_{11}$ possess complex crystal structures with vacancies or interstitials and thereby low thermal conductivities and excellent TE characteristics [1, 4].

Oxide materials, on the other hand, possess the apparent advantage of being environmentally benign and thermally stable up to appreciably high temperatures; the fact that they may also exhibit excellent TE characteristics (particularly for high-temperature applications) was first revealed already in the late 1990s for Na_xCoO_2 [7]. However, the TE performance of the currently available oxide materials is not yet up to the level of the best conventional TE materials: a common drawback of the simple oxide materials such as ZnO is their far too high thermal conductivity values, in particular the lattice thermal conductivity.

The lattice thermal conductivity is the product of heat capacity (C), phonon velocity (v) and phonon mean free path (l) and depends on – instead of n – the nature of chemical bonds and disorder/defects on different length scales. Accordingly, various nanoscale and/or multiscale engineering approaches have been searched for to manipulate phonon transport without significantly reducing the material's electrical conductivity [4, 8–10]. A remarkable demonstration of such an approach is the enhancement of the ZT value of PbTe up to 2.2 (at about 640 °C) through multiscale defect architecture from atomic-level lattice disorder to nanoscale endotaxial precipitates and mesoscale grain boundaries to achieve a full spectrum of phonon scattering [11].

Multilayered crystals – natural or artificial – provide another type of platform to manipulate electrons and phonons independently. Here, the so-called misfit-layered cobalt oxides such as $[CoCa_2O_{3\pm\delta}]_q CoO_2$ form an example of the first type. Due to their peculiar crystal structures consisting of mutually incommensurate layers of electrically conducting CoO_2 layers of hexagonal symmetry and rock-salt-structured oxygen-nonstoichiometric intervening layers, the misfit-layered phases may simultaneously exhibit high electrical conductivity, high Seebeck coefficient, and relatively low thermal conductivity.

9.1.3 Thin-Film Thermoelectric Devices

Thin-film thermoelectric devices are technologically appealing from both macroscopic and microscopic point of view [12, 13]. From the macroscopic-device-level perspective, thin-film thermoelectrics enable packaging solutions and usage scenarios that are radically different from those for bulk thermoelectric materials. For example, thin-film thermoelectrics with thicknesses in the range of 1–10 µm would greatly facilitate integration of thermoelectric cooling devices within microelectronics. Furthermore, the fabrication processes of thin-film thermoelectrics are often compatible with the standard tools of semiconductor device manufacturing, enabling scalable fabrication of TE modules. However, several major challenges related to thin-film thermoelectrics need to be solved to translate their high intrinsic ZT values into efficient TE devices. For example, the thermal management of the hot and cold sides of thin films remains a significant challenge because the temperature difference may vanish very quickly for high heat fluxes if the cooling of the cold side is not efficient enough.

From the microscopic perspective, the inherent properties of thin-film materials offer significant advantages for developing novel thermoelectrics. Nanostructuring is considered to be a key strategy for developing thermoelectrics with greatly reduced thermal conductivity values [4], and thin films as thin as 100–1000 nm offer a very good starting point for lowering the thermal conductivity in comparison to corresponding bulk materials. In particular, highly controllable thin-film fabrication techniques such as ALD open up the possibility for lowering the thermal conductivity via various superlattice approaches discussed in more detail in Section 9.3. For artificial superlattice thin-film materials consisting of alternating layers of dissimilar materials, the phonon transport is suppressed by internal interface scattering; by properly tuning the superlattice period, the electronic properties can be kept essentially unaffected or even enhanced due to, for example, quantum-confinement effects while decreasing the lattice thermal conductivity. Significant suppressions in thermal conductivity have been demonstrated, for example, for Bi_2Te_3–Sb_2Te_3 superlattices [14]. Interestingly, in superlattices, thermal conductivity may be reduced not only along the superlattice layer-piling direction but also in the in-plane direction [15].

Notably, thin-film thermoelectrics could be utilized either in cross-plane or in-plane configuration. The cross-plane configuration is analogous to a standard TE module configuration based on bulk materials, and the temperature gradient should arise in the cross-plane direction. This is the most relevant configuration for thermoelectric generator and cooling solutions. The in-plane configuration, where the temperature gradient arises parallel to the thin film and the substrate, is not that relevant for the generator or cooling solutions, but it can be very useful for thermoelectric sensor applications.

Thin-film thermoelectrics integrated with flexible substrates are a potential solution for powering wearable electronics with body heat [3]. Within this field, typical material solutions are organic thermoelectrics and hybrid inorganic–organic thermoelectrics. Highly controllable atomic-layer/molecular-layer deposition (ALD/MLD) processes are particularly suitable for fabricating the latter materials, enabling the high tunability of atomic-level structural features and properties of the hybrid thin films. Furthermore, ALD-based fabrication methods are applicable to depositing TE materials directly on flexible substrates such as polymers and textiles [16]. Atomic layer deposition (ALD) on textiles and polymers might require careful tuning of the deposition parameters to account for the increased complexity of the substrate, but after suitable parameters have been found, ALD could provide us with a practical solution for integrating TE materials with flexible substrates in a controlled manner.

9.2 ALD Processes for Thermoelectrics

The ALD technique uniquely possesses several features that could be beneficial in the TE technology, such as atomic-level thickness control of individual layers, large-area homogeneity, and multiple possibilities for nanostructure engineering and layer engineering including the easy deposition of precisely controlled nanolaminates and superlattices. Nevertheless, research in this field is just emerging [17, 18].

9.2.1 Thermoelectric Oxide Thin Films

So far, ALD processes have been developed for a few potential TE oxide materials and representative bismuth and lead chalcogenide TE materials. Among the ALD-fabricated thermoelectrics, the most widely studied is ZnO [19–21]. Even though ZnO is not the best TE material, it has several advantages. It consists of inexpensive earth-abundant, and nontoxic elements and remains stable up to relatively high temperatures. Also importantly, the ALD process to deposit ZnO from diethyl zinc $Zn(CH_2CH_3)_2$ (DEZ) and H_2O precursors is one of the prototype ALD processes; for a recent review, see [22]. The ALD process based on DEZ and H_2O typically yields polycrystalline films of the hexagonal wurtzite structure with a high degree of crystallinity even at relatively low deposition temperatures. The grain orientation then depends on the deposition temperature such that below 70 °C, c-axis orientation is preferred, between 70 and 200 °C, the a-axis orientation becomes gradually more and more prominent, and finally above 220 °C, the c-axis orientation again starts to dominate.

Zinc oxide is a wide-band-gap semiconductor and exhibits intrinsically n-type conductivity due to unintentionally formed defects. Accordingly, it is difficult to induce p-type conductivity through acceptor dopants. On the other hand, it is straightforward to enhance the n-type conductivity through trivalent cation substituents, in particular aluminum. For ALD-ZnO thin films, n-type doping is readily realized by replacing a fraction of the DEZ/H_2O cycles with TMA/H_2O (TMA = trimethyl aluminum Al(CH_3)$_3$) cycles. The optimal Al-for-Zn substitution level to maximize the electrical conductivity is around 2% (typically calculated from the ratio of the TMA/H_2O and DEZ/H_2O ALD cycles rather than the actual analyzed Al content), and for such optimally Al-substituted ALD-ZnO films, resistivity values as low as 10^{-4} Ω cm have been achieved [22]. The lightly Al-for-Zn substituted (Zn,Al)O films show, besides the low enough resistivity, decent (negative) values for the Seebeck coefficient (see Table 9.1) such that the so-called thermoelectric power factor, $PF \equiv S^2/\rho$, is optimized.

Another n-type semiconductor readily fabricated by ALD and exhibiting decent TE properties is TiO_2 with the anatase structure [23, 24, 29]; the most common ALD process for TiO_2 is based on titanium tetrachloride $TiCl_4$ and H_2O precursors; for a recent review, see [30]. Crystallinity of the TiO_2 films fabricated using the $TiCl_4$/H_2O process varies depending on the deposition

Table 9.1 ALD process parameters and resultant room-temperature thermoelectric characteristics for ALD-fabricated thermoelectric thin films.

Material	Precursors; deposition temperature	Requirement for crystallinity	Seebeck (μV K^{-1})	Resistivity (mΩ cm)	References
[$CoCa_2O_3$]$_q$$CoO_2$	Co(thd)$_2$, Ca(thd)$_2$, O$_3$; 275 °C	Air-750 °C Ar-600 °C	+113 +128	10	[17]
(Zn$_{0.98}$Al$_{0.02}$)O	DEZ, TMA, H_2O; 220 °C	As-deposited	−60	70	[22]
(Ti$_{0.75}$Nb$_{0.25}$)O$_2$	TiCl$_4$, Nb(OEt)$_5$, H_2O; 210 °C	H_2-600 °C	−12	1.4	[23]
(Ti$_{0.95}$Nb$_{0.05}$)O$_2$	TiCl$_4$, Nb(OEt)$_5$, H_2O; 160 °C	H_2-600 °C		1.4	[24]
CuCrO$_2$	Cu(thd)$_2$, Cr(acac)$_3$, O$_3$; 250 °C	Ar-800 °C	+330	10^3	[25]
Bi$_2$Te$_3$	BiCl$_3$, (Et$_3$Si)$_2$Te; 160 °C	As-deposited	−180	0.1–1	[26]
Bi$_2$Se$_3$	BiCl$_3$, (Et$_3$Si)$_2$Se; 160 °C	As-deposited	−180		[27]
Sb$_2$Te$_3$	SbCl$_3$, (Et$_3$Si)$_2$Te; 80 °C	As-deposited	+146	10	[28]

temperature: between 100 and 165 °C, the films are typically amorphous, from 165 to 350 °C, crystalline with the anatase structure, and above 350 °C, crystalline but with an increasing amount of the rutile phase with increasing deposition temperature. Further improvement of electrical properties is readily achieved using the aliovalent Nb-for-Ti cation substitution. In the ALD process for Nb-substituted (Ti,Nb)O_2 thin films, a fraction of the TiCl$_4$/H$_2$O cycles is replaced by Nb(OEt)$_5$/H$_2$O cycles. The Nb-substitution, however, considerably suppresses the film crystallinity such that, for example, at the deposition temperature of 210 °C, the $x = 0.05$ (Ti$_{1-x}$Nb$_x$)O_2 films are already essentially amorphous. Therefore, a postdeposition annealing treatment is typically carried out at 500–600 °C under strongly reductive conditions (in an H$_2$ gas flow to remove any excess oxygen) to crystallize the film into the anatase structure and to enhance the electrical conductivity. For the thus ALD-fabricated (Ti$_{1-x}$Nb$_x$)O_2 films, electron mobility is limited by grain boundary scattering in the low-doping regime ($x \leq 0.15$) and highly conducting ($\rho \approx 1$ mΩ cm) and notably c-axis-oriented films are only obtained for $x \geq 0.20$.

The drawback of the heavily Nb-substituted TiO$_2$ films is that the magnitude of Seebeck coefficient significantly decreases with increasing Nb content x to the level of -10 µV K^{-1} (Table 9.1). A smart solution to overcome this problem was recently discovered as it was found that the crystallinity in the as-deposited films actually hindered the grain growth during the postdeposition annealing treatment; in the initially amorphous (Ti,Nb)O_2 films deposited at 160–175 °C and then annealed in an H$_2$ gas flow, substantial grain growth was achieved during the annealing such that intragrain, rather than grain boundary, properties were found to govern the electron transport [24]. Particularly promising electron transport characteristics (see Table 9.1) were achieved for (Ti$_{0.95}$Nb$_{0.05}$)O_2 films deposited at 175 °C (as amorphous films) and then annealed in an H$_2$ gas flow at 600 °C.

Both ZnO and TiO$_2$ are n-type thermoelectrics. For p-type thermoelectric oxides, ALD processes have so far been developed for [Ca$_2$CoO$_3$]$_{0.62}$CoO$_2$ and CuCrO$_2$; in both cases, the drawback is that, to crystallize the as-deposited amorphous films, a postdeposition annealing at a relatively high temperature is required. The former films were prepared using Ca(thd)$_2$, Co(thd)$_2$, and O$_3$ as precursors at 275 °C [17]. A postdeposition heat treatment in an O$_2$ gas flow at 750 °C yielded well-crystallized highly c-axis-oriented [Ca$_2$CoO$_3$]$_{0.62}$CoO$_2$ films. Oxygen content of the O$_2$-annealed film could be further controlled through a reductive N$_2$ annealing. As the degree of reduction depended on the annealing temperature, the choice of the N$_2$-annealing temperature serves as a tool for the precise oxygen-content tuning. With decreasing oxygen content, the lattice parameter c and the Seebeck coefficient S were found to increase. The room-temperature S value was 113 and 128 µV K^{-1} for the oxygen-richest sample and the most reduced sample, respectively (Table 9.1).

The ALD process for CuCrO$_2$ is based on copper 2,2,6,6-tetramethyl-3,5-heptanedionate (Cu(thd)$_2$) and chromium acetylacetonate (Cr(acac)$_3$) as the metal precursors and ozone as the oxygen source [25]. Smooth and homogeneous thin films with an accurately controlled metal composition can be deposited in the temperature range of 240–270 °C; a postdeposition anneal at 700–950 °C in an Ar gas flow then yields well-crystalline films with the delafossite structure. Electrical transport measurements confirmed the p-type semiconducting behavior of the films (see Table 9.1).

9.2.2 Thermoelectric Selenide and Telluride Thin Films

Thin films of the archetype thermoelectric material Bi_2Te_3 have been grown by ALD from $BiCl_3$ and $(Et_3Si)_2Te$ precursors at 160 °C and at higher temperatures, although the GPC value decreased from 1.1 Å per cycle at 160 °C with increasing deposition temperature [26]. At 160 °C, the saturating growth behavior characteristic of an ALD-type growth was realized and the film thickness was easily controlled with the number of deposition cycles. The resultant films were found to be crystalline in the rhombohedral Bi_2Te_3 phase with a favored orientation along the (001) direction. Transport measurements revealed room-temperature values of 0.1 to 1 mΩ cm for resistivity and $-180\,\mu V\,K^{-1}$ for the Seebeck coefficient, somewhat inferior to but of the same magnitude as the values typically reported for Bi_2Te_3.

An ALD process parallel to that for Bi_2Te_3 has also been reported for Bi_2Se_3 based on the dechlorosilylation reactions between $BiCl_3$ and $(Et_3Si)_2Se$ precursors, yielding high-quality thin films with low impurity contents at a growth rate of 1.6 Å per cycle at 160 °C [27]. Similarly to other ALD processes utilizing alkylsilyl chalcogenide precursors, the growth rate decreased significantly with increasing deposition temperature. While the ALD-grown Bi_2Te_3 films showed metallic-type electrical conductivity versus temperature behavior, the Bi_2Se_3 films deposited with ALD showed semiconductor-type behavior.

While the ALD-Bi_2Te_3 and ALD-Bi_2Se_3 films were n-type conductors, p-type conducting Sb_2Te_3 films were grown with a dechlorosilylation ALD process from $SbCl_3$ and $(Et_3Si)_2Te$ at the rate of 0.16 Å per cycle at 80 °C [28]. The low deposition temperature moreover allowed the prepatterning of the films by standard lithography processes. Three different conductive regions in the temperature range of 50–400 K were found for these Sb_2Te_3 films; the room-temperature values were as follows: $n = 2.4 \times 10^{18}\,cm^{-3}$, $\mu = 270.5 \times 10^{-4}\,m^2\,V^{-1}\,s^{-1}$, $\sigma = 10^4\,S\,m^{-1}$, and $S = 146\,\mu V\,K^{-1}$.

9.3 Superlattices for Enhanced Thermoelectric Performance

In multilayers of two different materials with dissimilar electronic and vibrational properties, the material interfaces are anticipated to efficiently reduce the heat transport. The material interfaces may give rise to phonon boundary scattering and/or reduced transmission of thermal energy across the interface, and accordingly, the thermal boundary conductance may be controlled by ballistic and/or diffusive processes. Careful control of the structural and chemical properties of the interfaces and also the placement of these interfaces within the material provides us an exciting means to tailor the material's thermal conductivity preferably independent of its electrical conductivity.

9.3 Superlattices for Enhanced Thermoelectric Performance | 267

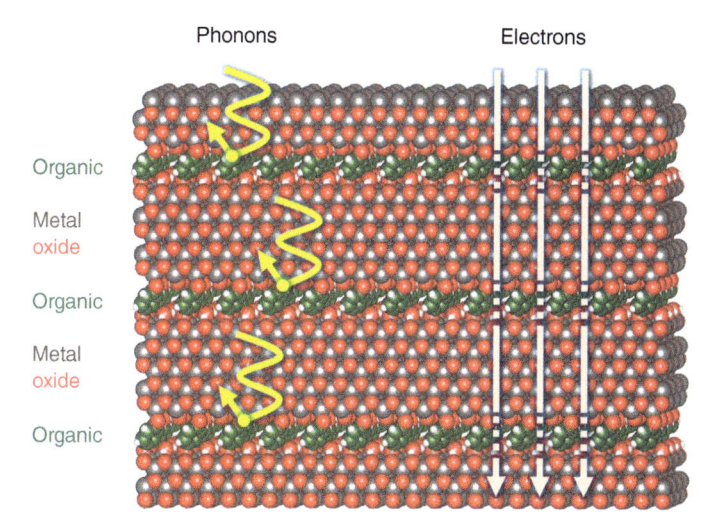

Figure 9.3 Scheme of the blocking of phonon transport while maintaining the electron transport in superlattice thin films consisting of thin organic layers regularly embedded between thicker thermoelectric oxide layers.

In thin-film superlattices with a high density of material interfaces, considerable reductions in thermal conductivity may be expected; the parameters to be controlled in these superlattices would include the individual properties of the constituent materials, the chemical bonding and structural mismatch between the different materials at the interface, and the frequency of the interfaces or the so-called superlattice period.

Inorganic and organic materials show in principle drastically different chemical and physical properties, and thus, we may expect that inorganic–organic superlattice structures where thin organic layers are regularly embedded within an electrically conducting inorganic matrix could show enhanced TE properties owing to the strongly reduced thermal boundary conductance over the inorganic–organic interface. A schematic illustration of the principle is presented in Figure 9.3.

Most importantly, a uniquely suited way to fabricate inorganic–organic superlattice thin films with sharp material interfaces in a highly controlled atomic/molecular layer-by-layer manner is to combine ALD cycles for nanometer-thick inorganic layers with single MLD cycles for very thin/monomolecular organic layers. The molecular layer deposition (MLD) technique for organics relies – similarly to the conventional ALD technique for inorganics – on self-saturating gas-to-surface reactions that guarantee the thin-film fabrication with a (sub)molecular-layer precision; for a recent comprehensive review for the MLD and the combined ALD/MLD techniques (see [31]).

The combined ALD/MLD technique has indeed already shown to allow the deposition of inorganic–organic superlattice thin films with the required precise control over the individual layer thicknesses and thereby the spacing between the

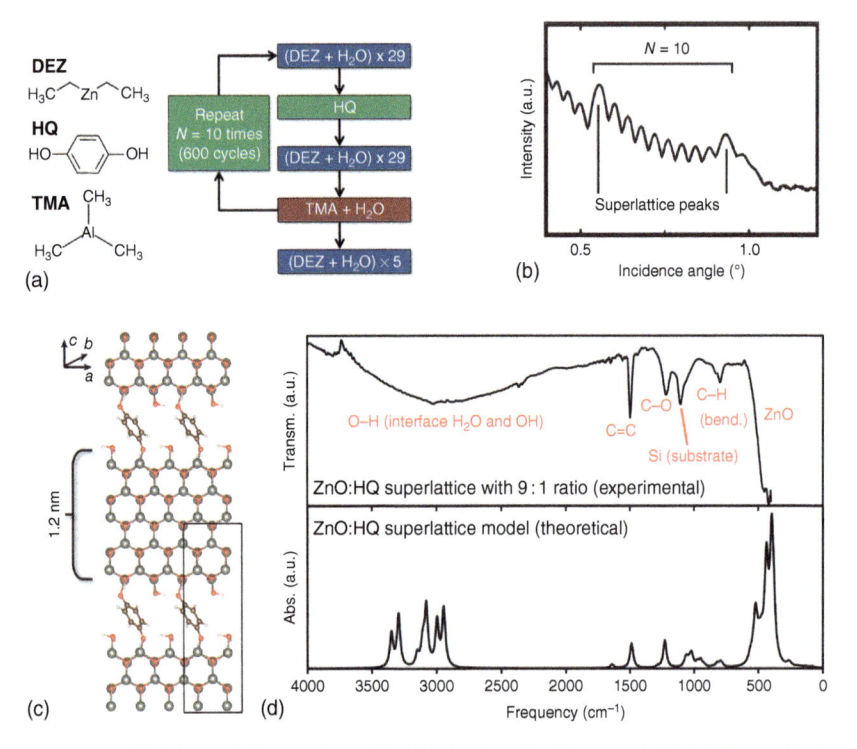

Figure 9.4 (a) Schematic illustration of the fabrication of oxide:organic ZnO:HQ superlattices by ALD/MLD; (b) verification of the superlattice structure by XRR technique; (c) atomic-level model of the ZnO:HQ superlattice derived from first-principles calculations; (d) experimental and theoretical IR spectra of ZnO:HQ superlattice.

organic layers within the thermoelectric inorganic matrix, that is, the superlattice period [32]. Figure 9.4 shows the fabrication principle and an illustrative example of the experimental verification of such ALD/MLD-grown inorganic–organic superlattice thin films using FTIR spectroscopy (Fourier transform infrared) and XRR (X-ray reflectivity) techniques. Moreover, state-of-the-art computational investigations based on *ab initio*/first-principles density functional theory (DFT) methods have been employed to study these novel inorganic–organic hybrid materials and also to elucidate their atomic-level structure–property correlations [33, 34].

So far, the experimental proof-of-the-concept data have been provided for the two oxide material systems, (Zn,Al)O and (Ti,Nb)O$_2$, combined with single organic hydroquinone-based (HQ) layers [4, 35–38]. In comparison to their parent constituents, the thermal conductivities in these superlattice thin films are considerably reduced, as clearly seen from Figure 9.5, where the room-temperature thermal conductivity values measured by a TDTR (time-domain thermoreflectance) technique are plotted against the superlattice period determined from XRR data for the nondoped ZnO:HQ and TiO$_2$:HQ systems. For both sample series, the reduction in thermal conductivity scales linearly with decreasing superlattice period, suggesting that incoherent phonon

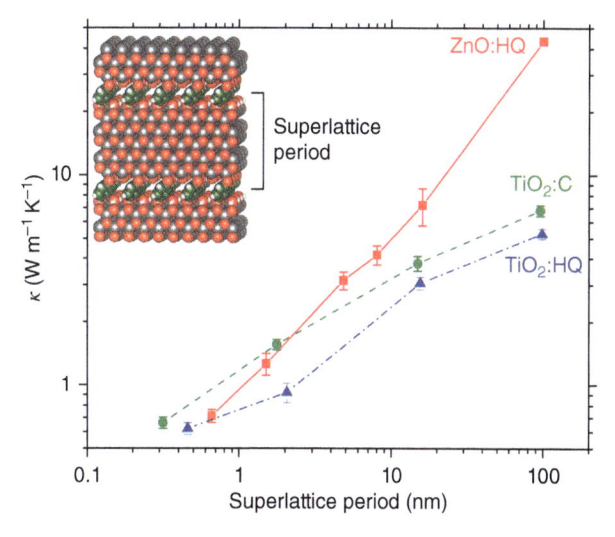

Figure 9.5 Thermal conductivity as a function of superlattice period for ZnO:HQ (as-deposited), TiO$_2$:HQ (as-deposited), and TiO$_2$:C (annealed) thin-film series fabricated by ALD/MLD. The data for about 100-nm superlattice periods refer to pure inorganic thin films.

boundary scattering at the inorganic–organic–inorganic interface dominates the heat transport in these multilayer structures [39]. Moreover, it was found that increasing the number of organic layers decreases the density and concomitantly the heat capacity of the superlattice film [40]. For both the ZnO:HQ and TiO$_2$:HQ systems, reductions in the thermal conductivity of more than an order of magnitude and thereby ultralow thermal conductivity values were obtained when the superlattice period was less than about 5 nm (Figure 9.5).

For the enhanced overall TE performance, the reduction in thermal conductivity should not come with a concomitant reduction in electrical transport properties. Fortunately, this seems to be the case as it has been shown for the (Zn,Al)O:HQ system through optical IR reflectivity, resistivity, and Seebeck coefficient measurements that in particular with a proper tuning of the Al content, the introduction of monomolecular organic layers within the (Zn,Al)O matrix does not considerably affect the electron doping level. Hence, as little change is seen in the power factor PF for the (Zn,Al)O:HQ superlattices upon the introduction of the organic layers, it would seem that the 10-fold or larger decrease in thermal conductivity observed for the superlattices would bring about a greatly increased thermoelectric figure of merit ZT for the material. It should, however, be noted that – before drawing more precise conclusions on this point – the state-of-the-art TDTR technique employed for the thermal conductivity measurements of TE thin films measures the cross-plane thermal conductivity of the films, while the electrical transport measurements reported so far have been made in-plane. Thus, the values may not be fully comparable. At the same time, it could be noted that experimental evidence exists for purely inorganic superlattices such that thermal conductivity may be reduced not only along the superlattice layer-piling direction but also in the in-plane direction [15]. Hence, the concept of enhancing the TE performance of inorganic materials by layering them with periodically intermittent organic layers using the ALD/MLD technique seems to be highly promising and should be more widely explored by expanding the selection of both the inorganic and organic constituents. Also importantly, in the case of the (Zn,Al)O:HQ system, the films including the superlattice structure were shown to remain stable upon heating in air up to the amazingly high temperature of about 500 °C.

From the high-temperature annealing experiments for the TiO_2:HQ superlattice films carried out in an H_2 gas flow at 600 °C, an exciting observation was made: this reductive annealing treatment results in the decomposition of the benzene rings to graphitic carbon, but the superlattice structure is preserved except for the slight contraction of the lattice (and the concomitant slight increase in the density and heat capacity of the films) tentatively assigned to the transformation of the out-of-plane-oriented benzene rings to the in-plane-oriented C_6 rings [38]. Most importantly, a highly parallel behavior of decreasing thermal conductivity with decreasing superlattice period as seen for the original TiO_2:HQ superlattice films was also confirmed for the TiO_2:C superlattice thin films (Figure 9.5) [38]. From the application point of view, this kind of oxide:carbon superlattice films may be even more attractive than the oxide:organic films owing to their even better thermal stabilities.

Studies have also been conducted to more deeply understand the mechanisms related to the reduction of heat transport across the inorganic–organic interphases. For these studies, ZnO:$(HQ-Zn)_k$ type superlattice thin films with thicker $(HQ-Zn)_k$ hybrid intermittent layers were investigated. It was found that, upon increasing the thickness of the hybrid (HQ-Zn) layer between the ZnO layers, the phonon transport across the interface changes from ballistic to diffusive [40].

Atomic-level structural models for the ZnO:HQ superlattice thin films have been derived using quantum chemical methods, enabling detailed investigation of their spectroscopic, electronic, and thermoelectric properties [33, 34]. The computational spectroscopic data have enabled detailed interpretation of experimentally measured infrared spectra, lending strong support for the presence of organic interfaces within the crystalline ZnO:HQ superlattices (Figure 9.4c,d). Band-structure calculations at the DFT-PBE0 level of theory suggest that the band structure of the ZnO:organic superlattices can be tailored by simple and experimentally feasible modifications of the organic constituent, providing guidelines for the band-structure engineering of ZnO:organic superlattices for improved TE efficiency. The predicted lattice thermal conductivities are also in agreement with the corresponding experimental values, the ZnO:organic superlattices showing significantly reduced thermal conductivities in comparison to bulk ZnO. The reduction in the thermal conductivity is limited by the period thickness of the ZnO blocks, and rough estimates on the effect of the superlattice can actually be obtained simply by evaluating the so-called cumulative lattice thermal conductivity of the parent oxide.

Finally, it should be mentioned that purely inorganic thermoelectric nanolaminate structures of alternating Bi_2Te_3 and Sb_2Te_3 layers have also been fabricated by ALD, using the corresponding dechlorosilylation processes [41].

Figure 9.6 Nanolaminate structure of alternating ALD Bi_2Te_3 and Sb_2Te_3 double layers exhibiting localized epitaxial growth as revealed by high-resolution TEM cross-sectional analysis [38]. (Nminibapiel et al. 2013 [41]. Reproduced with permission from The Electrochemical Society.)

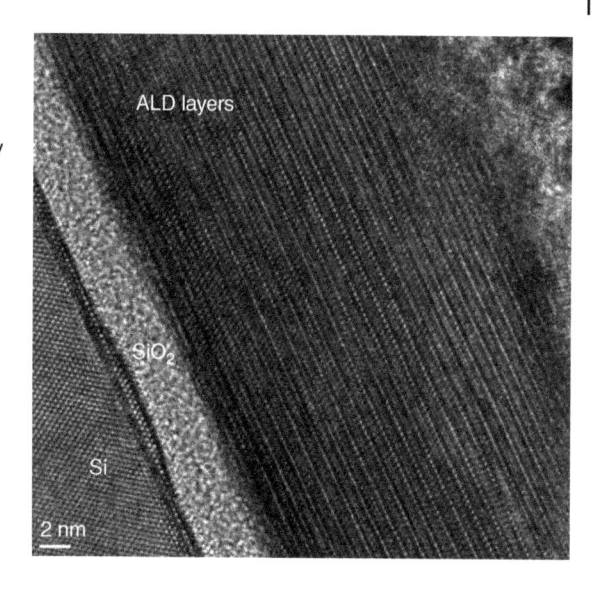

The nanolaminate concept for enhanced TE performance is not new, and Bi_2Te_3/Sb_2Te_3 nanolaminate structures have been obtained by various conventional techniques such as MBE and PLD; these techniques are, however, difficult to be transferred to industrial high-volume fabrication, and hence, efforts to obtain the same with ALD are highly warranted. From field-emission scanning electron microscopy (FE-SEM) images, an island-type growth with characteristic hexagonal crystallites was observed for both Bi_2Te_3 and Sb_2Te_3; then from high-resolution TEM cross-sectional analysis, localized epitaxial growth of alternating ALD Bi_2Te_3 and Sb_2Te_3 layers within the large islands was revealed (Figure 9.6). In addition, similar nanolaminates of PbTe and PbSe have been fabricated by ALD [42]. So far, these nanolaminate thin-film structures have not been characterized for the thermoelectric properties.

Finally, it should be emphasized that owing to the fact that both ALD and MLD inherently – based on their growth mechanism – yield highly conformal coatings, it should be straightforward to apply further nanostructuring of the described thermoelectric superlattice and nanolaminate materials using, for example, nanostructured substrate materials – natural or artificial – as sacrificial templates. Such nanostructuring approaches are yet to be attempted.

9.4 Prospects and Future Challenges

Thermoelectric power generation is a promising technology for energy harvesting or recovery from various excessively existing or unavoidably released waste heat sources, as thermoelectric materials in principle are capable of converting various types of heat flows directly into electricity. As an example of the area where ubiquitous thin-film-based thermoelectric devices would be in demand is the small wireless power sources for, for example, wearable electronics. Such devices could harvest, for example, human body heat to power medical sensors or other tiny portable devices. This is also the area where the ALD technique could become competitive as it allows for the fabrication of high-quality homogeneous coatings on large-area/nanostructured/flexible substrates.

The ALD technique, moreover, uniquely possesses other features that could be highly beneficial in the thermoelectric technology, that is, the atomic-level thickness control of individual layers and the ease of nanostructure engineering. In the former aspect and in particular when the ALD technique is combined with the emerging MLD technique for depositing organic layers with molecular-level accuracy, it is possible to fabricate precisely on-demand layer-engineered nanolaminates and superlattices of two different inorganic materials or inorganic–organic hybrids to suppress the thermal conductivity and thereby enhance the thermoelectric performance.

The research in the field is, however, in its infancy. So far, only a handful of ALD and ALD/MLD processes have been developed for prominent thermoelectric materials, and no efforts have been made to fabricate actual thermoelectric modules, even simple ones, where mutually n-type and p-type materials would be combined. The expansion of the variety of possible materials/processes is thus one of the urgent near-future tasks in the field. So far, the efforts have been directed toward the current Bi_2Te_3- and PbTe-based materials, on the one hand, and on oxide materials, on the other hand, but there would be other thermoelectric material families that could be considered as well.

References

1 Snyder, G.J. and Toberer, E.S. (2008) *Nat. Mater.*, **7**, 106–114.
2 DiSalvo, F.J. (1999) *Science*, **285**, 703–706.
3 Bahk, J.-E., Fang, H., Yazawa, K., and Shakouri, A. (2015) *J. Mater. Chem. C*, **3**, 10362–10374.
4 Sootsman, J.R., Chung, D.Y., and Kanatzidis, M.G. (2009) *Angew. Chem. Int. Ed.*, **48**, 8616–8639.
5 Slack, G.A. (1995) in *CRC Handbook of Thermoelectrics* (ed. D.M. Rowe), CRC Press, Boca Raton, FL, p. 407.
6 Pei, Y., LaLonde, A.D., Wang, H., and Snyder, G.J. (2012) *Energy Environ. Sci.*, **5**, 7963–7969.
7 Terasaki, I., Sasago, Y., and Uchinokura, K. (1997) *Phys. Rev. B*, **56**, R12685–R12687.
8 Dresselhaus, M..S., Chen, G., Tang, M..Y., Yang, R..G., Lee, H., Wang, D..Z., Ren, Z..F., Fleurial, J.-P., and Gogna, P. (2007) *Adv. Mater.*, **19**, 1043–1053.
9 Kanatzidis, M.G. (2010) *Chem. Mater.*, **22**, 648–659.
10 Nielsch, K., Bachmann, J., Kimling, J., and Böttner, H. (2011) *Adv. Energy Mater.*, **1**, 713–731.
11 Biswas, K., He, J., Blum, I.D., Wu, C.-I., Hogan, T.P., Seidman, D.N., Dravid, V.P., and Kanatzidis, M.G. (2012) *Nature*, **489**, 414–418.
12 Böttner, H., Chen, G., and Venkatasubramanian, R. (2006) *MRS Bull.*, **31**, 211–217.

13 Venkatasubramanian, R., Pierce, J., Colpitts, T., Bulman, G., Stokes, D., Posthill, J., Barletta, P., Koester, D., O'Quinn, B., and Siivola, E. (2012) in *Modules, Systems, and Applications in Thermoelectrics* (ed. D.M. Rowe), CRC Press, Boca Raton, FL, pp. 21-1–21-18.
14 Venkatasubramanian, R., Siivola, E., Colpitts, T., and O'Quinn, B. (2001) *Nature*, **413**, 597–602.
15 Yao, T. (1987) *Appl. Phys. Lett.*, **51**, 1798–1800.
16 Parsons, G.N., Atanasov, S.E., Dandley, E.C., Devine, C.K., Gong, B., Jur, J.S., Lee, K., Oldham, C.J., Peng, Q., Spagnola, J.C., and Williams, P.S. (2013) *Coord. Chem. Rev.*, **257**, 3323–3331.
17 Lybeck, J., Valkeapää, M., Shibasaki, S., Terasaki, I., Yamauchi, H., and Karppinen, M. (2010) *Chem. Mater.*, **22**, 5900–5904.
18 Niemelä, J.-P., Karttunen, A.J., and Karppinen, M. (2015) *J. Mater. Chem. C*, **3**, 10349–10361.
19 Tynell, T., Yamauchi, H., Karppinen, M., Okazaki, R., and Terasaki, I. (2013) *J. Vac. Sci. Technol., A*, **31**, 01A109.
20 Tynell, T., Okazaki, R., Terasaki, I., Yamauchi, H., and Karppinen, M. (2013) *J. Mater. Sci.*, **48**, 2806–2811.
21 Ruoho, M., Pale, V., Erdmanis, M., and Tittonen, I. (2013) *Appl. Phys. Lett.*, **103**, 203903.
22 Tynell, T. and Karppinen, M. (2014) *Semicond. Sci. Technol.*, **29**, 043001.
23 Niemelä, J.-P., Hirose, Y., Hasegawa, T., and Karppinen, M. (2015) *Appl. Phys. Lett.*, **106**, 042101.
24 Niemelä, J.-P., Hirose, Y., Shigematsu, K., Sano, M., Hasegawa, T., and Karppinen, M. (2015) *Appl. Phys. Lett.*, **107**, 192102.
25 Tripathi, T.S., Niemelä, J.-P., and Karppinen, M. (2015) *J. Mater. Chem. C*, **3**, 8364–8371.
26 Sarnet, T., Hatanpää, T., Puukilainen, E., Mattinen, M., Vehkamäki, M., Mizohata, K., Ritala, M., and Leskelä, M. (2015) *J. Phys. Chem. A*, **119**, 2298–2306.
27 Sarnet, T., Hatanpää, T., Vehkamäki, M., Flyktman, T., Ahopelto, J., Mizohata, K., Ritala, M., and Leskelä, M. (2015) *J. Mater. Chem. C*, **3**, 4820–4828.
28 Zastrow, S., Gooth, J., Boehnert, T., Heiderich, S., Toellner, W., Heimann, S., Schulz, S., and Nielsch, K. (2013) *Semicond. Sci. Technol.*, **28**, 035010.
29 Niemelä, J.-P., Yamauchi, H., and Karppinen, M. (2014) *Thin Solid Films*, **551**, 19–22.
30 Niemelä, J.-P., Marin, G., and Karppinen, M., manuscript (2017).
31 Sundberg, P. and Karppinen, M. (2014) *Beilstein J. Nanotechnol.*, **5**, 1104–1136.
32 Tynell, T. and Karppinen, M. (2014) *Thin Solid Films*, **551**, 23–26.
33 Karttunen, A.J., Tynell, T., and Karppinen, M. (2015) *J. Phys. Chem. C*, **119**, 13105–13114.
34 Karttunen, A.J., Tynell, T., and Karppinen, M. (2015) *Nano Energy*, **22**, 338–348.
35 Tynell, T., Terasaki, I., Yamauchi, H., and Karppinen, M. (2013) *J. Mater. Chem. A*, **1**, 13619–13624.

36 Tynell, T., Giri, A., Gaskins, J., Hopkins, P.E., Mele, P., Miyazaki, K., and Karppinen, M. (2014) *J. Mater. Chem. A*, **2**, 12150–12152.
37 Niemelä, J.-P. and Karppinen, M. (2015) *Dalton Trans.*, **44**, 591–597.
38 Niemelä, J.-P., Giri, A., Hopkins, P.E., and Karppinen, M. (2015) *J. Mater. Chem. A*, **3**, 11527–11532.
39 Giri, A., Niemelä, J.-P., Tynell, T., Gaskins, J., Donovan, B.F., Karppinen, M., and Hopkins, P.E. (2016) *Phys. Rev. B*, **93**, 115310.
40 Giri, A., Niemelä, J.-P., Szwejkowski, C.J., Karppinen, M., and Hopkins, P.E. (2016) *Phys. Rev. B*, **93**, 024201.
41 Nminibapiel, D., Zhang, K., Tangirala, M., Baumgart, H., Chakravadhanula, V.S.K., Kübel, C., and Kochergin, V. (2013) *ECS Trans.*, **58**, 59–66.
42 Zhang, K., Pillai, A.D.R., Bollenbach, K., Nminibapiel, D., Cao, W., Baumgart, H., Scherer, T., Chakravadhanula, V.S.K., Kübel, C., and Kochergin, V. (2014) *ECS J. Solid State Sci. Technol.*, **3**, P207–P212.

Index

a

accelerated degradation test (ADT)   165
ALD   212
  anode   198–199, 216
  cathode   216
  cycle   5, 16, 212
  doped ZnO   74–77
  electron-selective contacts   87–89
  FTIR   15
  growth characteristics   5–8
  in situ techniques   16
  In_2O_3 78–79
  Li-Al-O ($LiAlO_2$)   190
  Li-Al-Si-O   191
  Li-La-O   189
  Li_2CO_3 189
  Li_3N 192
  Li_3PO_4 192
  $Li_xSi_yO_z$ 191
  $LiCoO_2$ 195
  LiF   194
  $LiFePO_4$ 197
  $LiNbO_3$ 192
  linearity   5–8
  LiPON   193
  $LiTaO_3$ 192
  lithiation   196
  LLT   189–190
  metal chalcogenides   111–115
  $MnO_x/Li_2Mn_2O_4/LiMn_2O_4$ 196
  nanoporous materials   29–34
  OES   15–16
  passivating contacts   47

  plasma-enhanced ALD (PE-ALD)   8–11
  Pt deposits   215–216
  QCM   12–13
  QMS   13–14
  reactor   12
  saturation and ALD window   5–8
  SE   14–15
  Si hetrojunction solar cells   46–47
  Si homojunction solar cells   44–46
  sulphides   198
  supercycles   75
  TiO_2 films   124
  transition metal oxides
    electrocatalysts   174
  tunneling oxides   86
  V_2O_5 194
ALD metal chalcogenides
  bismuth sulfide (Bi_2S_3)   114
  cadmium sulfide (CdS)   113
  CIS   112
  Cu_2S 112–113
  CZTS   112
  indium sulfide (In_2S_3)   114
  PbS   113
  Sb_2S_3 113
  SnS   113
ALD process
  conformality of   16–21
  modellings   21–24
  nanoporous materials   29–34
  plasma-enhanced ALD   24–29
ALD-Pt cells   157
Al_2O_3 films   56, 57

Atomic Layer Deposition in Energy Conversion Applications, First Edition. Edited by Julien Bachmann.
© 2017 Wiley-VCH Verlag GmbH & Co. KGaA. Published 2017 by Wiley-VCH Verlag GmbH & Co. KGaA.

Index

aluminium back-surface field (Al-BSF) 44
anodic aluminum oxide (AAO) 17, 135, 165
anti-reflection coating (ARC) 44
antireflective coatings 102
artificial photosynthesis (APS) 225
aspect ratio (AR) 4, 17, 22
atmospheric pressure (AP) 59
atmospheric pressure chemical vapor deposition (APCVD) 238
atomic-layer/molecular-layer deposition (ALD/MLD) 263
Auger recombination 48, 82

b
band bending 52
band gap 69, 126, 229, 239
bismuth sulfide (Bi_2S_3) 114
blocking layer 126
Boltzmann constant 22, 48
Boron-doping 77
Bragg mirrors 68
Brillouin zone 102, 103
Burstein–Moss (BM) shift 71, 72, 74

c
cadmium sulfide (CdS) 113
carbon black (CC) 162
carbon monoxide (CO) 162
carbon nanotube (CNT) 136, 154
carbon spheres (CS) 164
carrier density 48, 69–73
carrier mobility 69, 71, 77, 260
catalytic probes 26
C–C bonds 216
chalcogenide materials 111
charge carriers 48
charge extraction length (L_{CE}) 105
chemical passivation 50
chemical spray pyrolysis (CSP) 212
chemical vapor deposition (CVD) 4, 59, 104, 108, 212
chronoamperometric tests 162
chronoamperometry (CA) 161
CIS 112

colloidal quantum dot solar cell (CQDSC) 120, 133
conduction band (CB) 227
conformality 3–34, 66, 115, 137
constant-voltage durability test 158
core-shell structures 238
corona oxide characterization of semiconductors (COCOS) 56
Coulombic scattering 69
crystalline silicon (Si) solar cells 43
Cu_2S 112
cyclic voltammetry 125
cyclopentadienyl indium (InCp) 135
CZTS 112

d
dechlorosilylation reactions 266
density functional theory (DFT) methods 268
dielectric constant 30, 88
dimethyl cadmium (DMCd) 134
direct ethanol fuel cells (DEFCs) 159
direct formic acid fuel cells (DFAFCs) 159
direct liquid-fed fuel cells (DLFCs) 159
direct methanol fuel cells (DMFC) 209
doping 51
Drude absorption 72
dual thermocouples 26
dye-sensitized solar cells (DSSCs) 113, 119, 244
dye/MOx structure 130

e
electrocatalytic tests 169, 171
electrochemical performance
  hydrogen evolution reaction (HER) 167
  hydrogen oxidation reaction (HOR) 157–159
  methanol/ethanol/formic acid oxidation reaction (MOR/EOR/FOR) 159–164
  oxygen reduction reaction (ORR) 164–167
electrochemical vapor deposition (EVD) 212

electrochemically active surface area (ECSA)   163
electron probe microanalysis (EPMA)   17
Eley–Rideal mechanism   10
ellipsometric measurement   14
ellipsometric porosimetry (EP)   31
energy-enhanced ALD   55
epitaxy   109
expanding thermal plasma (ETP)   73
extremely thin absorbers (ETAs)   107, 230, 233

f

Faradaic efficiency   246
Fermi–Dirac statistics   49, 53
Fermi level E_F 71
ferroelectric material   192
field-effect passivation   51, 54
field-emission scanning electron microscopy (FE-SEM)   271
fluorine-doped tin oxide (FTO)   234
Fourier transform infrared spectroscopy (FTIR)   15, 268
free carrier absorption (FCA)   72
free carrier reflection (FCR)   72
Fresnel equations   14
Fresnel's reflection laws   101
FTIR, *see* Fourier transform infrared spectroscopy (FTIR)

g

gadolinia-doped ceria (GDC)   214
gas conductance equations   22
generic coating technique   5
Gordon model   23
graphene nanosheets (GNS)   160
graphene oxide (GO)   164
grazing incidence small angle x-ray scattering (GISAXS)   16, 31
growth per cycle (GPC)   5, 12, 67
growth per super cycle (GPSC)   76

h

HAADF-STEM image   173
heat capacity (C)   261
Hf(NMeEt)$_4$ 64

highly-doped region (HDR)   51
high temperature fuel cells (HTFC)   209
high-volume manufacturing (HVM)   43, 58
hole transporter material (HTM)   120
HOMO   184
hybrid cell design   47
hybrid direct carbon fuel cells (HDCFC)   218
hydrofluoric acid (HF)   56
hydrogen evolution reaction (HER)   153, 167, 225
hydrogen oxidation reaction (HOR)   157–159
hydrogen sulfide (H_2S)   134
hydrophobic–hydrophilic repulsions   31
hydroquinone-based (HQ) layers   268

i

indium cyclopentadienyl (InCp)   79
indium oxide (IZO)   73
indium sulfide (In_2S_3)   114
indium tin oxide (ITO)   135, 234
infrared spectroscopy   10
infrared spectrum   101
in situ mass spectroscopy   10
interdigitated back-contact (IBC)   46, 65
intermediate band (IB)   109, 114
intermediate band photovoltaics (IBPV)   114
ionized impurity scattering (IIS)   70
IR surface spectroscopy   15
isoelectric points (EIP)   126

k

Knudsen diffusion   19, 31

l

Langmuir's law   24
layer-by-layer method   105
Li-Al-O (LiAlO$_2$)   190
Li-Al-Si-O   191
Li-La-O   189
Li$_2$CO$_3$ 189
Li$_3$N   192

Li_3PO_4 192
$Li_xSi_yO_z$ 191
$LiCoO_2$ 195
LiF   194
$LiFePO_4$ 197
light absorption efficiency (η_a)   101
$LiNbO_3$ 192
LiPON   193
$LiTaO_3$ 192
lithium lanthanum titanate (LLT)   189–190
low pressure CVD (LPCVD)   73, 77
low-temperature epitaxy   109
LUMO   184

m

macroscopic test structures   20
Masetti model   70
mass spectroscopy   26
membrane electrode assembly (MEA)   157
MEMS fabrication techniques   21
mesoporous silica membrane   31
metal oxides   229
metal-assisted chemical etching (MaCE)   232
methanol/ethanol/formic acid oxidation reaction (MOR/EOR/FOR)   153, 159–164
microwave plasma   27
mixed ionic-electronic (MIEC)   216
$MnO_x/Li_2Mn_2O_4/LiMn_2O_4$ 196
molecular layer deposition (MLD)   267
molten carbonate (MC)   209
molten carbonate fuel cells (MCFC)   210
molybdenum oxide (MoO_x)   89
Monte Carlo (MC) model   24, 27, 28

n

nanocylindrical arrays   238
nanonets (NNs)   235
nanoparticles
  architectures   129
  DSCC   127
  size distribution   105
nanoscale   262

nanosphere lithography   214
nanostructured solar cells
  absorber/HTM interface   130–132
  atomic layer deposition   132–134
  blocking layer, 126–130   134–138
  compact layer   121–126
nanotube/nanowires   134
nitric acid oxidation step (NAOS)   82
nitrides   230
NW arrays   135

o

octadecyltrichlorosilane (ODTS)   173
OES, see optical emission spectroscopy (OES)
Ohm's law   82
OPAL2   70
open-circuit-voltage (OCV)   214
optical emission spectroscopy (OES)   10, 15–16, 26
organic solar cells   120
oxygen evolution reaction (OER)   225
oxygen evolving complex (OEC)   225
oxygen reduction reaction (ORR)   153, 164–167
oxygen transmission rate (OTR)   132
oxygen vacancies (V_O)   241

p

partial metallization   84
passivated emitter rear contact (PERC)   44
passivating contacts   47, 82
passivation effect   19
PbS   113
perovskite solar cells   120
phonon mean free path (l)   261
phonon velocity (v)   261
phonon-boundary-scattering   266
phosphoric acid fuel cell (PAFC)   209
photoconductance (PC)   51
photocorrosion protection
  nanostructured photoelectrodes   246
  planar photoanodes   244
  planar photocathodes   246

photoelectrochemical cells (PEC)   120, 138
  catalyst deposition   239
  coating and functionalization   233
  micro-and nanostructuring   230
  nanostructured photoelectrodes   235
  passivation and modification   240
  photocorrosion protection   244
  photoelectrode materials   228
  principle of   227
  synthetize electrodes materials   234
photoelectrode architecture   237
photoelectrode materials
  elemental and compound semiconductors   229
  metal oxides   229
  nitrides   230
photoelectrode/electrolyte junction   241
photoluminescence (PL)   51
photonic band gap   30
photosensitizer   120
photovoltaic (PV)   43
photovoltaic conversion efficiency (PCE)   119
physical deposition methods   108
physical vapor deposition (PVD)   4, 59, 240
planar FTO   141
Planck's constant   70
plasma-enhanced (PE)   59
plasma-enhanced ALD (PE-ALD)
  advantages   10
  configurations   9
  conformality   24–29
  reactions   10
plasma-enhanced chemical vapor deposited (PECVD)   50
p-n junction model   232
point contact approach   106
poly-dimethyl-siloxane (PDMS)   137
polycarbonate filter   29
polycrystalline films   125
polymeric proton-conducting membrane (PEM)   151

polystyrene (PS)   235
polytetrafluoroethylene (PTFE)   155
post deposition anneal (PDA)   50
potentiostatic stability tests   245
power efficiency (η_p)   103
proton electrolyte fuel cell (PEFC)   218
proton exchange membrane fuel cell (PEMFC)   209
Pt alloy nanoparticle
  PtCo   170
  PtRu   170
Pt electrocatalysts, ALD of
  core/shell NPs   172–173
  electrochemical performance   157–167
  fabrication and microstructure   154–157
pulsed laser deposition (PLD)   70

q

QCM, *see* quartz crystal microbalance (QCM)
QD, *see* quantum dots (QD)
QD FET transistor   133
QD sensitizers   235
QDSSCs, *see* dye-sensitized solar cells (DSSCs)
quadrupole filter   13
quadrupole mass spectrometry (QMS)   13–14
quantum dot sensitized solar cells (QDSSC)   113, 120, 133
quantum dots (QD)   113, 132
quantum efficiency   232
quartz crystal microbalance (QCM)   12–13
quasi Fermi levels   48, 50

r

radical-assisted ALD   10
radio-frequency (RF) coil   10
random polarization   101
random-pyramid (RP)   64

reactive plasma deposition (RPD) 70
reversible hydrogen electrode (RHE) 161

s

Sb_2S_3 113
scanning electron microscopy (SEM) 17, 28
scanning transmission electron microscopy (STEM) 160
SE, *see* spectroscopic ellipsometry (SE)
Seebeck coefficient (S) 260, 266
self-assembled monolayers (SAMs) 173
Shockley–Queisser limit 103, 109
Shockley, Read, and Hall (SRH) equation 49, 50, 52
Si heterojunction solar cells
  compatibility 74
  lateral conductivity 69–71
  transparency 71–74
Si homojunction solar cells
  ALD Al_2O_3 54–59
  ALD passivation schemes 63–68
  solar cells manufacturing 59–63
  surface passivation 48–54
Si solar cell processing 58
signal/noise ratio 15
silicon heterojunction (SHJ) 46, 68
skutterudites 261
SnS 113
solar cells 44, 241
  efficiency 126
  nanostructuring of 142
  performance and stability of 143
  perovskite 142
  p-i-n design of 109
  third generation 127
solar fuels 225
  production efficiency 229
solar light absorbers
  ETA cells 107
  light harvesting and charge extraction 105–106
  low-temperature epitaxy 109
  pinhole-free ultrathin films 107
  stoichiometry and doping 107–109
  uniformity and precision 104–105
solid electrolyte (SO) 209
solid electrolyte interface (SEI) 183
solid oxide fuel cells (SOFC) 209
  ceria-based materials 214–215
  gallate materials 215
  zirconia-based materials 213–214
solid-state dye-sensitized solar cells (sDSSC) 230
space-charge region (SCR) 51
spatial ALD (S-ALD) 60, 80
spectroscopic ellipsometry (SE) 14–15, 19
Spiro-OMeTAD 134
stoichiometric semiconductor 108
stoichiometry 107
strontium titanate (STO) 83
subnanometer
  films 130
  scale 120
successive ionic layer adsorption and reaction (SILAR) 105
supercycle approach 75, 76, 80
superlattice period 267
surface recombination velocity (SRV) 51
surface-to-volume ratio 50, 132
synchrotron 16

t

technology readiness level (TRL) 226
tetrakis-dimethylamidotitanium (TDMAT) 31, 33, 245
tetrakisdimethylamido tin (TDMASn) 135
thermal ALD 8, 68, 87
thermal conductivity (κ) 260
thermodynamic equilibrium 48
thermoelectric (TE)
  designing and optimizing 260
  energy conversion and cooling 259
  oxide thin films 263
  performance 266
  power factor 264
  selenide and telluride thin films 266
thin film deposition 12

thin film electroluminescence (TFEL) 212
thin-film thermoelectric devices   262
TiCl$_4$ and H$_2$ plasma   10
time-domain thermoreflectance (TDTR)   268
titanium tetraisopropoxide (TTIP)   245
transition metal oxides (TMOs)   174
transmission electron microscopy (TEM)   17, 56
transparent conductive oxide (TCO)   43, 121, 134
traps   121, 126, 130, 133, 140, 242
trimethylaluminum (TMA)   10, 54, 156, 241
3D photonic crystals   30
tungsten trioxide   242
tunnel   47, 80–84, 86, 87, 133, 244
tunneling   83, 86–87, 121, 128, 132, 243, 245
tunnel oxide passivated contact (TOPCon)   47

u

ultraviolet (UV)
  radiation   55
  photons   10

v

valence band (VB)   227
vanadyltriisopropoxide (VTIP)   12
vapor transport method (VTM)   135, 238
V$_2$O$_5$ 194

w

water (photo)splitting   138–142, 233
water vapor transmittance rate (WVTR)   132
Wiedemann–Franz law   260
wireless cell   231

x

x-ray absorption near edge structure (XANES)   160
x-ray absorption spectroscopy (XAS)   241
x-ray diffraction (XRD)   16
x-ray fluorescence (XRF)   16, 31
x-ray photoelectron spectroscopy (XPS)   16
x-ray reflectivity (XRR)   16, 32, 268

y

yttria-doped ceria (YDC)   214
yttria-stabilized zirconia (YSZ) electrolyte   210

z

zintl phases   261
ZnO nanorod (ZnO-NR)   130, 162
ZrO$_2$ powder   15
Z-scheme   227